Strategic Approaches to Digital Platform Security Assurance

Yuri Bobbert
ON2IT BV, The Netherlands & Antwerp Management School, University of Antwerp, Belgium

Maria Chtepen
BNP Paribas Group, Belgium

Tapan Kumar
Cognizant, The Netherlands

Yves Vanderbeken
DXC, Belgium

Dennis Verslegers
Orange Cyberdefense, Belgium

A volume in the Advances in Information Security, Privacy, and Ethics (AISPE) Book Series

Published in the United States of America by
IGI Global
Information Science Reference (an imprint of IGI Global)
701 E. Chocolate Avenue
Hershey PA, USA 17033
Tel: 717-533-8845
Fax: 717-533-8661
E-mail: cust@igi-global.com
Web site: http://www.igi-global.com

Library of Congress Cataloging-in-Publication Data

Names: Bobbert, Yuri, author. | Chtepen, Maria, 1980- author. | Kumar,
 Tapan, 1983- author. | Vanderbekan, Yves, 1966- author. | Verslegers,
 Dennis, 1982- author.
Title: Strategic approaches to digital platform security assurance / Yuri
 Bobbert, Maria Chtepen, Tapan Kumar, Yves Vanderbekan, and Dennis
 Verslegers.
Description: Hershey PA : Information Science Reference, [2021] | Includes
 bibliographical references and index. | Summary: "With the field of
 digital transformation and the associated risk and security management
 rapidly changing due to emerging technologies and upcoming regulations
 this book offers extensive Design Science Research approaches to on one
 hand extensively examine the problem and on the other hand offer
 pragmatic solutions (artefacts) that can serve both academia and
 practitioners"-- Provided by publisher.
Identifiers: LCCN 2020050609 (print) | LCCN 2020050610 (ebook) | ISBN
 9781799873679 (h/c) | ISBN 9781799873686 (s/c) | ISBN 9781799873693
 (eISBN)
Subjects: LCSH: Computer security. | Application software--Development. |
 Business enterprises--Data processing.
Classification: LCC QA76.9.A25 B595 2021 (print) | LCC QA76.9.A25 (ebook)
 | DDC 005.8--dc23
LC record available at https://lccn.loc.gov/2020050609
LC ebook record available at https://lccn.loc.gov/2020050610

This book is published in the IGI Global book series Advances in Information Security, Privacy, and Ethics (AISPE) (ISSN: 1948-9730; eISSN: 1948-9749)

British Cataloguing in Publication Data
A Cataloguing in Publication record for this book is available from the British Library.

For electronic access to this publication, please contact: eresources@igi-global.com.

Advances in Information Security, Privacy, and Ethics (AISPE) Book Series

Manish Gupta
State University of New York, USA

ISSN:1948-9730
EISSN:1948-9749

MISSION

As digital technologies become more pervasive in everyday life and the Internet is utilized in ever in-creasing ways by both private and public entities, concern over digital threats becomes more prevalent.

The **Advances in Information Security, Privacy, & Ethics (AISPE) Book Series** provides cutting-edge research on the protection and misuse of information and technology across various industries and settings. Comprised of scholarly research on topics such as identity management, cryptography, system security, authentication, and data protection, this book series is ideal for reference by IT professionals, academicians, and upper-level students.

COVERAGE

- Risk Management
- Information Security Standards
- Security Information Management
- Data Storage of Minors
- Tracking Cookies
- Device Fingerprinting
- Access Control
- CIA Triad of Information Security
- IT Risk
- Global Privacy Concerns

IGI Global is currently accepting manuscripts for publication within this series. To submit a proposal for a volume in this series, please contact our Acquisition Editors at acquisitions@igi-global.com or visit: https://www.igi-global.com/publish/.

Titles in this Series

For a list of additional titles in this series, please visit: https://www.igi-global.com/book-series/advances-information-security-privacy-ethics/37157

Revolutionary Applications of Blockchain-Enabled Privacy and Access Control
Surjit Singh (Thapar Institute of Engineering and Technology, India) and Anca Delia Jurcut (University College Dublin, Ireland)
Information Science Reference • © 2021 • 297pp • H/C (ISBN: 9781799875895) • US $225.00

Multidisciplinary Approaches to Ethics in the Digital Era
Meliha Nurdan Taskiran (Istanbul Medipol University, Turkey) and Fatih Pinarbaşi (Istanbul Medipol University, Turkey)
Information Science Reference • © 2021 • 369pp • H/C (ISBN: 9781799841173) • US $195.00

Handbook of Research on Digital Transformation and Challenges to Data Security and Privacy
Pedro Fernandes Anunciação (Polytechnic Institute of Setúbal, Portugal) Cláudio Roberto Magalhães Pessoa (Escola de Engenharia de Minas Gerais, Brazil) and George Leal Jamil (Informações em Rede Consultoria e Treinamento, Brazil)
Information Science Reference • © 2021 • 529pp • H/C (ISBN: 9781799842019) • US $285.00

Limitations and Future Applications of Quantum Cryptography
Neeraj Kumar (Babasaheb Bhimrao Ambedkar University, Lucknow, India) Alka Agrawal (Babasaheb Bhimrao Ambedkar University, Lucknow, India) Brijesh K. Chaurasia (Indian Institute of Information Technology, India) and Raees Ahmad Khan (Indian Institute of Information Technology, India)
Information Science Reference • © 2021 • 305pp • H/C (ISBN: 9781799866770) • US $225.00

Advancements in Security and Privacy Initiatives for Multimedia Images
Ashwani Kumar (Vardhaman College of Engineering, India) and Seelam Sai Satyanarayana Reddy (Vardhaman College of Engineering, India)
Information Science Reference • © 2021 • 278pp • H/C (ISBN: 9781799827955) • US $215.00

Blockchain Applications in IoT Security
Harshita Patel (Vellore Institute of Technology, India) and Ghanshyam Singh Thakur (Maulana Azad National Institute of Technology, India)
Information Science Reference • © 2021 • 275pp • H/C (ISBN: 9781799824145) • US $215.00

701 East Chocolate Avenue, Hershey, PA 17033, USA
Tel: 717-533-8845 x100 • Fax: 717-533-8661
E-Mail: cust@igi-global.com • www.igi-global.com

Table of Contents

Preface

Nowadays, it is impossible to imagine a business without technology. Most industries are becoming "smarter" and more tech-driven (Desdemoustier, Crutzen, & Giffinger, 2019). We live in the era of the "platform society." We have become familiar with the new way of doing business that Uber, Airbnb, Amazon, Tencent, Alibaba, and many others have introduced to us and consider this the new normal. Digital platforms and eco-systems are not restricted by borders, locations, and industries (Nambisan, Zahra, & Luo, 2019). For example Coursera has become the world's most prominent online educator, collaborating with over 200 leading universities and companies. Pioneers ranging from Amazon to Lyft and Zillow and Airbnb to Zalando and ZBJ are disrupting the retail, healthcare, real estate, banking, lodging, steel industries, and labor markets.

In his book *Platforms Scale*, Choudary claimed that platforms would become increasingly powerful and concentrate on collecting more and more data to cross- or upsell services (Choudary, 2015). Ranging from small individual tech initiatives to complete business models with intertwined supply chains and "Platform" based business models. New ways of working, such as Agile and DevOps, are introduced, leading to opportunities and unknown risks. These risks do not restrict themselves to the technology domain; new challenges arise by teams working together in a distributed manner to deliver high paced value at a higher pace by reducing the time to market. We see smart-cities emerge, and society is taking a more holistic view of the regulation of such high-tech developments. Not only from a privacy perspective: who collects what, and for which purpose? Or from a human aspect: How can we give more autonomy to teams without losing "control" and pose compliance risks? New risks also emerge from a cybersecurity perspective: who protects our digital sovereignty and our "digital heritage"? Technology is no longer a domain that is shrouded in mystery; instead, it is an essential business discipline here to stay (Bobbert, 2018). Business schools worldwide include cybersecurity into their curriculum since roles emerge, and HR professionals need to equip themselves with new insights and understandings of these changing roles (Bobbert & Butterhoff, 2020). It is also a professional discipline that has got the attention of analysts and supervisory boards (ITGI, 2005). However, at the same time, organized crime has arrived on the scene in a big way (Stackpole & Oksendahl, 2011). Through hacks and denial-of-service attacks, all sorts of malicious actors are infiltrating our 'digital' society. They can easily take advantage of systems with poor design, implementation, and configurations. An alternative path consists of a wide range of advanced "social engineering" techniques to trick their way into organizations.

This research book aims to contribute in several ways. It addresses the significant problems when transforming an organization by embracing an API-based platform model and allowing an ecosystem of partners to take a central role in the (business) transactions. Certainly, when this model is applied by governments to improve citizen services. It goes in-depth into making use of small(er) DevOps teams and leveraging proven technological architectures as a means to release incremental features to the market fast. This technology is built and maintained through software-based production streets, referred to as Continuous Delivery Pipelines (Forsgren, 2018). This book aims to follow the thread of our business's function to the basement of the individual organization (construction) working in a eco-system of platforms (Hoogervorst, 2011). This function versus construction view is needed since CEO/CIOs need to provide reasonable assurance over this entire chain down to the nitty-gritty details of their "information age factories" (Bobbert & Ozkanli, 2020).

The field of digital transformation and the associated risk is rapidly changing due to emerging technologies and upcoming regulations. Organizations want to ensure speed and quality of technology delivery to serve customers, citizens, and other stakeholders (Forsgren, 2018). So far, little academic research is performed in this field, while the available research is rarely empirical (Bobbert, 2017). This book offers comprehensive Design Science Research approaches to, on the one hand, extensively examine the problem and, on the other hand, offer pragmatic solutions (artifacts) that can serve both academia and practitioners. Every section discusses the status quo and current challenges. It formulates core success factors and approaches that academic researchers and businesses can use. The book follows the structure below:

Chapter 1: Problems in the Area of Business Platform Models – How Are Governments Adapting the Platform Model to Improve Citizen Services

For both governments and enterprises, common obstacles block digital transformation progress, including legacy applications and processes that involve face-to-face visits to a government counter (with often irregular opening hours). Business models shift more towards technology-driven industries (OECD, 2019). In this first chapter, we elaborate on the multiple business models out there which rely on technology, how the technology contributes to the business goals of, in this case, governments. We specifically zoom in to Governmental platform services and the essential practices they should apply to get maximum value for citizens. We focus on this since we see tech-born companies already making the step and disrupt other business models.

Governments still need to make this step; therefore, examining what practices they should apply to become relevant for citizens in applying the business platform model appears to be a suitable research lens.

Research showed that many governments take a different route to the establishment of a platform model. Most are late in adoption technology and therefore will never realize the potential if the platform is not designed correctly. Citizens will be favoring to use a trusted and managed ecosystem for requesting and executing government services if the value is clear and governments vow for a transparent design of the platform where safety and privacy of the citizen' data is guaranteed. The author defines how a government can build up its technology stack to ensure citizen services are enabled on a platform model, using trusted providers' eco-system. Literature research is used to underpin the various levels and processes. Examples are included to showcase the best in class today.

Based on interviews and research, a stakeholder analysis is included to examine where the value is for politicians, government agencies and entrepreneurs wanting to be part of the platform model. This analysis is crucial to understand, as technology is not the only solution to put a successful business platform model out there.

Chapter 2: Research Findings in the Domain of Business Platform Models – Defining the Practices to Design a Perfect Government Business Model

The author starts by defining some of the best practices for defining a Government Business Platform model for citizen services. Based on research, practices were defined that will guide governments to the establishment of a successful platform model. One can assume that implementing a government business platform model does not come without several challenges. The author defines these challenges and gives recommendations. Some of the challenges can be related to the usage and institutionalization of the Agile approach to define, design but also release functionality via the platform to the citizens. This will be addressed in a further part of this book, where Agile Methodologies will be discussed in detail.

Next to the practices, the associated governance mechanisms were also defined to make sure the platform model is well management and delivers upon the defined vision and value. Governments, just like enterprises, are dynamic systems that need a standardized approach to be effective in organizing and executing work. Enter governance to make sure the organization effectively and efficiently strives to achieve its stated goals. A large section of this chapter is devoted to defining a governance model to support designing, building, rolling out and sustaining a platform model for governments.

At this point, the practices and governance model are considered foundational and theoretical. The author then provides a set of practical and reusable validation methods like a platform maturity model and a platform balanced score card. To proof that the validation methods are useful, effective, repeatable, and deliver consistent data, the author did a deep dive study at three case organizations in Flanders (Belgium) that are applying the business platform model as defined.

This chapter thus provides useful methods, guidance, and recommendations to all readers – based on thorough research – on designing and rolling out a success business platform model. The readers can apply these to understand what the status is of the organizations, how far they have come, how they can/ should apply the defined practices and governance mechanism and ultimately what the lessons learned are from reality.

Chapter 3: Findings and Core Practices in the Domain of Business Platform Models – Overall Evaluation of the Practices

This chapter concludes with critical success factors that we have turned into practices every organization can immediately apply. The reader will learn how the three case organizations all went through the classical norming, storming, performing stages to get the design right. The author provides a summary how it is possible to establish a platform model based on a trusted and managed ecosystem. Next, the author summarizes all these findings and converts them into a set of practical recommendations. As such, the reader can now confidently engage with their (government) organization to design, build and roll out a business platform model.

Chapter 4: Problems in the Area of Agile Methodologies

When establishing these platforms, this happens more and more in collaboration with multiple teams producing products. Agile has become the fastest growing IT development methodology, with most organizations doing agile implementations, as stated in Harvard Business Review. A challenge here is to lose efficiency and create waste due to distributed teams working with multiple Frameworks (Less/

Safe/Scaled Agile) and numerous regulatory requirements. Following the path of applying agile for small co-located teams, agile practices had also caught interest in larger organizations, as well as to large-scale software development projects (Paasivaara, Durasiewicz, & & Lassenius, 2008). However, it was observed that there had been some challenges in incorporating and practicing Agile to explore benefits as it was defined in the Agile Manifesto and was documented in the Standish Group Chaos Report 2020, which referred that big organizations were not as successful in practicing Agile for larger projects as compared to smaller projects (19% success in large size projects vs. 59% success in small size projects). This chapter will focus on the motivation for examining why Agile methodology doesn't work efficiently in large enterprises. The chapter aims to unearth with the help of several known statistics and see if Agile or Distributed Agile" is always a success or not and explore the cascading impact on organizations when the Agile implementation is not as per the expectation.

Chapter 5: Research Findings in the Domain of Agile Methodologies

Further on, the author explores all factors and characteristics associated with distributed agile software development with the literature's help. It intended to answer the risks and impact of distributed agile software development on team performance, which has a varying effect on organizations. First, literature is examined to get necessary information associated with Distributed Agile software development and Team performance and the models/tools/frameworks used to evaluate Agile teams' Team performance. It continuous to explain how the case study research design was used to get statistics using the data collection method and how the data has been analyzed to derive any conclusion and help answer the research question in a more quantified way. The chapter also elaborates with the literature analysis the characteristics, existing challenges, success factors, and practices associated with Distributed agile software development, Team performance, and existing Maturity models. The authors aim to explore the gaps surrounding distributed agile and team performance mentioned in the literature. In addition to this, the chapter highlights some key observations for the two selected maturity models. It highlights its strength and weakness which seems to have been ignored by the organization while using the models. For this research, a sizeable financial service company was examined as a case study.

Chapter 6: Findings and Core Practices in the Domain of Agile Methodologies

The chapter then concludes by drawing core strategic practices and approaches to organizations having similar distributed agile software development models. The chapter also leaves open discussion points for the readers to think upon and question themselves on the way Agile is being used within their organizations and how it contributes to improving the methods of working on digital systems and platforms.

Chapter 7: Problems of CI/CD and DevOps on Security Compliance

With the increasing importance of software development and IT systems for realizing governments and enterprises' goals, the need for specific governance between business and IT organizations has come to the forefront. At its most basic definition, IT governance is the process by which decisions are made around IT investments (De Haes, Van Grembergen, Anant, & Huygh, 2020). As defined by Calder and Moir (Calder, 2009), IT governance is a matter of optimizing the use of IT investments through strong collaboration and communication between the business, and IT's leaders and their strategies. Many frameworks became available that provide guidance on IT governance and set up the right processes and controls to give reasonable assurance that the intended goals and value are achieved to realize the organization's strategy (Gawer, 2014). IT Governance Frameworks, Security Governance Frameworks and Compliance aware Software-Defined Infrastructures that enable continuous compliance (McCarthy, 2014; Ozkanli, 2020) are on the rise and taken into account.

In this chapter, the authors define the main problems when working on technology products in DevOps Teams and on CI/CD pipelines about security and risk management. It focuses on the regulatory requirements and cyber threats that have an impact on organizations. Regulator requirements vary from industry and country. Working with multiple teams on products requires proper alignment in frameworks, controls, and architecture principles to be end-to-end protected throughout the connected platforms. This chapter examines the numerous compliance frameworks and architectural principles applied to an agile way of working and, more precisely, to CI/CD pipelines. It defines the main problem statement and questions the authors wanted to answer. The authors looked with a regulated industry lens since this industry suffers the most and therefore has the most significant benefit from this research project. The authors describe the main concepts of building software and sum up the main definitions of secure platforms.

Chapter 8: Research Findings in the Domain of CI/CD and DevOps on Security Compliance

It furthermore studies the mapping of governance and security control objectives impacted by DevOps to the corresponding DevOps control objectives. These DevOps objectives introduce either an opportunity or a risk for achieving the security & governance control objectives. Finally, the Artifact defines a list of SecDevOps controls that have proven to effectively combine the agility of the DevOps paradigm with security compliance assurance. The authors examine in collaboration with experts the multiple frameworks to be suitable. The authors define SecDevOps controls that have proven to be effective in combining the DevOps paradigm's agility with security compliance assurance. To design this Artefact, four widely-used frameworks/standards (COBIT 5, NIST Cybersecurity Framework, NIST SP 800-53, and ISO 27002) were reviewed for sufficiently detailed security and privacy control objectives and controls. Based on these criteria, NIST SP 800-53 and ISO 27002 standards were selected for comparison and mapping with (Sec)DevOps controls in this research. Additional validation was performed by comparing the findings of this study with high-level implementation and operational guidance of the Department of Defense DoD Enterprise DevSecOps Reference Design report (Lam, 2019). The information has as a purpose to describe the DevSecOps lifecycle and supporting pillars, in line with NIST Cybersecurity Framework, which is a high-level framework building upon specific controls and processes defined by NIST SP 800-53, COBIT 5, and ISO 27000 series.

Chapter 9: Findings and Core Practices in the Domain of CI/CD and DevOps on Security Compliance

This chapter concludes with a pragmatic set of core practices and strategic approaches academics and practitioners can use to ensure security compliance in CI/CD pipelines that ultimately enables teams to work agile on digital platforms.

Chapter 10: Challenges and Opportunities for Security Assurance in DevOps

The introduction of Agile methodologies in an organization brings, besides the numerous benefits, also some challenges. From a process perspective, some of these challenges include functional and non-functional requirements identification, tracking of projects, quality management, and risk management (Almeida, 2017). Traditional decision-making strategies such as hierarchical approval or review by a technical board no longer fit the increased speed with which development and operations are moving, resulting in the delegation of authority and end-to-end responsibility for the various teams (Moe, Dings, & Dyb, 2008). At the same time, technological advances such as infrastructure-as-code (Artač, Borovšak, Nitto, Guerriero, & Tamburri, 2017) containerization and micro-services require specific skills and knowledge to realize their full potential (Kang, Le, & Tao, 2016).

The close integration of the operations team with the development team, allowing them to collaborate early in the development cycle, is placed forward as an approach to meet these challenges and reduce friction (Ebert, Gallardo, Hernantes, & Serrano, 2016). This section of the book defines the main problems when working on DevOps Teams products and CI/CD pipelines concerning security and risk management. It focuses on the regulatory requirements and cyber threats that have an impact on organizations. Regulator requirements vary from industry and country.

Chapter 11: Research Findings in the Domain of Security Assurance in DevOps

Working with multiple teams on products requires proper alignment in frameworks, controls, and architecture principles to be end-to-end protected throughout the connected platforms.

This chapter examines the current practices in the field of DevOps security and proposes restorative practices and approaches that enable speed and quality without losing money and efficiency. These practices are plotted on the SecDevOps cycle so practitioners and scientists can work from that. The essential practices are highly technical and directly applicable in real-life environments.

Chapter 12: Findings and Core Practices in the Domain of Security Assurance in DevOps

This chapter concludes with a pragmatic set of core practices and strategic approaches academics and practitioners can use to ensure security compliance DevOps that ultimately enables teams to work agile on digital platforms. It concludes with core practices and strategic approaches that we have turned into practices every organization can immediately apply.

This book offers comprehensive Design Science Research-based approaches to, on the one hand, an examination of the problems via extensive literature study, and on the other hand, offer pragmatic solutions (artifacts) that can serve both academia and practitioners. It concludes with core practices and approaches that academic researchers, as well as business leaders, can use.

Yuri Bobbert
ON2IT BV, The Netherlands & Antwerp Management School, University of Antwerp, Belgium

REFERENCES

Almeida, F. (2017). Challenges in migration from waterfall to agile environments. *World Journal of Computer Application and Technology, 5*(3), 39–49.

Artač, M., Borovšak, T., Nitto, E. D., Guerriero, M., & Tamburri, D. A. (2017). DevOps: Introducing Infrastructure-as-Code. *39th International Conference on Software Engineering Companion (ICSE-C)*, 497–498. doi:10.1109/icse-c.2017.162

Bobbert, Y. (2017). Defining a research method for engineering a Business Information Security artefact. *Proceedings of the Enterprise Engineering Working Conference (EEWC) Forum*.

Bobbert, Y. (2018). *Improving the Maturity of Business Information Security: On the Design and Engineering of a Business Information Security Administrative tool*. Radboud University.

Bobbert, Y., & Butterhoff, M. (2020). *Leading Digital Security: 12 ways to combat the silent enemy*. https://12ways.net/blogs/emerging-roles-in-digital-security/

Bobbert, Y., & Ozkanli, N. (2020). LockChain technology as one source of truth for Cyber, Information Security and Privacy). In *Computing Conference*. London: Computing Conference.

Choudary S. P. (2015). *PlatformScale*. Retrieved from https://www.amazon.com/Platform-Scale-emerging-business-investment-ebook/dp/B015FAOKJ6

De Haes, S., Van Grembergen, W., Anant, J., & Huygh, T. (2020). Enterprise Governance of Information Technology. *Enterprise Governance of Information Technology*. 387-84882-2 doi:10.1007/978-0-

Desdemoustier, J., Crutzen, N., & Giffinger, R. (2019). *Municipalities' understanding of the Smart City concept: An exploratory analysis in Belgium*. Academic Press.

Ebert, C., Gallardo, G., Hernantes, J., & Serrano, N. (2016). DevOps. IEEE Software. https://doi.org/10.1109/ms.2016.68.

Forsgren, N. (2018). *Accelerate: The Science of Lean Software and Devops: Building and Scaling High Performing Technology Organisations*. IT Revolution Press.

Gawer, A. (2014, March). Bridging differing perspectives on technological platforms: Toward an integrative framework. *Research Policy, 43*(7), 1239–1249.

Hoogervorst, J. (2011). A Framework for enterprise engineering. *International Journal Internet and Enterprise Management, 7*(1), 5-40.

ITGI. (2005). *Information Risks; Who's Business are they?* IT Governance Institute.

Kang, H., Le, M., & Tao, S. (2016). Container and Microservice Driven Design for Cloud Infrastructure DevOps. In IEEE International Conference on Cloud Engineering (IC2E). IEEE. https://doi.org/10.1109/ic2e.2016.26

Lam, T. C. (2019). *Enterprise DevSecOps Reference Design*. Department of Defense.

McCarthy, M. A. (2014). A Compliance Aware Software Defined Infrastructure. *Proceedings of IEEE International Conference on Services Computing*, 560-567.

Moe, N. B., Dings, T., & Dyb, T. (2008). Understanding Self-Organizing Teams in Agile Software Development. *19th Australian Conference on Software Engineering*, 76-85. doi:10.1109/aswec.2008.4483195

Nambisan, S., Zahra, S. A., & Luo, Y. (2019). *Global platforms and ecosystems: Implications for international business theories.* doi:10.105741267-019-00262-4

OECD. (2019). *Strengthening Digital Government (Issue March).* doi:10.1787/9789264307636-en

Ozkanli, N. (2020). *Implementation of Continuous Compliance; Automation of Information Security Measures in the software development process to ensure Continuous Compliance.* Utrecht: Open University Press.

Paasivaara, M., Durasiewicz, S., & Lassenius, C. (2008). *Distributed Agile Development: Using Scrum in a large project.* doi:10.1109/icgse.2008.38

Stackpole, B., & Oksendahl, E. (2011). *Security Strategy.* Auerbach Publications.

Acknowledgment

Noteworthy accomplishments are rarely achieved alone. A tremendous group of people have, directly or indirectly, contributed to our research projects. We would therefore like to acknowledge the contributions made by:

Prof. Dr. Wouter Van Dooren, Mr. Floris Ampe, Mr. Ivan Stuer, Dr. Nils Walravens, Mr. Eddy Vander Stock, Mrs. Annabel De Craene, Mr. Andreas Nikolakopoulos, Mrs. Gigi De Schryver, Mr. Tobias Verbist, Mr. Frederic Coene, Mr. Steven Bradley, Mr. Sanne Hoekstra, Mr. Leon Kortekaas, Mr. Danny Onwezen, Mr. Karthik Rajopalan, Mr. Wiebe de Roos, Mr. Dominik De Smit, Mr. Tim Hemel, Mr. Sangam Gupta, Mr. Francois Raynaud and the ex-chief of cabinet of the Belgian Federal government (who preferred to remain anonymous)

Our special thanks also go out to Prof dr. Barry Derksen and Mr. Mark Butterhoff, for sharing their insights and challenging us through critical reviews.

Our gratitude goes out to professors Prof Dr. Yuri Bobbert and Prof. dr. Hans Mulder for their mentorship and support during our research, to Prof. dr. Steven de Haes for his sponsorship and guidance and to the entire Antwerp Management Team, especially Danny Lauwers and Celine Jansen.

We also want to thank our colleagues from Antwerp Management School including:

Edzo A Botjes and Ashish Ranjan for their suggestions, support and insightful comments,

Vincent Pattijn for the lengthy debates on the foundations of agile and flow.

Last but certainly not least we want to thank our families for supporting us during this journey and putting up with the time and effort spent to complete our research projects and contributions to this publication.

Introduction

These days it's impossible to imagine business without technology. Most industries are becoming "smarter" and more tech-driven—ranging from small individual tech initiatives to complete business models with intertwined supply chains and "Platform" based business models (Betz, 2016). New ways of working such as Agile and DevOps are introduced and thereby new risks arise (Bobbert & Ozkanli, 2020; CROForum, 2018). Not only technology risks, but also risks that are caused by teams working together at high pace and autonomy (Kumar, 2020; Lencioni, 2002). Where decisions on risk acceptances or security measures most of the time take place in the team itself rather than looking at the bigger picture of accumulated risks (Ozkanli, 2020). According to CROForum[1] this is an increasing "silent risk". This autonomous way of working in agile teams – in most case in a distributed manner- is needed to enable speed, quality and craftmanship and there is a quicker time to market (Forsgren, 2018). For policymakers and business leaders, technology is no longer a domain that is shrouded in mystery; rather it's an essential business discipline that is here to stay, and it's taught at business schools all over the world. It's also a professional discipline that has won the attention of analysts and supervisory boards. However, at the same time, nefarious nation -state activity and organized crime have arrived on the scene in a big way. Through hacks and denial-of-service attacks, all sorts of malicious actors are infiltrating our 'digital' society. They can easily take advantage of systems that are sloppily designed, built and configured and they frequently use advanced "socially engineering" techniques to trick their way into organizations. Platform oriented businesses are typically built on api-based-ecosystems of data, assets, applications and services (DAAS). These hybrid technology landscapes, most of the time built-in software defined networks in clouds (McCarthy, 2014), lack real-time visibility and control when it comes to their operations (Bobbert, 2017; Hilton, 2017). This makes it hard for boards to take ownership and accountability over cyber (ITGI, 2005), IT, or HR risks.

Digital Platforms and Frameworks

Frameworks such as the ISO27000 are being applied in order to implement Information Security troughout the value chain. Each individual provider applying there individual practice framework. According to Siponen (Siponen & Willison, 2009) "these frameworks are generic or universal in scope and thus do not pay enough attention to the differences between organizations and their information security requirements". In practice, we have seen the application of security and privacy frameworks falter because they tend to become a goal on their own rather than a supporting frame of reference to start dialogues with key stakeholders (Kuijper, 2020). Kluge et al. (Kluge & Sambasivam, 2008) for example also noted that the use of frameworks as a goal on its own does not support the intrinsic willingness and commitment to improve. This is especially the case for mid-market organizations that lack dedicated security staff, capabilities and/or sufficient budgets. Puhakainen and Siponen (2010) noted that information security approaches are lacking not only theoretically grounded methods, but also empirical evidence of their effectiveness. Many other researchers (Workman, Bommer, & Straub, 2008; Lebek, Uffen, Neumann, Hohler, & Breitner, 2014; Yaokumah & Brown, 2014) have also pointed out the necessity of empirical research into practical interventions and preconditions in order to support organizations improve the effectiveness of their security. These theoretical voids, as well as the practical observation of failing compliant-oriented approaches, widen the knowledge gap (Flores, Antonsen, & Ekstedt, 2014). This "knowing-doing gap" is also perceived in the current approaches which predominantly aim at the technology or by the technology industry but seem to lack a holistic approach.

From IT Security to Business Information Security to Digital Assurance

Back in the years 2000 security was mainly IT-oriented and the main focus was on using IT controls to mitigate or detect security vulnerabilities. Research has shown that the number of security incidents has increased (Ponemon, 2016) over the years, as has the financial impact per data breach (Ponemon, 2016). Mastering emerging technologies such as big data, Internet of Things (Conti, Dehghantanha, Franke, & Watson, 2018), social media and combating cybercrime (Cashell, Jackson, Jickling, & Webel, 2004), while protecting critical business data, requires a team instead of a single IT person (ITGI, 2005). To protect this data, security professionals need to know about the value of information and the impact if it is threatened (ITGI, 2005). IT risk management requires different capabilities, knowledge and expertise from the skills of IT security professionals (Hubbard, 2009). Hubbard (Hubbard, 2009) refers to the failure of 'expert knowledge' in impact estimations and to the importance of experience beyond risk and IT security, such as collaboration and reflection (Bobbert, 2018).

In the past (Yaokumah & Brown, 2014), IT security controls were implemented based on best practices prescribed by vendors, without a direct link to risks or business objectives (Yaokumah & Brown, 2014). These controls depended on technology and the audits and assessments were used to prove their effectiveness (Zitting, 2015). The problem with this approach lay in the limitations of mainly IT-focused security and security experts working in silos with limited, subjective views of the world (Flores, Antonsen, & Ekstedt, 2014). Let alone of the entire value chain they are working in. This is important, as information security is subject to many different interpretations, meanings, and viewpoints (Van Niekerk & Von Solms, 2010) from all parties involved. In the case of IS, this refers to interactions and reflection between actors, e.g., the business, data owners and industry peers on the appropriate level of risk appetite and security maturity (Flores, Antonsen, & Ekstedt, 2014). Thus, objectivity relates to reality, 'truth reliability', testability and reproducibility, while subjectivity refers to the quality of personal opinions. Intersubjectivity involves the agreements between social entities and the sharing of subjective states by two or more individuals (Seale, 2004). Providing objective assurance over the entire value chain of suppliers applying different methods, frameworks and controls becomes more and more cumbersome.

The state of security in 2010 shifted towards 'information security'. ISO specifies information security as "protecting information assets from a wide range of threats in order to ensure business continuity, minimise business risk and maximise return on investment and business opportunities" (ISO/IEC27001:2013, 2013). Its core principles are Confidentiality, Integrity and Availability (CIA) (ISO/IEC27001:2013, 2013). Later non-repudiation and auditability were added to comply with audit and compliance regulations when working in a more collective manner and to demark the responsibility of each party involved in the assurance process. Thus, Information Security should ensure a certain level of system quality and assurance (Cherdantseva & Hilton, 2013).

The scope of Information Security was then expanded to other disciplines in the enterprise since digital became more and more common in our way of doing business (ISF). In their book *Information Security Governance*, Von Solms and Von Solms describe the growing number of disciplines involved in IS (von Solms & von Solms, 2009). By 2011 IT managers and IT security managers were increasingly urged to engage with business to determine risk appetite and the desired state of security. In 2005 ITGI proposed to co-develop IS together with the business (ITGI, 2005). Since 2011, the role of culture (Van Niekerk & Von Solms, 2010), awareness (Al-Omari, El-Gayar, & Deokar, 2012b), compliance (Al-Omaria, El-Gayar, & Deokar, 2012) and knowledge sharing (Flores, Antonsen, & Ekstedt, 2014) has also been included in security strategy frameworks (Stackpole & Oksendahl, 2011). Due to research on IT governance at the Antwerp Management School (AMS) (Van Grembergen, De Haes, & Guldentops, 2004), relational mechanisms such as culture, behaviour and knowledge were incorporated in the COBIT 5 Information Security Framework (ISACA, 2012) in 2012.

IT staff still find it difficult translating security controls into concrete actions in the initial phase of a design and build of software (Visser, 2016) (Ozkanli, 2020). Because of these complex processes, employees focus on continuous maintenance of documentation to please internal and external regulators, instead of value creation for customers. Khan (2018) states in his paper "Due to constantly shifting regulations, businesses today are having to audit their IT compliance requirements on average four and a half times per year. Now more than ever, the act of adhering to regulatory requirements requires an ongoing commitment." Without an automated process security & privacy by design and continuous delivery will not be possible (Forsgren & Humble, 2018). Compliance processes are complex and time consuming, often manual and the evidence has to be found numerous times for different audits, reviews and different regulators (Khan, 2018).

Up to 2016, the subjective silo approach to IS was designed, maintained and reported via spreadsheets (Zitting, 2015). Experts mapped multiple control frameworks (ITGI, 2007) from ISO, ISF, COBIT 5 in spreadsheets and these are still used by regulators such as the Dutch Central Bank (Koning, 2014). Powell et all (Powell, Baker, & Lawson, 2009) discovered in 483 error instances in 50 spreadsheets. The Powell research is one of the largest examinations into spreadsheet errors. Volchkov stated that collecting evidence of effectiveness of the controls via spreadsheets has limitations (Volchkov, 2013) and pose a risk on its own. This "Assurance" problem increases when throughout the value chain information assurance needs to be established with multiple technology providers delivering services in and to the platform.

According to Richardson (2015)

Information assurance (IA) is the process of getting the right information to the right people at the right time. IA benefits business through the use of information risk management, trust management, resilience, appropriate architecture, system safety, and security, which increases the utility of information to authorized users and reduces the utility of information to those unauthorized.

This relates to the field of business information security; with business level engagement and strategic risk management of information and related systems, rather than the creation and application of security controls (Bobbert, 2018).

Considering the issues mentioned above we summarize the most important ones:

- Platforms are extended enterprises with high dependency of each other
- This requires expert knowledge of risk and security professionals as well as tech-business-leaders
- IT-focused security and security experts working in silos
- Providing objective assurance over the entire value chain is cumbersome
- This problem increases when teams work in distributed environments (across geographical and cultural boundaries)
- Difficult translating security controls into concrete actions in design and build of software
- Compliance processes are often manual and the evidence has to be found numerous times for different audits, reviews and different regulators
- This "Assurance" problem increases when throughout the value chain assurance needs to be established with multiple technology providers delivering services in and to the platform

Taking this into consideration there is a need to establish a more collaborative way of working among stakeholders when addressing the dynamics of the environment and the organization, gain a more qualitative and integral view based on facts related to tactical and operational data, to secure an increase in awareness at board level, to cultivate a certain level of reflection and self-learning to achieve continuous improvement and to use accepted best-practice frameworks produced and maintained by existing security communities and bodies.

REFERENCES

Al-Omari, A., El-Gayar, O., & Deokar, A. (2012a). Security policy compliance: user acceptance perspective. *Proceedings of the 45th Hawaii International Conference on System Sciences.*

Al-Omari, A., El-Gayar, O., & Deokar, A. (2012b). Information security policy compliance: the role of information security awareness. *Proceedings of the American Conference on Information Systems.*

Betz, C. (2016). *The Impact of Digital Transformation. Agile, and DevOps on Future IT Curricula.* Academic Press. doi:10.1145/2978192.2978205

Bobbert, Y. (2017). Defining a research method for engineering a Business Information Security artefact. *Proceedings of the Enterprise Engineering Working Conference (EEWC) Forum.*

Bobbert, Y. (2018). *Improving the Maturity of Business Information Security: On the Design and Engineering of a Business Information Security Administrative tool.* Radboud University.

Bobbert, Y., & Ozkanli, N. (2020). *LockChain technology as one source of truth for Cyber, Information Security and Privacy.* Computing Conference, London, UK.

Cashell, B., Jackson, W., Jickling, M., & Webel, B. (2004). *The Economic Impact of Cyber-Attacks.* Congressional Research Service, The Library of Congress.

Cherdantseva, Y., & Hilton, J. (2013). A Reference Model of Information Assurance & Security. *IEEE Proceedings of ARES, SecOnt Workshop.*

Conti, M., Dehghantanha, A., Franke, K., & Watson, S. (2018). Internet of Things security and forensics: Challenges and opportunities. *Future Generation Computer Systems-The International Journal of Escience, 78,* 544-546.

CROForum. (2018). *Understanding and managing the IT risk landscape: A practitioner's guide.* Retrieved from https://www.thecroforum.org/2018/12/20/understanding-and-managing-the-it-risk-land- scape-a-practitioners-guide/

Deloitte. (2009). *Spreadsheet Management, Not what you figured.* Deloitte.

Flores, W., Antonsen, E., & Ekstedt, M. (2014). Information security knowledge sharing in organizations: Investigating the effect of behavioral information security governance and national culture. *Computers & Security, 2014*(43), 90-110.

Forsgren, N. (2018). *Accelerate: The Science of Lean Software and Devops: Building and Scaling High Performing Technology Organisations.* IT Revolution Press.

Forsgren, N., & Humble, J. K. (2018). *Accelerate: The Science of Lean Software and DevOps: Building and Scaling High Performing Technology Organizations.* Lean IT Strategies LLC.

gov.uk. (n.d.). *The Security Policy Framework (SPF).* Author.

Halkyn. (2015, February 19). *ISO27001 Self Assessment Checklist hits record downloads.* Author.

Hilton, M. N. (2017). Trade-offs in continuous integration: assurance, security, and flexibility. *Proceedings of the 2017 11th Joint Meeting on Foundations of Software Engineering.* 10.1145/3106237.3106270

Hubbard, D. (2009). *The Failure of Risk Management.* John Wiley & Sons.

ISACA. (2012). *COBIT5 for Information Security.* Information Systems Audit and Control Association, ISACA.

ISF. (n.d.). *Corporate Governance Requirements for Information Risk Management.* Information Security Forum.

ISO/IEC27001:2013. (2013). *ISO/IEC 27001:2013 Information technology -- Security techniques -- Information security management systems -- Requirements.* Geneva: ISO/IEC.

ITGI. (2005). *Information Risks; Who's Business are they?* IT Governance Institute.

ITGI. (2007). *COBIT Mapping: Mapping of CMMI for Development V1.2 With COBIT.* IT Governance Institute.

Khan, J. (2018, June). *The need for continuous compliance.* Academic Press.

Kluge, D., & Sambasivam, S. (2008). *Formal Information Security Standards in German Medium Enterprise.* Edsig.

Koning, E. (2014). *Assessment Framework for DNB Information Security Examination.* Amsterdam: De Nederlandsche Bank. Retrieved 7 29, 2015, from https://www.toezicht.dnb.nl/3/50-203304.jsp

Kuijper, N. (2020). *Effective privacy governance and (change) management practices (limited to GDPR article 32) A view on GDPR ambiguity, non-compliancy risks and effectiveness of ISO 27701 as Privacy Management System*. Antwerp Management School.

Kumar, T. (2020). *What is the impact of distributed agile softWare development on team performance?* Antwerp Management School.

Lebek, B., Uffen, J., Neumann, M., Hohler, B., & Breitner, M. (2014). Information security awareness and behavior: A theory-based literature review. *Management Research Review*, *12*(37), 1049–1092. doi:10.1108/MRR-04-2013-0085

Lencioni, P. (2002). *The Five Dysfunctions of a Team; a leadership fable*. Wiley Imprint Jossey Bass.

McCarthy, M. A. (2014). A Compliance Aware Software Defined Infrastructure. *Proceedings of IEEE International Conference on Services Computing*, 560-567.

Ozkanli, N. (2020). *Implementation of Continuous Compliance; Automation of Information Security Measures in the software development process to ensure Continuous Compliance*. Open University Press.

Papazafeiropoulou, A. (2016). Understanding Governance, Risk and Compliance Information Systems the experts view. *InfoSyst Front*, (18), 1251-1263.

Pfeffer, J., & Sutton, R. (2001). *The Knowing-Doing Gap: How Smart Companies Turn Knowledge into Action*. Harvard Business School Press.

Ponemon. (2016). *Cost of Data Breach Study: Global Analysis*. Ponemon Institute LLC.

Powell, S., Baker, K., & Lawson, B. (2009). Errors in Operational spreadsheets. *Journal of Organizational and End User Computing*, *21*(3), 24–36. doi:10.4018/joeuc.2009070102

Puhakainen, P., & Siponen. (2010). Improving employees compliance through information systems security training; an action research study. *Management Information Systems Quarterly*, *34*(4), 757–778. doi:10.2307/25750704

Richardson, C. (2015). *Bridging the air gap: an information assurance perspective*. University of Southampton.

Seale, C. (2004). *Researching Society and Culture* (2nd ed.). Sage Publications.

Siponen, M., & Willison, R. (2009). Information Security management standards: Problems and solutions. *Information & Management*, *46*(5), 46. doi:10.1016/j.im.2008.12.007

Stackpole, B., & Oksendahl, E. (2011). *Security Strategy*. Auerbach Publications.

Van Grembergen, W., De Haes, S., & Guldentops, E. (2004). Structures, Processes and Relational Mechnisms for IT Governance. In *Strategies for Information Technology Governance* (pp. 1–36). Idea Group Publishing. doi:10.4018/978-1-59140-140-7.ch001

Van Niekerk, J., & Von Solms, R. (2010). Information security culture; A management perspective. Elsevier.

Visser, J. (2016). *Building Maintainable Software*. O'Reilly Media Inc.

Volchkov, A. (2013). How to Measure Security From a Governance Perspective. *ISACA Journal, 5*.

von Solms, S., & von Solms, R. (2009). Information Security Governance. New York: Springer Science. doi:10.1007/978-0-387-79984-1

Workman, M., Bommer, W., & Straub, D. (2008). Security lapses and the omission of information security measures: A threat control model and empirical test. *Computers in Human Behavior, 24*(6), 2799–2816. doi:10.1016/j.chb.2008.04.005

Yaokumah, W., & Brown, S. (2014). An Empirical Examination of the relationship between Information Security / Business strategic alignment and Information Security Governance. Journal of Business Systems. *Governance and Ethics, 2*(9), 50–65.

Zitting, D. (2015). Are You Still Auditing in Excel? *Sarbanes Oxley Compliance Journal*. Retrieved from http://www.s-ox.com/dsp_getFeaturesDetails.cfm?CID=4156

ENDNOTE

[1]   Chief Risk Officer Forum; The CRO Forum's Emerging Risk Initiative continually scans the horizon to identify and communicate emerging risks

Chapter 1
Problems in the Area of Business Platform Models:
How Are Governments Adapting the Platform Model to Improve Citizen Services

Yves Vanderbeken
DXC, Belgium

ABSTRACT

Platform models are all around us and here to stay. However, governments are generally late in adopting new technologies and also late in strategizing how a platform model can improve citizen services. This chapter will define how a government can build up its technology stack to ensure citizen services are enabled on a platform model, using an ecosystem of trusted providers. Literature research is used to underpin the various levels and processes. Examples are included to showcase the best in class today.

1 WHAT IS "GOVERNMENT AS A PLATFORM"?

1.1 Introduction

We live in the era of the "platform society." We get up in the morning and read our news or catch up with friends on social media platforms. Our journey to the office may involve a platform-enabled lift in a stranger's car, using Uber. We stay in other people's homes or rooms when we travel (Airbnb). We purchase everything online, even groceries via Amazon, never thinking to consider where the actual seller is located or who they employ, for how long.

We have become familiar with the new way of doing business that Uber, Airbnb, Amazon, Tencent, Alibaba, and many others have introduced to us and consider this the new normal. Digital platforms and ecosystems are not restricted by borders, locations, and industries (Nambisan et al., 2019). Pioneers ranging from Amazon to Lyft and Zillow and from Airbnb to Zalando and ZBJ are disrupting the retail, healthcare, real estate, banking, lodging, and steel industries, and labour markets (Jacobides et al., 2019).

DOI: 10.4018/978-1-7998-7367-9.ch001

We turn to casual task platforms like Helpling or TaskRabbit whenever we want to hire a person for a specific job.

As such, *"there is no doubt that platform models are here to stay"* (Choudary, 2015). Choudary claimed in his book "Platforms Scale" that platforms will even become increasingly powerful and concentrate on collecting more and more data to cross- or upsell services. For example, data captured by commerce platforms like Alibaba also served as input to a financial credit rating system. The author argued that platforms that facilitate interactions and capture data will have an increasingly larger role to play in the future of global trade.

OECD: "By 2022, over 60% of global GDP will be digitized. An estimated 70% of new value created in the economy over the next decade will be based on digitally enabled platforms."

We find supporting evidence at the World Economic Forum (World Economic Forum, 2019) when they state that platform-oriented companies are seriously disrupting traditional business. Even though traditional players are fast in embracing the platform model to safeguard or grow their products and/or services portfolio they risk being overtaken by new players (often start-ups).

Andersson Schwarz (Andersson Schwarz, 2017) warned us that although consumers or users may see digital platforms as technologies that enable us to share, communicate, or transact freely, in reality, they are governing systems that control, interact, and accumulate.

Although more people and things are becoming connected through networks, it is clear that the digital transformation of governments is still only at an early stage and that governments are late in adopting the platform model (Accenture, 2018).

Platforms also change the way we interact with government authorities (using, e.g., online portals and digital registries). However, OECD stated that the policy response of governments to the digital transformation has been mixed. Some are developing a strategic and pro-active approach to leveraging its benefits. In contrast, others have made piecemeal decisions to contain or roll back the consequences of specific incidents (e.g., security breaches) or the impacts of new technologies, applications, or business models (OECD, 2017).

Many policies, and public sectors' internal processes and dynamics, are a legacy of an analogue era that assumed a physical context and are ill-adapted to the digital era. OECD advises that policymakers must work to ensure that the opportunity offered by the digital transformation is used to improve the well-being of all citizens.

The lack of an integrated, whole-of-government approach increases the risk that policies in one area will have unintended, possibly adverse, impacts on another, or that opportunities for synergies that enhance positive effects are missed.

OECD already recommended in 2014 (OECD, 2014) to optimize public service delivery, whereby the government is no longer necessarily the provider of public services but increasingly acts as a broker that allows for the right solutions to a specific citizen's problem to be delivered by the provider best fit to do it. This will enable the government to benefit from the effects of scale and network. This is, however, leading to complex trade-offs, such as balancing privacy and convenience.

An example where scale and network effects play, is the rapid rise of Internet Hospital Platforms in China (Wu et al., 2019), sponsored by the government. The Chinese government views digital medicine using the platform model as a solution to address several challenges like the overprescribing of profitable drugs and diagnostic tests (leading to a waste of health resources). China's fast increase in internet users (from 22.7% in 2008 to 59.6% in 2018) provides an opportunity for the development of digital medicine platforms. The online platform supports the health-care provider at the community level in diagnosis and treatment decisions. The provider can link the patient directly to the platform for a video consultation with a physician. If no diagnosis can be made, the patient is referred to hospitals associated with the platform. As good as it sounds, this way of working also has its challenges, including patient safety, data security, and lack of surveillance and evaluation frameworks.

In summary, adopting the platform model is relevant today and will continue to be the state-of-the-art in public sector innovation (Smedlund et al., 2018).

1.2 Broad Definition of Business Platform Model

As this research is about business platform models, it is necessary to first provide a good definition what exactly this model is about. As there is a lot of literature found, this section will keep it short as we will refer the reader to the most interesting books and authors around platform models.

A platform is defined as a **business model**, *not a technology infrastructure that brings together producers and consumers for a certain product or service (Van Alstyne et al., 2016).*

The purpose is to facilitate interactions across many participants and establish a business transaction once a connection is established.

Figure 1. Platform Business Model visualized - Source: Van Alstyne et al.

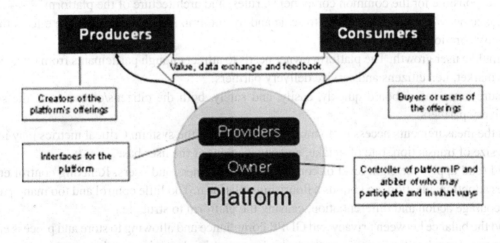

The platform business model itself does not own the products or even the means of production, but rather creates and facilitates the means of a connection between the producer and the consumer. Establishing the connection (i.e., matching demand and supply) is the key value driver as the two stakeholders can then create an interaction. These interactions could take the form of short-term transactions like connecting buyers and sellers for a taxi drive (e.g., Uber) or renting an apartment (e.g., Airbnb). However, the interactions could also focus on creating a longer-term social relationship (e.g., Facebook) or longer-term collaboration to achieve a shared outcome or learn faster together (e.g., Coursera).

Parker et al. (Parker et al., 2016) stated that platforms must perform three key functions to encourage a high volume of valuable interactions, which they summarized as **pull, facilitate, and match**. The platform must **pull** both enough producers and consumers to the platform to enable interactions between them. Next, the platform must **facilitate** interactions by providing them with tools and rules that make it easy for producers and consumers to connect and encourage the establishment of a transaction or service request. Finally, it must **match** producers and consumers effectively by using information about each to connect them in ways they will find mutually rewarding. If there is value in the match, then the transaction will be initiated.

The product is often dematerialized as it becomes a service (e.g., a virtual training, a taxi drive) that a consumer buys via the platform. After the transaction is finished, the material component (e.g., the taxi, the apartment) is not transferred to the consumer but stays with the producer.

In a platform model, value is created by the partners and the users rather than by the platform owner itself. The interaction between the participants is facilitated and managed by an intermediary or platform owner (Belleflamme, 2016). Platform owners have several key roles to perform. Some important ones below were adapted to a government context based an MIT article focusing on Financial Sector (Kansu & Parker, 2018):

- Setting up and managing the platform - make sure the architecture and functionality is open and transparent to the intended usage. Provide the infrastructure and make it performant, open, and scalable.
- Be the point of contact for all users - both the citizens, the developers of content, or the service provider.
- Be responsible for the common components, rules, and architecture of the platform.
- Manage network effects to be able to scale and for minimizing potential negative effects through quality curation.
- Stimulate user growth. The platform owner needs to attract enough participants from both sides of the market, i.e., citizens and service delivery partners.
- Ensure users can onboard quickly, easily, and safely, both the citizens/customer as the service delivery partners.
- Set the measurements necessary to monitor the health of the system. Critical metrics may include the size of transaction, rate of uptake, and engagement of the user base.
- Find (and apply) the right level of control over the partners and users. Relaxing control encourages adoption, ideation, and builds a flourishing platform. Too little control and too many partners discourage action and differentiation, causing the platform to struggle.
- Find the balance between privacy and GDPR compliance and allowing to store and process enough data to personalize the experience through - for example - behaviour analysis.
- Measure the success and failure of matches to ensure that users can find valuable interactions. Platform owners should use the data to understand what is missing or not working well on the platform and adjust.

The concept of a platform business is not a new phenomenon. According to the Oxford English Dictionary, the word "platform" has been used since the 16th century to denote *"a raised level surface on which people or things can stand, usually a discrete structure intended for a particular activity or operation."*

Somewhat surprisingly, the word has been used in an abstract sense for nearly as long. The dictionary cites examples from as early as 1574 in which "platform" refers to "a design, a concept, an idea, (something serving as) a pattern or model".

When we project this into the 20th century, one can think about the massive American shopping malls or even exhibition centres. The difference is that nowadays, platforms are increasingly global and supported by digital technology infrastructures that help to scale participation and collaboration.

In summary, the concept of using a platform to do business is not new. However, due to the fast technology advances, platform companies embracing technology are now disrupting and redefining the way goods are distributed and sold to customers as services instead of as a product.

1.3 Evolution of Government Service Models Towards the Platform Model

Lately, the term "**Government as a Platform**" emerged whereby technology is embraced, and government agencies make the transition to becoming a broker of services. Using technology for a new business model thus. However, the term is used in a variety of ways across the world. To understand the true meaning of Government as a platform and come to a standard baseline definition, we must first explain the history of the evolution of government services, mainly referred to as e-government.

Until the 1990s, most governments, like much of the business world, used information technology to automate back-office operations, with little emphasis on automating customer-facing functions such as information dissemination or service delivery.

Starting in the mid-1990s and with the rapid rise of the internet and computing power, the terms e-government emerged and referred to the application of the Internet and other information technology (IT) to provide government information and services electronically. It offered increased convenience to the public by making such services available 24 hours a day, seven days a week, coupled with the

Figure 2. Historical overview of government transformation – Source: Adapted from Gartner

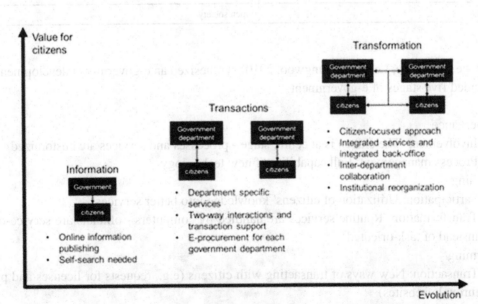

advantages of improved accuracy and reduced cost to the government, deriving from its requiring little or no direct interaction with a government employee (e.g., through the usage of websites).

Early 2000, Gartner (Christopher H. Baum | Andrea Di Maio, 2000) already published an overview of the desired evolution of governments service. This evolution depicted the transformation from one-way communication towards two-way interactions:

Behind these models are clear definitions (Kutlu & Sevinc, 2010) what the characteristics of e-government are, as shown in Table 1.

Table 1. Characteristics of e-government - Source: Kutlu et al.

Traditional State	E-State
Passive Citizen	Active Customer-Citizen
Paper-based communication	Electronic communication
Vertical/Hierarchical Structuring	Horizontal/Coordinated Network Structuring
Administration uploads data	Citizen uploads data
Workers response	Automated voice mail, call centre, etc.
Workers support	Self-support/expert support
Worker-centred control	Automated data update control
Cash flow/cheque	Electronic Fund Transfer
Standard service	Individualized/differentiated service
Fragmented/disrupted service	Inclusive/permanent/one stop service
High transaction costs	Low transaction costs
Inefficient growth	Efficient management
One-way communication	Multi-level communication
Subordinate relations	Participation relations
Closed state	Open state
Closed society	Open society

Around the same time, Lee et al. (Jungwoo, 2010) synthesized an e-government development model which provided five stages of e-government:

- e-governance
  - Involvement: The citizen is at centre stage - processes and services are customized
  - Process management: Full capability of new technology
- Morphing
  - Participation: Utilization of citizens' knowledge into better services
  - Transformation: Routine services are handled by computers - officials are service-oriented instead of task-oriented
- Reforming
  - Transaction: New ways of transacting with citizens (e.g., requests for licenses and payment through websites)

Figure 3. e-government development model – Source: Lee et al.

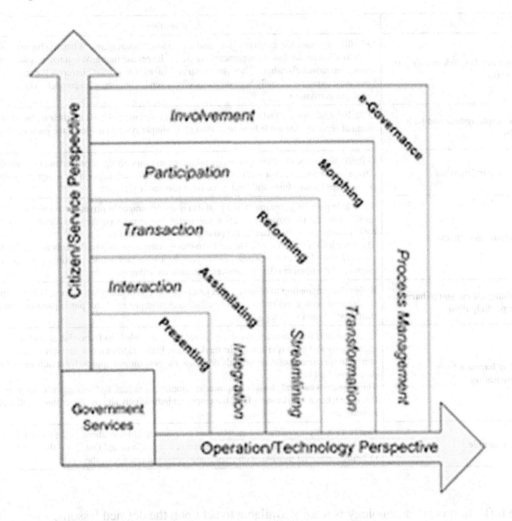

- ○ Streamlining: Technology makes processes more efficient
- Assimilating
  - ○ Interaction: Opportunity of interaction-based services (e.g., downloading forms)
  - ○ Integration: The information and processes in the organization are integrated with technology
- Presenting
  - ○ Simple presentation of information (e.g., on a website)

The topic of this research - platform business models - aligns well with the characteristics of **e-governance** and implies a further transformation of government organisation and services delivery.

Tim O'Reilly went one step further when he wrote his vision for **Government-as-a-Platform** or government 2.0 already in 2011 (O'Reilly, 2011). This vision was again driven by the rapid expansion of the internet, but also the availability of emerging platforms like Facebook and Amazon. To embrace technology and let governments make the most of technology, he defined seven (7) key lessons (Table 2).

Table 2. Key lessons for governments to embrace technology - Source: O'Reilly

Lesson	Explanation
Open standards spark innovation and growth	O'Reilly advocates the usage of open standards to foster innovations. When the barriers to entry to a market are low, entrepreneurs are free to invent the future. When barriers are high, innovation moves elsewhere. Then the author reinforces that vibrant platforms become less generative over time, usually because the platform vendor has begun to compete with its developer ecosystem.
Build a simple system and let it evolve in	O'Reilly pleads for the end of large, feature-filled programs, and calls for their replacement by minimal services extensible by others. Designing simple systems is one of the great challenges of Government 2.0.
Design for participation	O'Reilly states that the goal is to design programs and supporting infrastructure that enable citizen self-service and even citizen self-organization. The key question for designing a platform is then what architectures will lead to the most generative outcome.
Learn from your "hackers"	The whole point of government as a platform is to encourage the private sector to build applications that the government did not consider or does not have the resources to create. Open data is a powerful way to enable the private sector to do just that. The same lesson can be applied to platform-based systems when governments accept that the most creative ideas for how a new platform can be used would not necessarily come from the creators of the platform (i.e., the government) but the industry.
Data mining allows you to harness implicit participation	Rather than attempting to enforce better practices through detailed regulations, O'Reilly suggests the usage of open government data to enable innovative private sector participants to improve their products and services.
Lower the barriers to experimentation	O'Reilly states that governments should strive for open-ended platforms that allow for extensibility and even revision by the marketplace. Platform thinking is an antidote to the complete specifications that currently dominate the government approach not only to IT but to programs of all kinds. This requires a cultural change. Empowering employees to "fail fast" accepts and acknowledges that even when an experiment fails, the employee individually and the government will still learn something.
Lead by example	Finally, O'Reilly states that a great platform provider should stay ahead of the curve and allow time for the market to catch up to. If government is a platform and Gov 2.0 is the next release, O'Reilly states, "let us make it one that shakes up—and reshapes—the world."

The reflection is that technology is readily available to act upon the defined lessons.

OECD (OECD, 2019) provided a simple overview of the typical transitions that governments go through when using more technology. These are:

- **m-government** or mobile-enabled government. This is the period where most governments made their website accessible to mobile devices or where they create apps to be used primarily on mobile devices. The goal here is to adopt technology for easing access to citizen services.
- **t-government,** or transformational government is to realize public sector reform. Efforts are aimed to move beyond the e-government efforts of creating better service delivery for citizens and companies and realize public sector reform (Van Veenstra & Janssen, 2012). The goal here is to reform bureaucracy and start thinking from an outside-in perspective.
- **i-government** is all about integration and information sharing with an ecosystem that provides new functionality to improve services to citizens (given respect for privacy). Until now, this model was dominated by the supply of Open Data and the creation of an Open Data Portal, where the government provides access to free and machine-readable datasets. Lately, new technologies are

added to integrate functionality between different parties. Given this, governments can focus on their core tasks and provide reusable blocks of functionality to third parties to interact with the government functionality (e.g., ID verification). The goal is here is efficiency and doing more with less.

Like most companies, public administrations constantly seek to enhance their relationships with their customers or citizens. The motto for such efforts today is "**digital-first**". This implies that governments must undergo a paradigm shift in their support and uptake of digital opportunities. The shift happening across governments worldwide is from using technology to digitize existing procedures and services in search of efficiency gains (e-government) to using data and digital technologies to rethink and re-organize how governments deliver public value (digital government) to foster open, innovative and collaborative governance, as also mentioned above (Welby, 2019).

What about **e-inclusion**? Will a platform model not be restricted to those that are well connected to the Internet? By promoting a "digital-first" approach, governments may inadvertently create new digital divides by excluding those who cannot use online services. Muller researched this further and found evidence of a digital divide that reflected demographic and social differences between citizen groups (Müller & Skau, 2015). When implementing e-government services, citizens' acceptance is key to success. Age, level of education, attitudes, beliefs, and norms influence citizens' acceptance and adoption of e-government services according to the research of Muller.

Looking back, e-government was mainly a support instrument for analogous processes. To citizens, this was disappointing. There is an increase in citizens' expectations for effective, fair, and citizen-centric services. This requires a shift from inward, disjointed and process-oriented organizational structures to highly collaborative teams that seamlessly deliver the services (United Nations, 2018).

The 2018 United Nations E-Government Survey reported steady progress in improving e-government and public services provision online (United Nations, 2018). The European countries lead e-government development, with the Americas and Asia following closely. Belgium was classified in the highest category.

The UN also stated that the target is moving. In a digital government, governments now also need to incorporate the simultaneous proliferation of big data, artificial intelligence, data science, blockchain, robotics, and other emerging technologies. The challenge is that the speed with which technology is evolving, surpasses the speed with which governments can respond to and use technology to their advantage.

The report concluded that while e-government began with bringing services online, the future will be about the power of digital government to leverage technology to transform citizen services.

In 2019, OECD (OECD, 2019) gave the strong advice that **governments should build supportive ecosystems that support and equip public servants to design effective policy and deliver quality services**. That ecosystem enables collaboration with and between citizens, businesses, civil society, and others to harness their creativity, knowledge, and skills in addressing challenges facing a country.

However, let us not forget why governments are embarking on a platform roadmap, what is the value this model brings? Using a platform model is a way to enhance the citizens' well-being in the state they live and function, using technology as a basis to communicate and deliver government services. OECD (Welby, 2019) recommended that the focus therefore should be on benefits that are not only material, but that also reflected benefits derived from a different approach to government interactions with its citizens. This requires:

- **A responsive government**, considering the redesign of government and its end-to-end services model - using appropriate new technology - rather than being content with the existing architecture of the public sector.
- **A protective government,** encouraging efforts to create and distribute trust throughout social networks and the political discourse, certainly when using the platform model as the prime interface between governments and citizens
- **A trustworthy government**, delivering high quality and reliable services characterized by a desire to understanding their users and being open to challenges and feedback. This also requires transparency how the model operates and can be consumed by citizens

When we look back at the core functions and features of a platform, it was necessary to refine the list by adding core capabilities that are specific to a digital government service. Based on the work of Andrews (Andrews, 2019), the following additional features were proposed:

- **Identity and authentication**, for verifying the citizen, but also to log into systems that may be provided from outside of government but are connected to delivering (part of) the service.
- **Service analytics** across digital and non-digital channels to baseline the user experience and journey with government and identify what works through evidence. This is the basis for providing personalization service.
- A **consistent government web platform/experience** to manage and keep together relevant government information for service delivery, even in a personalized way (i.e., every citizen sees his unique page).
- A **Services register** in the form of a consumable catalogue of government services (e.g., grant allocation API).
- A **public eligibility and calculation engine**, to make the rules of eligibility and calculation available as an API for anyone to consume, including government agencies themselves. This will contribute to the transparency of the government.
- **Centralized Payment, notification, and verification services**, bringing together services that are common but traditionally duplicated across government.
- Support the "**Only Once**" principle whereby the citizen gives consent for his data to be reused inside the government over and above the current service request. This reduces the need for data-sharing and dramatically reduce the cost, processing, time, and indignity around many services today.

1.4 What Should a Government Business Platform Enable?

An important lesson governments can learn from a platform company is that successful platform companies are **created with intent** (IBM Government Industry, 2017). Platforms are not supposed to be created by just opening out what you have available, although many of these were found during this project. The current approach of applying platform thinking in government seems just to take data and open it out to the world. The difference with industry is that successful platform companies first create a strategy and a vision for the ecosystem in terms of what they can achieve by being on the platform. The next step is to provide the right tools and services that enable success for the ecosystem.

Translating the above into a government context, the following questions were asked:
What if citizens could opt-in for:

- Arranging all your request for healthcare assistance from home.
- Participate in enriching the government vision and influence spending budgets.
- A personalized platform that connects you seamlessly to all government services, regardless of origin (local, provincial, regional, federal, etc.) - one entry point - one dashboard.
- Personalized career opportunities with a bit of advice on enhancing your skills.
- Taking the next course or diploma in enhancing your (digital) skillset.
- Targeted home improvement features including a personalized quote from private companies that are certified by the government.
- Maximum self-service of ordering government services like permits (e.g., fishing permit) that were redesigned to provide you a good experience.
- Logging on to private companies e-services catalog as a citizen reusing the government log on credentials.
- Controlling who has access to what data and where all your data is used securely and under control of the government.
- Reporting any event in your neighbourhood (abandoned waste, pothole, safety) via one channel, even alerting your neighbours to a potential situation.

What if you -as a citizen - have full control over…

- Selecting that private partner that governments have certified to provide you personalized and fitting services.
- Providing feedback to those private partners on how well they performed.
- Self-servicing all your permits, subsidies, etc. in a way that is considering your case data (e.g., housing, electricity, waste, etc.).

What if the government provides you with…?

- Options to choose your healthcare providers yourself, based on a budget and the right to spend your allocated budget pending your insights.
- Optimized traffic conditions using the latest technology to steer the flow and provide you with alternative means to
- Targeted alerts on potential weather situations

Reponen (Reponen, 2017) stated that in a government business platform model, it is all about the citizen in the middle, as depicted in Figure 4.

So, by redefining government services across the silos, one comes to an integrated set of citizen services that is abstracted from the government organizational setup. Using the business platform model, we truly come with an outside-in driven approach, whereby governments design processes and workflows from the citizen's view, not the organizational view behind it. In platform ecosystems, the focus shifts from internal optimization to external interaction.

Figure 4. Gov. platforms models put citizens in the middle (Reponen, 2017)

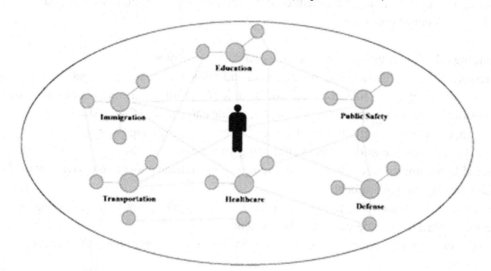

Table 3. Characteristics of a government service broker – Source: ISS & PWC

From	To
Citizen Under Control	Citizen in Control
Governing for citizens	Governing with citizens
Organization silos	Organization networks
Public sector organizations as big, all-in-one giants	Public sector organizations as small, flexible, purpose-driven entities
Government owning inputs and processes	Government and citizens owning outcomes
Defining and controlling input	Measuring outcomes
Forced cooperation based on enforcement	Mutual collaboration based on trust
Trust in the "strong leader"	Trust in each other, the "servant leader"

First, we must acknowledge that governments need to change from being a service provider to becoming a service broker or service facilitator(ISS & PWC, 2019). This means governments need to become a director of services, being the broker that connects the needs of the citizens to the right government and industry partners that together can help solve the need. Governments can then commission a part of the service delivery to the ecosystem it has helped set up (via certifications or audits). The article states some of the transition characteristics governments need to accept, shown in Table 3.

Given these insights for a fundamental transition of the services model of a government, it is now possible to determine how such a model should like and why it is different than current digitization programs. That was the key focus of this research.

As a summary, let us end with an example of a government / country that is applying all the above. Reference is made to Dubai and their vision of Smart Dubai 2021 (Smart Dubai Office, 2018), where the objective is to establish a digital, lean connected government build on technology. Its key features will be:

- **A government with zero visits**—eliminating the need to commute to and physically interact with the Government by providing 100% of eligible public services through digital channels and targeting full digital adoption.
- **Paperless, cashless government**, driven by cutting-edge, disruptive technologies, defining the Government of the future now.
- **Delivering optimized experiences** by connecting public services that target residents' and visitors' critical needs and major life events, saving time, and simplifying life.
- A Government powered by world-class city-wide **shared services and infrastructure** that drive significant efficiencies for the Emirate.

1.5 Problem Statement

Governments are struggling to adapt to the fast-changing realities of a digital age (Fenwick et al., 2019). Most governments are aware of digital disruption and have prioritized embarking on a platform journey, but scaling best practices remains a challenge. Even the top countries adopting Government as a Business Platform model still have significant room to improve and lag industry progress (Accenture, 2018).

Platforms represent the largest transformation since the Industrial Revolution (Accenture, 2016a; Kenney & Zysman, 2016), revolutionizing industry after another. However, until now, the platform revolution has transformed only a handful of industries, leaving some of the most important areas of our society, such as education, government, and healthcare, still unaffected, meaning that the revolution in these sectors is only beginning (Parker et al., 2016).

Even when governments are deploying a platform model, most countries are taking different routes along their journey(Accenture, 2018), leveraging the strengths and ecosystem of their nation (i.e., very local) first. Even the most advanced countries seem to struggle evolving to the role of ecosystem moderator from their traditional role of government as a provider.

In the article "State of the art in the use of emerging technologies in the public sector"(Enzo et al., 2019), the authors listed the lack of common IT standards, the (digital) skills shortage in the public sector, an unsuitable legal framework and the general absence of political support or strategic vision as inhibitors to use more technology in the public sector service delivery model. This resulted in the observation that (too) many countries have their national strategies for digital government.

In the EU, 'Digital Strategy' refers to the design and implementation of eGovernment strategies, roadmaps, or action plans needed to modernize public administrations at different levels that respond to the evolving societal needs and technological trends (DIGIT, 2019). Although the EU has an ambitious plan with the forthcoming adoption of a new Multi-Annual Financial Framework (MFF), which will have a large budget dedicated to digital transformation, most of the investments are still for laying the basics of (internet) connectivity. Few initiatives seem to be focused on adopting new business (platform) models.

If a government is to embark on establishing a platform model, there is a need for

a) Defining the best practices for opening government services on a platform.

b) Defining how governments can create value for all stakeholders in general, but citizens, in particular, whom will be using the platform business model to find relevant services

c) Defining how governments will define and manage an ecosystem of trusted partners that will provide a fair service to citizens

As a summary, research indicated that platform models are here to stay, and industry has taken a frontrunner role. However, governments are late in adopting technology in general and the platform model specifically. Governments also use different approaches, even inside one nation. If the needs, as mentioned above, are not well defined and adhered to, it will lead to ineffective and disparate execution, hampering the creation of value for (digital) citizen services and not leveraging the power of the ecosystem at all.

Based on this literature, the problem statement of this research was formulated as follows:

If governments do not design, run, and maintain a platform model according to a clear vision, supported by (political) leadership, using the latest technologies, and with a new culture in mind to innovating services, the resulting value will not be achieved, and citizens will refrain from using the platform model for government services.

Using the platform model cannot be a hit and miss for a government. Hence the importance of this research of defining the right practices to make it right the first time.

1.6 Research Questions

This research centred around the fact that governments are embarking on a similar transition as industry, i.e.,

- From a linear provider of services (i.e., you ask, you might get), over e-government (i.e., you ask online, sometimes governments decide in real-time) …

Figure 5. Linear way of working for government services - Adapted from Choudary

- Towards acting as the orchestrator of services, not the grantor of money and executioner of the service request.

Figure 6. Platform way of working for government services - Adapted from Choudary

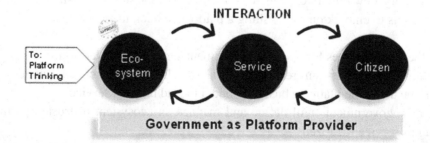

This implies setting up an ecosystem of partners that provide the right services to citizens, giving these partners the mandate to execute (part of) the services and allow citizens to drive their service requests online, anytime, anyplace.

Using the *hypothetico-deductive method* from the Design Science method(Johanesson & Perjons, 2014), the following key question emerged:

Will citizens be favouring to use a trusted and managed ecosystem for requesting and executing government services if the value is clear and governments vow for a transparent design of the platform where safety and privacy of the citizen' data is guaranteed?

This question was then turned into a hypothesis:

Citizens will be favouring to use a trusted and managed ecosystem for requesting and executing government services if the value is clear and governments vow for a transparent design of the platform where safety and privacy of the citizen' data is guaranteed.

In this hypothesis, the emphasize is on defining the practices for a platform model for governments to be successful, like:

- Setting up a trusted and managed ecosystem
- Being transparent in the design of the services and
- Creating value for all stakeholders
- Ensuring safety and privacy of the citizens' data

To design a properly functioning government business platform, we needed to answer the following research questions that act as predictions of the hypothesis:

- **Research Question 1 (RQ1):** What are some of the best practices for defining a Government Business Platform model for citizen services?
- **Research Question 2 (RQ2):** What are the criteria that constitute a well-governed design for a Government Business Platform model for citizen services?

In the next section, the approach to answering the research questions is detailed.

1.7 How Was Research Conducted?

This research project was the result of work done in a descriptive, exploratory, analytical, and creative way. Hence it was following a qualitative approach. A deductive process was followed, meaning that the problem statement resulted in the hypothesis, which was theoretic in the beginning. The hypothesis was translated in two research questions. Based on data collected in case organizations, findings were documented that led to an answer to the research questions and thus a conclusion on the hypothesis.

To obtain the necessary insights into the matter of platform models, the following methods were applied:

- Literature Research
- Interviews with domain experts (government, platform models, governance, etc.)
- Case studies

The definition of research in general as "*something that people undertake in order to find out things in a systematic way, thereby increasing their knowledge*" (Sauders et al., 2009) is reflected in the artifacts that were created as a result of the material studied.

In the next sections, more details are provided about each method used. The reason for this section is to let the reader understand how the necessary information and insights were obtained to come a reasonable underpinned answer to the research questions.

Based on the work of Aspers & Corte(Aspers & Corte, 2019), the research approach included the collection of empirical material like case studies, personal experiences from case owners, interviews with politicians, observational, historical, interactional, and visual texts obtained from various sources.

This qualitative research followed the definition of Aspers & Corte(Aspers & Corte, 2019), i.e., "*the iterative process in which improved insight and understanding to the scientific community is achieved by making new significant distinctions resulting from getting closer to the phenomenon studied*". In other words, as more insights became available, a next iteration of searching for relevant material was sometimes needed to come to a level of detail deep enough to answer the research questions.

1.7.1 Approach to Literature Research

Much of the research involved searching for and reading through literature. Literature formed the foundation upon which the artifacts are based.

This research project was not the first to discuss government business platforms and certainly not the first related to platform models in general and in commercial companies. It was therefore essential to collect and interpret literature in a systematic way to find out what the issues and opportunities for platform models were in general and in a government context.

The main references to books, papers and articles were found through the library search functionality from the AMS / University of Antwerp(Universiteit Antwerpen, 2020). The search functionality provided access to the Web of Science where articles were searched and read according to a specific keyword. A list of keywords that were used to search can be found in Table 4.

Next to the formal search through the Web (of Science) and library of the University of Antwerp, an additional type of literature was added. This included a selection of conference papers, reports from governments and the EU, reports from reputed consultancy companies, or expert opinions via blogs. These were found either via Open Access repositories like ResearchGate, Google Scholar, via Google Search or by recommendations from the experts that were interviewed.

As literature was processed, the references led to other papers. This is referred to as snowballing. Snowballing means using the reference list of a paper or the citations of the paper to identify additional papers(Wohlin, 2014). The above table was the starting point and was therefore not exhaustive. The Reference Section at the end of this research project represents all material used.

During the search, many articles were found that were not related to the subject of this research project but contained some of the keywords. The definition of platform also applies in other fields of science like biology, healthcare, nature, and others. These articles were manually separated from the selection.

Table 4. Keywords used in literature research - own work

Keywords used to Literature Selection (alphabetic order)		
Concept-Centric Search Terms	**Author-Centric**	**Supporting keywords**

General
• Platform Architecture
• Platform Brokering of demand and supply
• Platform Business model
• Platform Definitions, Types, and Characteristics
• Platform Infrastructure
• Platform Governance
• Platform Matching Process
• Platform Network Effects
• Platform Owner
• Platform Portal
• Platform Risk
• Platform Thinking
• Platform transition for traditional companies
• Platform Value
 	
Government Specific
• Government and Open Data Platform
• Government as a platform
• Government as Platform Provider
• Government Platform and improving citizen services
• Government Services Design
• Governments and New Technologies
	Books & Author Centric
	• Tim O'Reilly on government platform models
	• The Standish Group on reasons for the failure of IT projects
	• Geoffrey G Parker, Marshall W Van Alstyne and/or Choudary, Paul Sangeet with various articles and books like
	   o Platform Revolution,
	   o Platform Scale,
	   o Pipelines, Platforms, and the New Rules of Strategy
	• George Westerman, Didier Bonnet and Andrew McAfee in "Leading Digital"
	• Michael A Cusumano, Annabelle Gawer and David Yoffie, "The Business of Platforms"
	• COBIT2019 handbooks by ISACA
	• Steven De Haes, Wim Van Grembergen, Joshi Anant, Tim Huygh, "Enterprise Governance of Information Technology"
	
	Thesis material from
	• Sara Reponen, "Government as a Platform: Enabling Participation in a Government Service Innovation Ecosystem"
	• Lieselot Danneels, "Transforming Government: The way towards digital era governance."
	
	Note: exact references are added in the Reference Section
		• Agility
		• API Strategy and Standardization
		• Behaviour Analytics or Analysis
		• Best Practices
		• Capabilities
		• Case Study
		• Citizen Value
		• Critical Success Factors
		• Critical Success Factors for e-government projects
		• Critical Success Factors for IT Projects
		• Critical Success Factors for platform companies
		• Culture
		• Customer or Citizen Experience
		• Data-Driven
		• Ecosystem
		• Ecosystem Design
		• Ecosystem Moderator
		• Governance
		• Government Declaration
		• Government Policy
		• Innovation
		• IT Governance Frameworks
		• KPI
		• Leadership
		• Orchestration
		• Politics
		• Privacy
		• Processes
		• Risk Management
		• Self-Service
		• SLA
		• Strategy
		• Transparency
		• Value
		• Vision

1.7.1.1 Approach to Refining the Problem Statement

The first purpose of this literature research was to fully understand what the definition and characteristics are of a platform model, how governments were dealing with it, what the challenges and opportunities were, and where we could find examples of already operating platform models. This understanding was needed to formulate an exact problem statement.

The literature research followed the objectives that Budgen & Brereton defined(Budgen & Brereton, 2006), but were adapted to suit the specific nature of research towards platform models. These objectives were:

• To summarize existing evidence concerning platform models in general and related to platform models used in governments specifically.

- To identify what the current issues and opportunities were in current research about government platform models, to help determine where further investigation was needed.
- To help position the research activities that would be addressed in this research project.
- To examine how far the practices and design criteria for a government platform model were supported or contradicted by the available empirical evidence.

The reason for this approach was to ensure that the problem statement and research questions were based on objective analysis of reported issues and opportunities in literature. The insights resulted in the problem statement and research questions.

1.7.1.2 Approach to Answering the Research Questions

Literature research was specifically used to define an answer to research question 1:

- **What are some of the best practices for defining a Government Business Platform model for citizen services?** The purpose of the literature review was to get an overview of practices that influence success (or failure) of government platform projects. However, as there were few references found, the search criteria were extended with a combination of the supporting keywords, as mentioned in the previous table. For example, a specific filter was applied to search for criteria for success in IT projects, e-government projects, or platform related projects. Extra material from The Standish Group was used to find the common causes for failures or success for IT projects. The practices for the research topics were extrapolated using all this material.

For Research Question 2:

- **What are the criteria that constitute a well-governed design for a Government Business Platform model for citizen services?** To design the right governance model for business platform models for government, first research was done what IT Governance frameworks could be used. Weill and Ross (Weill & Ross, 2011) argued that in order to be effective and get the best value from IT investments, IT governance must be actively designed to the type of organisation. IT Governance cannot be the result of isolated mechanisms (e.g., steering committee, office of IT architecture, service level agreements) implemented at different times to address the challenge of the moment. De Haes & Van Grembergen(De Haes et al., 2020) build upon Weill and Ross in stating that value is the prime driver to define a governance model that supports the strategy. For that to happen, business and IT goals must be well understood and interlocked before defining a suitable governance model. In this research, the objective was to define a governance model that is fit for a government organization that is embracing the business platform model for citizen services. A one-size-fits-all approach was not possible, hence the assumptions on how the organization brings value to the citizen were the prime drivers for selecting the right processes. As the business platform model is relatively new to government, a specific set was to be designed.

1.7.2 Approach to Getting Expert Opinions Through Interviews

After creating the artifacts for the research questions, interviews were held with experts in applying platform models, in government and academics. There were two goals in doing this: first, to get an even better understanding of the problem statement. Second, to record the opinion of the experts versus the defined artifacts, i.e., both the practices as the prioritized list of processes to confirm that the research was solid enough to take it to case organisations.

People representing the following roles were interviewed as subject matter experts:

- PWC - Government Offering lead & COBIT expert – member of COBIT2019 review committee
- Ex-Chief of cabinet Belgium Federal Government
- Professor of Public Administration at the Department of Political Science of the University of Antwerp
- CTO Agency of Information – Flemish Government
- Senior Researcher IMEC - SMIT - Vrije Universiteit Brussel (specialty: Smart Cities)
- Chairman of Digipolis – the IT organization of Antwerp, Member of the City Counsel of Antwerp, Chairman of the commission for housing in Antwerp, Member of commission on Economy that includes digitalization

The interviews also captured soft elements and subjective opinions on politics, leadership, and perceived value to anticipate the success of platform models in government organizations. These helped set the priorities for the practices.

1.7.3 Approach to Case Studies

The artifacts created based on research were applied in several government organizations in Flanders (Belgium) to validate whether the defined practices were applicable and were – in some form or another - used in the organization itself.

Case research was done through interviews with leading representatives of the case organisations. First a good understanding of the vision, mission, and objectives of using the platform model was acquired. Then, a deep dive into the defined practices was done to understand why and how the platform model was established. An extensive elaboration on the organization' leadership, vision, strategy, and objectives in general was discussed, focusing how the platform model fitted in realizing the strategy. A series of validation methods were used to remain consistent across the organisations. This also allowed to baseline where an organisation stood and where it wanted to be in terms of applying the platform model.

2 Problems in government platform models

2.1 Introduction

Today, many definitions and interpretations for platform models exist for governments in literature. Some refer to a technology platform, whereby infrastructure is made available as a service to all entities of that governments. Some refer to a platform whereby citizen services are reinvented, and the execution is outsourced to an ecosystem of trusted third parties. In between these 2, there are many variations.

Platform models are primarily an industry given, whereby the commercial transaction is at the core of the interaction between the stakeholders. Governments have a different business model and thus it is confusing to copy the platform model as such to a government entity. Hence, many interpretations can be found in literature.

A baseline on government business platform models was defined as part of this research. The objective was to define a set of logically but interconnected building blocks that enable the improvement of citizen services. This meant connecting the technology, data, and functionality features of a platform to allowing a citizen service to be executed via a business platform model. This way, it was possible to compare the various government interpretations, map them on the established baseline and indicate what was needed to mature the roadmap to a citizen services platform model.

For both governments and enterprises, there are common obstacles blocking progress on digital transformation, including legacy applications and processes that involve face-2-face visits to a government counter (with often irregular opening hours). These hurdles are real and pose a challenge for transforming traditional business models and information legacy into a digital enterprise.

Some governments are beginning to engage in promising digital transformation implementation efforts that bring information, analytics, and devices together to persistently apply them to support their goals. Among which are to be faster, smarter, sensing, responding, and analytical in requesting and providing citizen services.

However, there is much work to be done. McKinsey (McKinsey, 2018) noted that digitization efforts in U.S. government agencies have typically fallen short of their potential. Despite their desire to improve performance with digital transformation, many federal and state agencies have been unable to realize the full potential of digital technology.

Governments' typical response is that the government is not a business and, therefore, cannot be run like one. However, technology is agnostic when it comes to organizations and industries, and citizens are digital. Citizen expectations are rapidly changing based on their everyday experiences that are rapidly becoming more fluid across devices and commercial entities (IDC, 2018). Governments must become digital natives. Hence, the increasing pressure on governments to adopt to the technology reality and expectations of digital citizen.

2.2 Levels towards a citizen services oriented "Government as A Platform"

As demonstrated before, governments evolved from using analogue means over e-government to digital government. In this topic, research was done how governments defined and implemented "**Government as a Platform**" (GaaP) in their nation. This research was used to define a baseline of levels towards a full citizen services platform model.

Many governments are still in the early stages of formulating their response to the platform economy. The reason is that governments are still in the stage of attempting to fully grasp the phenomenon and its implications for all parties involved (Lenaerts et al., 2017). This was confirmed by Accenture (Accenture, 2018) in their GaaP Readiness Index, where it was demonstrated that most governments were taking a different road to implement the platform model.

A study from the European Economic and Social Committee (de Groen et al., 2017) found that government responses to the platform economy are generally narrow in scope, reactive, and concentrated on dealing with side effects rather than attempting to reap potential benefits.

To define common or universal practices, it was first needed to standardize the many definitions used for "Government as a Platform". Based on a synthesis of literature and research into many government's approaches, the following logical view was created to map most governments against (Figure 7).

Figure 7. Functional view of government as a platform – Own Model

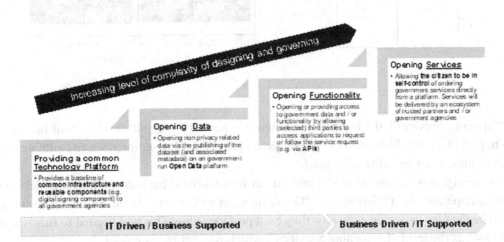

The above picture was inspired by the categorization proposed by Smedlund et al. (Smedlund et al., 2018) on the types of service platforms. The authors stated that the type of platform depends on the IT investments and the openness of the platform. Smedlund stated that IT investments differentiate the platforms from each other. The more investments are done in front-end technology to enable a good User Experience (i.e., UX), the more attractive the platform will become. Others focus on investing in back-end technology to optimize processes to match producers and consumers. Examples here include Amazon, or Alibaba.

The other dimensions Smedlund considered were the open or restricted entry to the platform. An open platform is likely to attract more transactions, which increases the complexity of the network of relationships in the platform ecosystem. Restricting entry to the platform lowers the need for communication and coordination. This results in Figure 8.

From these definitions, the "Providing a common technology platform" model fits the left below quadrant as this is internally focused on government agencies, and usage of the infrastructure is restricted to government agencies.

The "Opening Data" usually refers to the provision of Open Government Data and is always open to anyone. Open Data platforms provide access to the data and are typically requiring the IT department to set up back end platform to collect, prepare, and publish the datasets.

As one can see, "Opening Functionality" was split in 2. For functionality that is only available to government agencies, the internal label was given. All functionality that is opened to the public was labelled externally.

Figure 8. Four types of service platforms – Source: Smedlund et al. & Own work

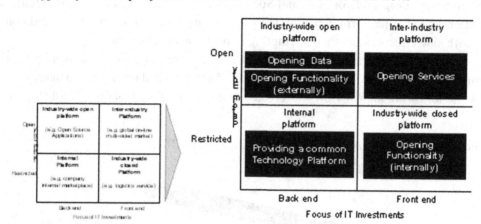

For "Opening Services", the top right quadrant is a perfect fit, as the services will be open to the ecosystem providers. The IT investment is focused on providing the user interface to facilitate the transactions and, thus, front-end citizen focused.

As a summary, this section defined a baseline of four levels of business platform models that governments can operate, often referred to as "Government as a Platform". In the following sections, these levels will be detailed to demonstrate how they each provide value. This was needed to firmly scope the research towards the level of "Opening Services", which was our target model.

2.2.1 Level 1: Providing a Technology Platform

2.2.1.1 Definition

Research indicated that most governments are using the platform model internally to standardize and optimize the usage of common IT Services (e.g., Cloud, Application Development, IT advisory, etc.). In the early days, this was often referred to as a Shared Services platform. However, with the addition of extra services like advisory or business expertise, the offering has been extended to include a wide range of both business and IT services.

The platform provides access to a concentration or consolidation of functions, activities, services, or resources into one stand-alone unit that can be ordered and delivered directly to the requesting agency.

The provider and clients enter partnerships based on service-level agreements (SLAs) on a fee basis, in much the same way a company agrees with its customers. This arrangement implies that the client could potentially go elsewhere for the service and thus pushes the provider to offer the best possible product at best possible cost, much like a competitive open market (Burns, 2008).

The main drivers are standardization and cost optimization for all stakeholders, while the consumers (i.e., government agencies) are being able to get access to professional services from preselected third-party private partners to realize the agency's IT goals. This facilitates the procurement and usage of external expertise and technology.

Figure 9. Technology-focused platform model – Own view

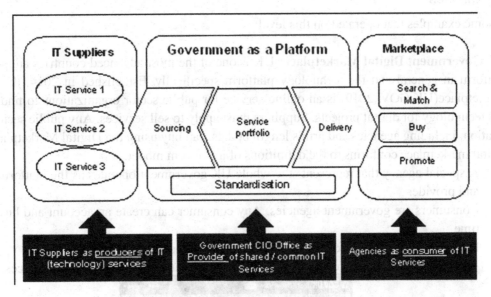

This model is well established, and success was reported as high. Already in 2008, 74% of governments investigated by Burns (Burns, 2008) reported success and - at least - achieving the goal of cost savings and efficiency.

2.2.1.2 Platform Roles

If we map this approach to the defined platform model, the following platform roles were defined in Table 5.

Table 5. Roles in the Technology Platform Model – Own View

Role	Who	What
Producers	IT Suppliers	• Act as producers of services • Offer their services to the government agencies • Usually went to a compliance process to meet standards and government regulations
Platform Provider (and regulator)	The Government - usually through a centralized department like the CIO office	• Provides the platform where the service portfolio is displayed. • Manages the ecosystem of IT suppliers and have a procedure in place to select the right IT Suppliers (e.g., procedures to negotiate prices or SLA or compliance to specific government regulations) • Guarantees to the agency a correct and transparent compliance to the overall government standards (e.g., architecture, usage of methods, SLA, penalties, etc.) • Let competition play through sourcing similar vendors for the same services, although this is never a free market because of the specific imposed rules to comply
Consumers	Government agencies	• Act as consumers of the services offered on the platform • Can order and consume directly • Cost & Value are clear • Can provide feedback through formal channels (usually not available as a platform feature) • Provide input to central CIO office what new services would be desirable or needed

2.2.1.3 Examples

Hereby some examples that operated on this level:

- **UK Government Digital Marketplace:** UK is one of the most advanced countries adopting the platform in general and the technology platform specifically. Established in 2014, the Digital Marketplace(UK.GOV, 2019) is an online service for public sector organizations to find people and technology for digital projects. Suppliers must apply to sell services. Any public sector organization, including agencies and arm's length bodies, can buy using the Digital Marketplace. The digital marketplace conforms to the definitions of a platform model:
  - o A special agency that represents the whole UK government operates as the platform owner and provider
  - o Consumers are government agencies. Any consumer can create an account and buy at any time
  - o Producers are the selected suppliers to sell (IT) services
  - o There is a well-defined governance model on how to apply and how to buy services.

- **Belgium Federal G-Cloud Model:** At the Belgium Federal Level, the G-Cloud uses a similar approach as in the UK but is, at present, focusing on providing Infrastructure as a Service (SMALLS, 2019).

- **Estonia Digital Government Platform:** Estonia is seen as an overall example of a digital government. As part of this overall strategy, a technology platform has been rolled out recently (RIKS, 2019).

- **Flanders:** Flanders is using an external supplier to manage a predefined set of IT Services that can be ordered by any agency in Flanders, local, or regional (HFB, 2019).

2.2.1.4 Governance Basics

Table 6 provides a summary of the processes that were reported in the literature to successfully set up and manage this type of model (Cheung, 2014) versus an own mapping of corresponding COBIT2019 processes.

This mapping will be reused further when we define the practices and map these to the set of corresponding management control processes.

2.2.1.5 Relevance to this Research Project

This model is added for completeness but will not be further investigated in detail. The model was relevant to understand the specific dynamics when the complexity of the services increases. Most other services will rely on robust IT infrastructure services to build upon as otherwise, the costs for creating a platform per agency would be enormous.

2.2.2 Level 2: Opening Data

2.2.2.1 Definition

We now concentrate on the first level that externalizes government information. Data is the product that is being offered by government agencies to public. Providing data required more than just technology to deliver, as we will explain below.

Table 6. "Technology Platform" - Mapping of processes reported in literature versus COBIT2019 - Own work

Reported process	Corresponding COBIT2019 Process
• Customer-focused strategic leadership • Strategic alignment aided by separating strategic from operational concerns and avoiding charge-back models of cost recovery	• APO02 - Managed Strategy
• Risk management by clearly defining service delivery expectations across all participants	• APO12 - Managed Risk
• Performance management via metrics to prove the cost savings gained by Shared service delivery as well as issue resolution	• MEA01 - Manage Performance and Conformance Reporting
• Control and legislative compliance, for issues such as privacy and accountability	• MEA03 - Managed compliance with external requirements
• Relationship management, which addresses transparency, communications to participants, and attentiveness to the culture.	• APO08 - Managed relations
• Transformation management, including skilled staffing, defining transition timelines, and managing the growth and scope of the SSC	• BAI05 - Managed Organizational Change
• Value management, to deal with on-going funding for services (who pays for what and in what proportion?)	• EDM02 - Managed Benefits delivery
• Not Reported - added	• APO05 - Managed Portfolio • APO10 - Managed Vendors • APO03 - Managed Enterprise Architecture

The background and reason for adding this level was that governments are working on programs to be more transparent by - amongst others - making their data freely available in a machine-processable format for other parties to reuse and build applications. This is referred to as **Open (Government) Data**. The main idea behind this initiative is to stimulate a new data-driven economy that lets commercial companies use these datasets (often for free, sometimes for a fee) to develop applications and make money out of these.

To achieve this goal, governments all over the world created data portals where one can search and find the datasets. These portals bring producers and consumers together to obtain (access to) the dataset. An open data portal uses software that includes features such as searching, metadata support and management, basic visualizations, user management tools, data publishing, data storage capabilities, and natively exposed API support.

Behind the scenes, government agencies set up procedures to produce quality datasets and publish these on the portal. Governments also set up portals to publish the datasets and even build mechanisms to push the datasets to higher levels of governments, thus aggregating all datasets. For example, in Flanders, datasets from cities or local municipalities are automatically promoted from the local level to the Flanders Open Data portal, from there on the Belgium portal and ultimately to the EU Open Data portal.

The main drivers for this level were compliance to EU standards that ordered all non-privacy related data to be, by default, made available through an Open Data portal, either via direct download of the dataset or via API (EU, 2019).

In this research project, the features of the open data process and portal were mapped to the definition of "Opening Data" in the platform model.

Figure 10. Illustrating the basics of Open Data - Own work

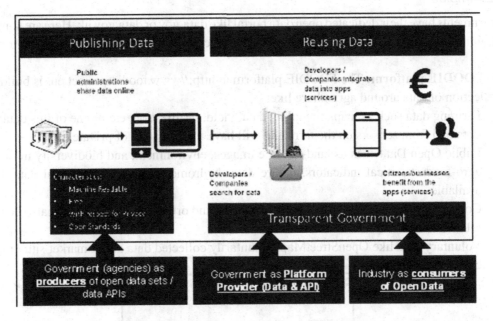

2.2.2.2 Platform Roles

If we outline the "Opening Data" approach to the defined platform model, the platform roles were defined in Table 7.

Table 7. Roles in the "Opening Data" Platform Model – Own View

Role	Who	What
Producers	Government agencies	• Act as producers of data sets that are registered on the Open Data portal • Offer either direct access to the data or via an API • Data meets the criteria of Open Data (i.e., Machine-readable, including metadata, etc.) • Data is of high and consistent quality
Platform Provider (and regulator)	The Government - usually through a centralized department like Open Data or Information Department	• Provides the platform where the Open Data sets can be searched and obtained • Publishes and forwards data catalog entries (with the associated metadata) to other aggregating portals or platforms • Ensures the data published is of good quality and has the right metadata • Ensures there is a feedback loop to contact the data publishing agencies
Consumers	Industry or individual consumer	• Search the data on the government provided generic portal or specific thematic focused portal (e.g., food data, health-related data, weather-related, etc.) • Adhere to the license model that is linked to the dataset • Access or retrieve the data via direct download of the dataset or using the provided API • Are free to use and reuse the data in any application and sell this off to consumers

2.2.2.3 Examples

Most governments have their dedicated open data portal like data.gov or data.gov.uk. Here are some cases of less straightforward open data portals that use a thematic approach to bring the right data together:

- **The FOODIE platform:** The FOODIE platform at http://www.foodie-project.eu/ is building out a collection of data around agriculture like:
  - Farming data such as maps, sampling data, yield, fertilization, etc. Some of this data will be obtained from sensors on the farm and will have the character of private data.
  - Public Open Data such as land satellite images, environment, and biodiversity information, agro-food statistical indicators, nature data, hydrometeorological data, soil data, etc. are available
  - Commercial data, mainly from satellite images, and orthophotos, but it could also be market-related data.
  - Voluntary data like OpenStreetMap, voluntarily collected data about market situation, agriculture production.

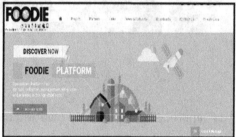

- **ENERGYDATA.INFO:** ENERGYDATA.INFO is an open data platform providing access to datasets and data analytics that are relevant to the energy sector. ENERGYDATA.INFO has been developed as a public good available to governments, development organizations, private industry, non-governmental organizations, academia, civil society, and individuals to share data and analytics that can help to achieve the United Nations' Sustainable Development Goal 7 of ensuring access to affordable, reliable, sustainable, and modern energy for all.

- **Global Health Observatory:** The Global Health Observatory at http://apps.who.int/gho/data/ node.home is the World Health Organization's gateway to health-related statistics for more than 1000 indicators for its 194 Member States. Data are organized to monitor progress towards the Sustainable Development Goals (SDGs), including health status indicators to monitor progress towards the overall health goal, indicators to track equity in health indicators, and indicators for the specific health and health-related targets of the SDGs.

- **World Bank:** The World Bank's goal is to decrease the percentage of people living with less than $1.90 a day to no more than 3 percent by 2030. The goal is to promote income growth of the bottom 40 percent of the population in each country. The databank at https://databank.worldbank.org/home.aspx provides access to indicators and projects that under the World Bank's supervision.

2.2.2.4 Governance Basics

Few research articles provided insight into the practices or preferred processes for a government open data portal. Zuiderwijk et al. (Zuiderwijk, A., Susha, I., Charalabidis, Y., Parycek, P., Janssen, 2015)

Table 8. "Opening Data" - Mapping of processes reported in literature versus COBIT2019 - Own work

Reported process	Corresponding COBIT2019 Process
• Legislation, regulation, and licenses	• MEA03 - Managed compliance with external requirements
• Strategy and political support	• APO02 - Managed Strategy • EDM05 - Ensured stakeholder engagement
• Management support and publication processes within governmental agencies	• APO09 - Managed Service Agreements
• Training of and support for civil servant	• BAI08 - Managed Knowledge
• Evaluation of the open data initiative - define and report on metrics of success	• MEA01 - Manage Performance and Conformance Reporting
• Sustainability of the open data initiative	• APO14 - Managed Data
• Collaboration	• BAI08 - Managed Knowledge
• **(Architecture of) Open data platforms, tools, and services**	• APO03 - Managed Enterprise Architecture
• **Accessibility, interoperability, and standards (e.g., data modelling standards, metadata, API)**	• APO14 - Managed Data

provide empirical data from several case studies that fit this research. This data was extended with an own mapping of corresponding COBIT2019 processes (Table 8).

2.2.2.5 Relevance to this Research Project

This model was relevant as it was the basis for interacting with an ecosystem of partners that directly uses the data provided by the government. In other words, this is the first level of "Government as a Platform" whereby data is provided to an ecosystem to create new insights and/ or applications. This ecosystem can consist of internal stakeholders to the government, but also external to the world. This last model is referred to as open data.

An open data portal adheres to almost all features defined earlier, except that the default is free delivery of the data set. Only when well documented and covered by the right license agreements, governments are allowed to ask a nominal fee - i.e., compensation - for the production of a specific dataset, as defined in the EU PSI (EU, 2019).

Providing data sets for download or access via API is the most basic way whereby government externalize data for reuse. The lessons learned from this model are thus directly applicable when we move up the stack towards opening functionality and services.

2.2.3 Level 3: Opening Functionality

2.2.3.1 Definition

This is the level whereby governments are opening (part of their) functionality to the external world. Opening functionality can be in the form of basic processes like identity verification or providing access to specific government applications. In both cases, this is done through the provision of APIs.

Figure 11. "Opening Functionality" – illustration of platform roles – Adapted from Internet Sources

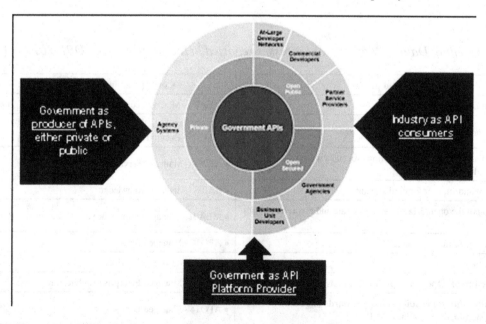

APIs are small parts of functionality (e.g., calculating a benefit payment or checking the status of an application) that are loosely joined by open standards. It is a way for an organization to expose defined assets, data, or services for consumption by a selected audience of developers, either inside or outside the organization. APIs are simple for application developers to use, access, understand, and invoke. APIs can thus make it easy to use government services or access information. Private-sector developers can access government APIs to provide value to citizens and potentially earn some revenue, although monetizing APIs is not a standard benefit in the government sector.

APIs represent a break with the traditional ways the work of government has been organized. Traditionally, governments are organized into vertically integrated thematic agencies (environment, benefits, tax, driving, etc.). Next comes the layer of hierarchically and geographically organized government (local, municipal, provincial, regional, federal government). APIs mean agencies at any level of government may become responsible for operating APIs used by services outside the walls of their organization, and potentially outside of government altogether. This approach requires a change of perspective, from organizations thinking about their mission, towards thinking about where they fit in the broader purpose of government and their role in supporting society. Nevertheless, it also creates new power dynamics(Pope, 2019a).

Governments started providing APIs to standard functionality like citizen identification (e.g., e-ID). Both governments and private companies can use this functionality to identify the person logging into the system. Next in line was providing access to open data via APIs as we have seen in the previous level. From there on, governments provided (select) access to functionality that resides in applications.

However, APIs — and the investment required to support them — must be positioned as an enabler of outcomes, not as a result. Gartner (Lacheca, 2018) predicted failure if strategies or business cases for investment in APIs or ecosystem platforms are perceived as abstract technology investments.

2.2.3.2 Platform Roles

If we map the "Opening Functionality" approach to the defined platform model, the following platform roles were defined:

Table 9. Roles in the "Opening Functionality" Platform Model – Own View

Role	Who	What
Producers	Government agencies	• Create and make APIs available, either for internal or external reuse.
Platform Provider (and regulator)	The Government - usually through a centralized department like Open Data or Information Department	• Provides the platform where the APIs can be searched, and instructions on how to use them • Provide standards to the producers how an API needs to be built and deployed • Ensures there is a feedback loop to contact the API publishing agencies
Consumers	Industry or individual consumer	• Find, select and integrate the API in their business process • Are free to use and reuse the API in any application and sell this off to consumers

Example Cases of Opening Functionality via APIs

Below are a few examples where governments are providing support functionality via APIs:

- **Gov.UK Verify me:** GOV.UK Verify is a secure way to prove who you are online. When you use GOV.UK Verify to access a government service, you choose from a list of identity providers (also called 'certified companies') that the government has approved to verify your identity. This component is an example where the UK Gov has opened functionality (via an API) to a selected ecosystem of trusted partners. You can choose which partner will do the verification on behalf of the government. It is offered as functionality and is often embedded in other applications.

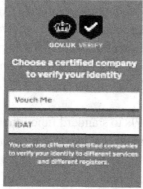

- **Belgium e-ID:** The eID is an electronic proof of identity (with chip) with which you can carry out electronic transactions. You can use the eID for:
  o Identification of different authorities
  o Signing electronic documents
  o Securely logging in to online public services
  See https://eid.belgium.be/en/what-eid

- **IndiaStack:** India Stack describes itself as "a set of APIs that allows governments, businesses, start-ups, and developers to utilize a unique digital Infrastructure to solve India's hard problems towards presence-less, paperless, and cashless service delivery.".
  See https://indiastack.org/about/

2.2.3.3 Specific Examples: Smart City Platforms

When we talk about APIs and Governments, one of the most visible areas where this is already applied, is the area of Smart Cities, often also referred to as **City Platforms**.

Desdemoustier and others (Desdemoustier et al., 2019) state that today the definition of Smart Cities is still fuzzy. They proposed to define it as the promotion of technologies and mostly IT into city infrastructure to increase effectiveness and efficiency in the city. The construction of Smart Cities is summarized by the deployment of IoT, sensors, networks, and intelligent systems to improve the efficiency of local territories.

However, the concept is broader than applying technology. In 2011, Nam and Pardo (Nam & Pardo, 2011) stated that a Smart City is layered into three categories of core components. These components stand for:

- **Technology** (infrastructures of hardware and software),
- **Human** (creativity, diversity, and education) and
- **Institution** (governance and policy)

Smart City platforms are being developed extensively in sectors such as transportation (including traffic management), waste, and energy. Lately, a new layer (or focus area) has been added: **Sustainability**, whereby Smart Cities promote a green economy and a high social awareness, including quality of life and place.

One of the drivers behind smart cities is that the world population is concentrating in cities all over the world. This puts enormous stress on public services like traffic, housing, energy, waste management, and others. With the use of new technologies, smart cities aim to improve the citizen's life in the cities and make cities ready for more growth. However, most of the solutions currently being developed, focus

Figure 12. API Maturity State - Source: IDC

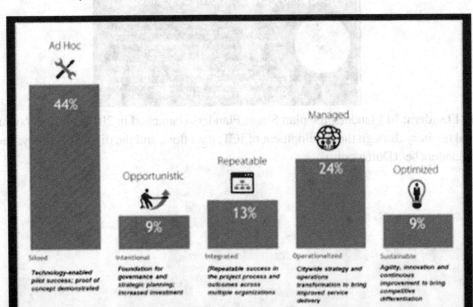

on a specific domain, target a particular problem, or were developed from scratch, with little software reuse. Since they do not interoperate, they lead to duplication of work, incompatible solutions, and non-optimized resource use (Santana et al., 2017).

Most Western European Smart City initiatives are still in a low maturity mode. Based on the IDC maturity model, a study in 2019 revealed the following evaluation (IDC, 2019) (Figure 12):

With 44% evaluated as (still in) ad-hoc status, IDC stated we have a long way to go before reaching a mature status. This is due to the lack of standards and convergence of initiatives, even in one country, despite the efforts of the EU to standardize the architecture. This is worrying as we will link this model as a cornerstone for "Opening Services", which will be the subject of this research project.

Often, city platforms link to Open Data platforms as introduced in the "Opening Data" section before. Providing Open Data (via APIs) is indeed a large part of a smart city platform. Still, we extend the scope to APIs that directly link to government functionality or applications.

For the sake of this research, we will define a city platform as **a marketplace where a collection of APIs is made available by governments to developers who then turn these into applications**. As APIs link to functionality, this fits the defined label of "Opening Functionality".

Examples include:

- **City Platform as a Service (CPaaS):** The main goal of this project is to develop a City Platform as a Service (CPaaS) that can be federated to support regional or even global applications, and that forms the basis for a smart city data infrastructure. Technical challenges that need to be addressed include data provenance, data quality, adaptive privacy levels, policies, and adaptive processes for distributing and deploying processing intelligence to the cloud or the edge. The CPaaS.io project is jointly funded by the European Commission (grant agreement n° 723076) and NICT from Japan (management number 18302) See cpaas.bfh.ch/

- **Smart Flanders:** In Flanders, the plan Smart Flanders (launched in 2017) aims to "smarten" the regional territory through the development of ICT, data flow, and the digital economy. See https://smart.flanders.be/ (Dutch only)

- **API.Store Brussels:** This site provides several APIs to connect to government services in Brussels like list and search POI venues, fixmystreet, GeoLocalization, and schedule of events. See https://api.brussels/store/ The site also explains well how the strategy has been set, how this has been translated in a governance mechanism, and what the objectives are. See http://www.smartcity.brussels/the-project-strategy#

- **Flemish Government - Agency of Information:** The Flemish Government is in its early days to provide APIs to external parties. The Agency of Information is leading the pack with delivering a set of standard APIs into basic registers, geolocation, and others. Both the various agencies of the Flemish Government as external parties are encouraged to use these standardized APIs. See https://overheid.vlaanderen.be/webdiensten-ons-api-aanbod (Dutch Only)

- **MAGDA: Maximum Data Sharing between administrations (Flemish Government):** The MAGDA (Maximum Data Sharing between Administrations and Agencies) platform provides a service-oriented data exchange infrastructure for accessing base registries of citizen and enterprise data, both at the regional and the local level and increasingly at the federal level. The platform takes care of authentication & authorization of data users and filtering & logging of data access, to comply with privacy protection regulations with regards to personally identifiable information. A service-oriented infrastructure is the only feasible technical means to interconnect different base registries and to allow for the exchange of authentic data between these base registries and the applications that want to use that data. See https://overheid.vlaanderen.be/magda (Dutch Only)
  Note: This example could also have been classified under the "Opening Data" label, but as it also processes data before returning an answer (e.g., anonymizing specific fields), this fits better the "Opening Functionality" section.

- **ACPAAS: Antwerp City Platform as a Service:** The City of Antwerp has the ambition to become a top player among the smart cities and even set the tone at the European level. ACPAAS is the name of the project to create a platform with a set of components (amongst them a set of APIs) that can be used internally to optimize the use of technology, but also externally to improve citizen services. See https://acpaas.digipolis.be/nl/about (Dutch only)

  Note: this case will be further investigated in the "Opening Service" category as well, as it is an excellent example of where the strategy is to go beyond offering APIs but also includes optimizing services to citizens (of Antwerp).

The Antwerp City Platform-as-a-Service

- **Traffic Management - Talking Traffic platform (NL):** Traffic and transport are becoming increasingly intelligent. Information technology is already able to improve the use of road capacity. By approaching mobility issues ever more smartly, journey times can be further shortened, and traffic flows will be enhanced, resulting in lower government spending. Talking Traffic is a step towards fewer road traffic incidents; an adequately warned, alert road user can focus more on the traffic and make the right choices at critical moments. Agreements have been reached with all partners to ensure that Talking Traffic services make the maximum possible contribution to reducing the number and improving anticipation of unsafe traffic situations See https://www.talking-traffic.com/nl/

2.2.3.4 Governance Basics

Own research on API platforms revealed a strong focus on technical aspects. That is why we also focused on researching governance basics on Smart City Platforms, as there was an abundance in reported success factors and focus process areas. Table 10 is a selection of this research:

Table 10. "Opening Functionality" - Summary of reported Success Factors / processes – Own work

Source	Success Factor / Processes reported
IBI (IBI, 2017)	• Stakeholder benefits • Engagement & buy-in • Regional alignment with a community focus • Strategy momentum and foundational initiatives • Clarity • "Dust-proofing" the strategy • Lessons learned • Urban integration • Performance indicators • Creating a lasting smart city culture
Harms (Harms, 2016)	• Define a clear vision • Focus on humans instead of technology • Focus on a specific topic • Develop a city-wide smart strategy • Make use of a smart city strategy framework • Bring Local government, businesses, knowledge institutes and citizens together
IDC (Yesner & Ozdemir, 2017)	IDC created a maturity model for Smart Cities. In there they defined the following process measures across five dimensions and 19 success areas, ad depicted below:
API Governance and Management **(IBM, 2018)**	Business Governance • Defining and managing the API portfolio & Roadmap • Privacy - ensuring Consumer / citizen privacy • Legal considerations - ensuring proper usage of government assets, especially when assets are being used by the ecosystem of (non-government) partners • Measurements - establishing success criteria, metrics to be collected and implementation of the gathering and communication of these metrics • Entitlement and Enforcement - who can use what • Monetization - although one would think this is not relevant to governments, at least the policy for users must be clear Technical Governance • API Architecture & Standards • API Policy enforcement - A central point where policies are created and enforced. Usually, a team is created that handles API governance across all government agencies • API Taxonomy • API Lifecycle Management • API Operations • API Technology (e.g., Gateway) • API Register / Catalogues • API Documentation (especially for external users) • API Security (authentication, encryption, anonymization, thread protection) Why? • API governance helps save time and money because it enables consistency across APIs, allows components to be reused, and ensures that APIs are built proactively to achieve specific goals and bring value to the business. • API governance also helps companies make intelligent decisions regarding API programs and establish best practices for building, deploying, and consuming APIs
API governance in public and private sectors **(U.S. Department of Veterans Affairs, 2018)**	• Leadership - Providing leadership for teams when it comes to APIs. • Innovation - A focus on innovation using APIs across the organization. • Communication - Facilitating communication across all teams and projects. • Advisory - Acting as an advisor to existing leadership and management. • Strategy - Helping existing teams develop, evolve, and realize their plan. • Success - Focusing on helping existing teams be successful when it comes to APIs. • Architect - Bringing a wide variety of software architectural skills to the table. • Coaching - Being a coach to existing teams and decision-makers across the organization.

When we add a mapping of the above to COBIT2019 processes, this results in Table 11.

Table 11. "Opening Functionality" - Mapping of processes reported in literature versus COBIT2019 - Own work

Reported process / Success Factor	Corresponding COBIT2019 Process
• Define a clear vision • Strategy • Leadership	• APO02 - Managed Strategy
• Culture	• APO07 - Managed Human Resources • BAI05 - Managed Organizational Change
• Technology Architecture • API Management	• APO03 - Managed Enterprise Architecture
• Defining and managing the API portfolio & Roadmap	• APO05 - Managed Portfolio
• Privacy - ensuring Consumer / citizen privacy • Entitlement and Enforcement - who can use what	• APO12 - Managed Risk • APO13 - Managed Security
• Legal considerations - ensuring proper usage of government assets, primarily when assets are being used by the ecosystem of (non-government) partners	• APO12 - Managed Risk • APO13 - Managed Security
• Measurements - establishing success criteria, metrics to be collected and implementation of the gathering and communication of these metrics	• APO09 - Managed Service Level Agreements
• Monetization	• APO06 - Managed Budgets and Costs
• Data	• APO14 - Managed Data

2.2.3.5 Relevance to this Research Project

Defining and setting the basis for a government API strategy is the fundament for having third parties interact with or use government functionality in a controlled way. Standardization may be well established for supporting functionality like e-ID. Still, research showed that the standardization of APIs in governments (or smart city projects) has a long way to go.

Status reports from IDC (Yesner & Ozdemir, 2017) showed that governments are struggling with setting this up correctly and are often not aligned in the roll-out of APIs and platforms, resulting in many disperse local initiatives and using individual protocols. Governments argued they cannot wait for the standardization to happen but need to book results now. This is a catch-22 situation.

The EU realized something must be done about standardization and commissioned the "APIs4D-Gov —digital government APIs: the road to value-added open API-driven services' project in 2018 to tackle this. In the interim report (Santoro et al., 2019), the authors provide an overview, but this project is far from finished yet.

When we move up the stack and open government applications even broader through APIs (e.g., requesting a grant or scholarship), governments will have to standardize the outward and inward architecture to make this efficient and effective. If the industry does not find the APIs attractive, they will not incorporate them in the services that they will develop to assist in the service execution.

Governments need to play a leading role in standardizing the technology and set up the platform to attract both citizens and industry partners. So, getting the functionality layer correct (through the provision of well-architected and monitored APIs), is seen as a prerequisite for the next tier up, i.e., opening services.

2.2.4 Level 4: Opening Services

2.2.4.1 Definition

As governments are familiar with technology platforms, have successfully opened data via a platform, and demonstrate how to offer access to functionality (via APIs) to external users, it is time to raise the stakes one more level.

In this next and ultimate level, the government will now externalize the execution of the services to an ecosystem of partners that use (parts of) the functionality offered by governments. Governments will focus on facilitating the transaction rather than executing it.

This is where the true definition of a platform business model kicks in, where the government is the platform owner, where industry delivers (parts of) the service and citizens self-service themselves via the catalog on the platform. A platform that is designed for self-service is one that users can easily find, understand what it does, and start using without intervention (Pope, 2019b).

Figure 13. "Opening Services" – illustration of platform roles - Own Work

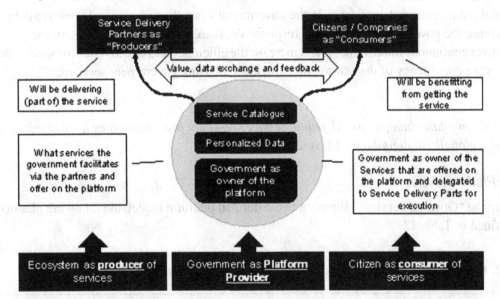

In this model, governments will be acting as an intermediary. Instead of executing the service request themselves, governments will focus on facilitating collaboration, connecting people and providers, and coordinating ground-breaking public service delivery models of the future. It is the foundation that allows government and non-governmental organizations to deliver next-generation public services.

When we talk about services, we refer to public-facing services that allow citizens or their representatives to achieve a desired personal outcome. With appropriate governance, services can be provided by any layer of government, and by commercial or third sector organizations. Services are the things that are built on top of platforms (Pope, 2019b).

On a well-designed platform, citizens can request the delivery of transactional services offered by the government themselves. Services are easy to find and use. Identification is simple but secure. Citizens have two-way communication with the government about their needs for information and services (Rantanen et al., 2019). The execution of these services is delegated to and handled by third-party partners.

The most critical determinant of any platform's success is its ability to attract participants to join and contribute to it as a platform cannot create value on its own. Attracting participants will be explored in the practice section. In any case, citizens demand for digital services continues to rise, which increases the pressure on governments to respond appropriately.

Many governments are embarking on a roadmap to define and roll out this type of model. An excellent example of this model can be found in Australia. The Australian government has recently published a strategy with the vision that Digital platforms will help shift the delivery of government services to faster and better digital channels that are easily accessible, making government more responsive, convenient, and user-focused (Australian Digital Transformation Agency, 2019). Improving digital services will also help boost the effectiveness and efficiency of service delivery. They state the following benefits:

- Simplify and reduce unnecessary burden on people and businesses who access government information and services.
- Encourage innovation and interoperability, creating a highly interoperable ecosystem of government and market.
- Build public trust and confidence in the government's handling of user and sensitive data.
- Increase the government's agility and responsiveness to meet changing expectations.
- Deliver continuous improvements and increase the efficiency and effectiveness of service delivery.
- Improve consistency of the end-to-end experience in using government services.

"To succeed in a fast-changing world, public services must not just adapt. They have to reinvent themselves for a globalized, digital age" (Atos, 2018).

2.2.4.2 Platform Roles

If we map the "Opening Services" approach to the defined platform model, the following platform roles were defined in Table 12.

Table 12. Roles in the "Opening Services" Platform Model – Own View

Role	Who	What
Producers	The ecosystem of third parties, not even government agencies	• Offer a service that integrates with the provided functionality of the government • Execute a service that governments delegate
Platform Provider (and regulator)	The Government - usually through a centralized department	• Provides the platform where the services can be searched, where the partners are listed that can execute them and where citizens can request them • Provides certification to the ecosystem • Ensures there is a feedback loop to contact the government and ecosystem partners • Regulates what data can be used to tailor or personalize the services • Facilitates the matching of the request versus what the providers offer
Consumers	Citizens	• Find, select, and request the service via the platform in a self-service approach

2.2.4.3 Generic Examples

Below are some thematic examples in and around Flanders, where governments use the "Opening Services" approach to better serve their citizens:

- Welfare - Surviving in Brussels: "Surviving in Brussels" focuses on people in difficulty (who may or may not be homeless, have little or no resources, with or without papers) but also to social workers. They can contact them for information about the region of Brussels. This case has all the characteristics of an "Opening Services" platform:
- The Brussels Government acts as a broker to services that are provided by an ecosystem
- The Ecosystem delivers the services (Food, Shelter, Homes, etc.)
- The homeless can tap into the platform to find the most suitable service, either via the website or the app, but also via selected kiosk, which makes the approach very inclusive for all

  There is an ecosystem Steering Committee that consists of 70 representatives of all organizations for homeless people in Brussels. Together with homeless people, they have identified 23 needs, from nutrition, shelter, medical care, and relaxation options to legal assistance.

  See https://www.survivinginbrussels.be/app/nl/

- **BXLRefugess – Citizen Platform: BXLRefugees** aims to mobilize and unite citizens to transform societal attitudes towards people in migration and to help provide them with a human response to their needs. It does that by offering a sleeping place or a home to stay temporarily, sourced from citizens directly. It is a compelling case, not only because of the nature of the services they are trying to match but also because they use Facebook as an existing platform to connect providers and potential consumers. See http://www.bxlrefugees.be/nl/

- **Employment - VDAB Matching Platform:** The VDAB is the public employment service of Flanders. VDAB employment services offer a wide range of services to job seekers, employers, and employees through various channels. VDAB wants to continue expanding the competency-based matching of employees and employers using a **platform model** whereby an ecosystem of partners has access to the vacancies and can do the matching themselves and directly. To help realize this, VDAB expressed the desire in 2018 to create a new organization that will operate as the platform provider to the ecosystem. This program is known as "**Matching 2.0**". The idea was converted into a vision note in 2018 that was approved by the VDAB board and transformed into an assignment to set up the new organization to host and exploit the platform model, including managing the ecosystem. VDAB is looking outside its walls to provide better services and information jointly to citizens and stakeholders with partners from other sectors. They choose deliberately for a partner model with the industry and opt for a co-creation model, which fosters natural partnerships and encourages unusual ones to cultivate new value and innovation. The implication is that VDAB must undergo an organizational change, hence the creation of the Matching organization team. VDAB is working closely with its French counterpart, Pôle Emploi, which is pursuing a similar approach. See https://www.vdab.be/

VDAB
samen sterk voor werk

matching 2.0.

producten van de toekomst

- **Citizen Services – ACPAAS:** We repeat the example of the "Antwerp City Platform as a Service" as the platform also allows the creation of new citizen services. The focus here is on the interaction with an ecosystem of start-ups. Start-ups are invited by the City of Antwerp to use the platform and its components to write new functionality and thus optimize the delivery of services to citizens. Innovation and creativity are stimulated through specific agile projects, and the ecosystem is encouraged to reuse the existing services. The value for citizens is that more innovation in services is brought through a consistent experience. See https://acpaas.digipolis.be/nl/about (Dutch only)

- **Education - Learning and teaching for Wales:** Also, in Education, many platforms have been created. One might think about commercial platforms like Coursera, Lynda.com, etc., but during the research, examples were found where the government has set up and manages an education supply to its citizens. In this example, Wales determined what the skills and competencies are for its citizens to make the best of the digital world. The platform enables learners and teachers to access online resources anywhere, at any time, from a range of devices. It also provides tools to help teachers create and share their resources and assignments. So, instead of competing with industry, Wales differentiates by providing the necessary training for all its citizens to be ready for the digital world. See https://hwb.gov.wales/

- **Transportation - Connected Citizens Platform (Waze):** The Connected Citizens Platform is an example in where collaboration between public and private industry is critical to deliver an optimized service to commuters and travellers. Over a thousand cities and other public sector partners are already working with Waze to gain better data and insights, reduce driving alone with Waze Carpool, and to make their communities safer with Waze's Beacons and Crisis programs. See https://www.waze.com/ccp

☺ **waze**
for cities

**Working together
for a smarter,
more sustainable
future**

- **Healthcare - Self-Directed Services:** Moving to a patient-centric model for health care means moving from a system based around closed formal organizations to one that takes the form of an open network. This means breaking down the silos, unbundling capacities, and building platforms where those resources can be made available on-demand via information-based networks. This requires taking individual capabilities that are currently closed off behind organizational walls and opening them up to usages external to organizational boundaries within common shared platforms. Examples include:

  ○ **Buurtzorg Nederland:** In the Netherlands of the mid-2000s, the healthcare sector was characterized by a lack of continuity of care and was faced with declining quality, rising costs, and a disillusioned nursing workforce. This inspired a change from a publicly subsidized institutional care system to a more privatized home-based care regime, introducing private sector management tools in the Dutch long-term care sector. Buurtzorg pioneered healthcare services with a nurse-led model of holistic care that has revolutionized community care in the Netherlands. A platform unites patients and caretakers while governments provide oversight and trust. See https://www.buurtzorg.com/or the case description at https://www.centreforpublicimpact.org/case-study/buurtzorg-revolutionising-home-care-netherlands/

  ○ **Medicaid USA:** Self-directed Medicaid services means that participants, or their representatives, if applicable, have decision-making authority over certain services and take direct responsibility to manage their services with the assistance of a system of available supports. The self-directed service delivery model is an alternative to traditionally delivered and managed services, such as an agency delivery model. Self-direction of services allows participants to have the responsibility for managing all aspects of service delivery in a person-centred planning process See https://www.medicaid.gov/medicaid/ltss/self-directed/index.html

  ○ **Whoog: Whoog's** value proposition is to immediately identify available and competent people within a community of professionals (e.g., home healthcare assistance) and then facilitate the coordination of the intervention while taking care of resource management services and automation of the generation of administrative and accounting documents versus the government. The service has been co-designed with the French government so that the providers are certified and can be trusted. See https://whoog.com/

○ **Doctolib: Doctolib** is the largest e-health service in Europe, supported and endorsed by the government. It brings healthcare providers (e.g., doctors) and patients together to facilitate a healthcare transaction (e.g., online doctor's consultation). It provides a new generation of software to transform patient and consultation management. For patients, this translates into a comprehensive online service to improve their experience in healthcare. See https://about.doctolib.com/

○ **Flemish Agency for people with a handicap:** The **Flemish Agency for people with a handicap (i.e., VAPH)** offers a platform for specific support when you are disabled. On the platform, you can search for a wide variety of care providers that are selected from the Flemish agency and are close to your neighbourhood. You make a choice and can order the service. This is linked to the overarching "personal handicap care budget" whereby people with a handicap get a budget they can spend with the service provider they want. See https://isis.vaph.be/wegwijzer/pvb

Many other examples exist, but the above cases give a good view of the breadth of services that governments are already positioning using the platform model, whereby the execution of the service is handled through an ecosystem of selected and trusted partners, realizing the matching process.

2.2.4.4 COVID-19 Examples in Flanders

At the time of writing this research project, the first wave of COVID-19 pandemic was causing a lockdown of public life in most of Europe, including Belgium. Amazingly, this triggered a surge in quickly established and launched platforms operated by the Flemish Government to match the demand and needs of several groups.

For schools, the Flemish government created a platform to make sure students were still able to enjoy e-learning facilities. Companies (as producers) jumped on the wagon to offer their services, often free, sometimes at reduced rates to

One example is the platform of the Department of Education, *Klascement*, where extra classes and producers were uploaded and made available to all students, parents, or even anyone interested in further education:

Figure 14. Klascement Screen - Source: klascement.be

Another example here is the WeZooz Academy, which is a collaboration between commercial companies like Microsoft and the government, through the Department of Education of the Flemish Government:

Figure 15. WeZooz Academy Screen - Source: wezooracademy.be

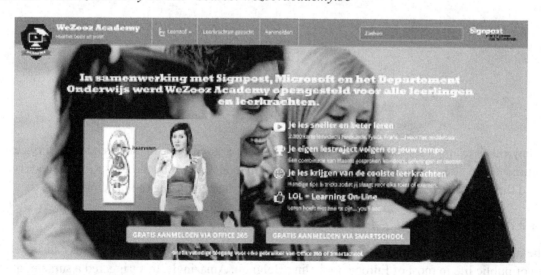

The "Help Heroes" platform turns the model around. Citizens can volunteer their services like helping to run errands for sick and elderly, medication, or just a visit. The caretakers can select the right person in the right neighbourhood (hence the matching) and offload some of the work that is usually done by professionals to volunteers.

Figure 16. Help Heroes Screen - Source: hulpvoorhelden.be

We hereby found extra empirical evidence that the "Opening Services" level of our baseline is becoming popular and that circumstances like COVID-19 can act as a catalyst for using the business platform model for citizen services. But we need to add some caution, as these solutions are often triggered by enthusiasm rather than strategy.

2.2.4.5 Governance Basics

When we translate this into governance mechanisms, we found good examples in literature. For example, the Australian roadmap for a digital platform (Australian Digital Transformation Agency, 2019) has identified six keys to success for "Opening Services":

- Build **trust and confidence** - we will need to earn the trust and confidence of the public, governments, and other users of digital platforms
- **Transform our culture, skills, and capabilities** - we will need a culture that cultivates innovation, and access to the right people with the right skills and capabilities to make digital platforms successful
- **Use technology and data** to connect and unify government services - we will develop digital platforms that are interoperable, flexible, and extensible
- Strengthen **digital leadership, governance, and accountabilities** - we will define clear roles and accountabilities to govern digital platforms
- Address **funding** and **legislative barriers** - we will work across government to remove barriers that stop or hinder cross-government collaboration or adoption of digital platforms

- Foster **collaboration** and innovation across government and beyond - we will collaborate and engage with a much broader audience to fully realize the benefits of digital platforms

Australia acknowledges that **strong leadership and governance** are at the heart of any successful transformation.

South Korea is also seen as an early adopter of transforming its government services model to embrace technology. Korea's e-Government experience provides an abundance of lessons for countries pursuing Digital Government reform (Karippacheril et al., 2016). Using the platform model is a part of the technology used to improve the services model. This case provided good insight into the practices to make sure such a transformation can be done correctly, which were summarize here:

- Sustained high-level leadership support
- Cultivation of technical capacity within government
- Establishment of a body at the highest level of government to manage cross-agency coordination
- Appropriate sequencing of citizen service innovations to help stimulate demand for the core technology infrastructure needed to sustain reforms
- Employee motivation, innovation, and commitment to deliver better quality and more efficient services to citizens
- Adoption of a national level enterprise architecture approach
- A strong partnership between the public and private sector

As citizens will engage in a transaction, they are entitled to some guarantees they will get what they ordered. Therefore, Service Level Agreements are important to set at business and technical level:

- **Business Level - between users and providers** - the service provider needs to be accountable, but so does the user. The provider may be accountable for a price and service quality, but the user needs to be accountable for using the service appropriately (Burns, 2008)
- **Technical Level - between the platform provider and the producers**, where the performance of applications is critically dependent on the resilience and scalability of the platform, which a platform owner ought to be able to guarantee in the form of service-level agreements and uptime promises (Tiwana, 2014)

Based on the above-investigated countries, Table 13 was created of reported processes and Success Factors, including a link to the corresponding COBIT2019 process (own view).

2.2.4.6 Relevance to this Research Project

This research looked explicitly into the model that opens the execution and delivery of services to third parties, which is the subject of this research project. Most of the literature was however referring to possible future use cases, expressing opinions and thoughts on how governments should change. Few articles referenced and / or commented actual working platform models. There were indeed quite a lot of individual initiatives mentioned that came close, built with good intentions, but not backed up with a strategy and /or a coordinated set of structures.

Table 13. "Opening Services" - Mapping of processes reported in literature versus COBIT2019 - Own work

Reported process / Success Factor	Corresponding COBIT2019 Process
• Strong Leadership • Digital Leadership • Sustained high-level leadership support	• APO02 - Managed Strategy
• Governance • Establishment of a body at the highest level of government to manage cross-agency coordination	• EDM01 - Ensured Governance Framework Setting and Maintenance
• Trust and confidence	• APO12 - Managed Risk • APO13 - Managed Security
• Culture, Skills, and Capabilities • Cultivation of technical capacity within government • Employee motivation, innovation and commitment to deliver better quality and more efficient services to citizens	• APO07 - Managed Human Resources • BAI05 - Managed Organizational Change • BAI08 - Managed Knowledge
• Adoption of a national level enterprise architecture approach • Develop digital platforms that are interoperable, flexible and extensible	• APO03 - Managed Enterprise Architecture • APO14 - Managed Data
• Service Level Agreements, both on the business and technical level	• APO09 - Managed Service Level Agreements
• Funding • Appropriate sequencing of citizen service innovations to help stimulate demand for the core technology infrastructure needed to sustain reforms	• APO05 - Managed Portfolio • APO06 - Managed Budget and Cost
• Legislation - regulate both the platform as the ecosystem providers that use the platform. Also, be transparent to citizens what protection or trust the government provides when engaging	• MEA01 - Manage Performance and Conformance Reporting • MEA03 - Managed compliance with external requirements
• Collaboration - building the ecosystem, manage the platform (internally)	• APO08 - Managed relations • APO10 - Managed Vendors
• Collaboration - building a strong partnership between the public and private sector (externally)	• EDM05 - Ensured stakeholder engagement

• Value, in terms of
   o Ensuring citizens get the benefits they are entitled to and requested via the platform
   o Ensure the government provides benefits to the citizens via the ecosystem
   o Ensure the investment of the government produces a better service delivery or more efficiency for all stakeholders
   o Ensure the ecosystem providers deliver the benefits as requested and is commercially sound
	• EDM02 - Managed Benefits delivery

The need to look specifically for the right success factors and design criteria to make it a success for the whole government, not just for one agency, was hereby confirmed.

2.3 Summary

The above set of levels describe an own view of increasing functionality-rich government platform models. As literature research indicated, these models are also increasingly more sophisticated in technology, governance and management as governments move up the value chain towards the "Opening Services" model, which is the target of this research.

This overview was needed to establish a baseline on what model is the only correct one to answer the research questions. The hypothesis to validate is that the model of "Opening Services" would likely be able to provide valuable services to the citizens when they were executed by a trusted and managed ecosystem of partners and under the ownership of a government organization that used the platform model to search and find these services. By using this baselined approach, we can now define what is necessary in term of practices to stipulate the roadmap towards improved citizen services.

3. STAKEHOLDER ANALYSIS: WHO IS WAITING FOR A GOVERNMENT PLATFORM MODEL IN FLANDERS?

3.1. Introduction

Via interview sessions with government experts, a broad stakeholder analysis was done focused on Government of Flanders (Belgium). Feedback was collected via interviews. These sessions were used to find out how Flanders' politicians and citizens looked forward to using the government business platform model ("Opening Services").

To get a 360° view, the interviews focused on three aspects of the business platform model:

- A view from the politicians – is the platform model – and associated change in citizen service approach - on the political agenda?
- Looking from the citizen point of view – is this what citizens are waiting for?
- Are commercial companies interested in offering their services on the government platform and under what conditions?

3.2 Are Politicians Waiting for the Platform Model?

First, politicians will have to agree and support the idea that in the digital era, the private sector, NGOs, citizens, and even machines can produce public services—often faster, more organically, and with better user experiences than the government itself. Politicians must act to foster the platform revolution. Or will they stand by, watch new players from other sectors take the lead, and risk future relevance? (Accenture, 2016b).

Driven by an increasing digitalization of our society, it is the citizens that provide the political leadership with authority to react appropriately and - for example - adopt the platform model. Citizens expect, in return, effective governance for public services (UNDP - Global Centre for Public Service Excellence, 2016).

The rise of government as a platform – incorporating the availability of data and functionality to external users - means that government service delivery will face unprecedented interest from organizations looking to invent new approaches to serve the public. Will politicians help orchestrate or facilitate this, or avoid pushing the platform model? Digital changes all aspects of government, including operating models, processes, skills and culture, and stakeholder relationships, so there is a significant risk associated with enforcing platform models without a good strategy, leadership, and technology advancement.

As the benefits of digitization are dramatically underestimated by politicians, decisions and policies are being made with a poor understanding of reality. Effective management of the digital economy depends on the politicians ability to accurately assess the value of free digital goods and services (HBR, 2019). HBR suggested looking into alternatives measures like the OECD Better Life Index (OECD, 2019) or United Nations' Human Development Index (United Nations, 2019) to measure the return on investment of rolling out a platform model.

According to Accenture(Accenture, 2016b), politicians have three options to deal with this phenomenon:

- **Lead**: politicians can opt to create a vision fostering an environment that supports collaboration among all stakeholders using the platform business model. Leading the vision means politicians have active control over it. Still, it requires addressing regulatory issues, providing the technology foundation, determining how providers integrate with the platform, and managing transparent and centralized processes and governance.
- **Partner**: politicians can opt for a co-creation model with government, citizens, and stakeholders as partners to encourage new ways of delivering government services, cultivate new value and innovation using the platform business model. The VDAB is an excellent example in Flanders that has chosen to go with a partnership model.
- **Let Happen:** politicians can choose to have another entity provide a given public service but retain responsibility for specific aspects of its provision, such as regulatory oversight, funding, or data management. The industry partner might overtake the government in speed or might want more say in the execution, so sound operating principles must be agreed upfront.

Translated to Flanders, the experts interviewed expressed quite some scepticism when asked if the "Opening Services" model is on the political agenda of the regional government. The interviewed saw politicians acting in the "Let Happen" approach as stated above.

In Flanders, a new government declaration was published after the elections of 2019 that holds quite some hook-on points to define a business platform model. Technology is seen as a significant game-changer. Flanders launched its "Radical Digital" Program in 2015, with the ambitious goal of having all interactions between citizens and administrations done electronically by 2020. This goal was only partially achieved, and the new Flemish Government has defined the second wave in this program to catch up and make the original goals(Vlaamse Overheid, 2015).

The experts testified that Flanders has little experience with platform models and is hesitant to engage in a top-down approach. Nevertheless, there are Flemish government agencies that got political support to start building an "Opening Services" model. The VDAB case with the Matching Platform is one of them and will be further detailed out in the cases section. There is action on Antwerp city level with ACPAAS as a prime example.

Here and there, we find a local initiative like Theatre Finder in Antwerp, that - probably unintentionally - gets close to the defined characteristics of an "Opening Services" model. This application allows owners of festivity rooms to maximize idle time, i.e., time the rooms are not used by renting them out to organizations that want to meet. More of these cases exist in local communities, like renting out sport facilities. So, somewhere, the idea lives, but not yet consolidated in a strategy or brought together in one platform.

In summary, individual initiatives do pop up, but do not seem to be coordinated from the regional level and lack top-down political ownership. As there seems to be no overarching vision defined in Flanders like in other countries (e.g., Australia, UK, Singapore), platform initiatives are likely to remain isolated and success concentrated to individual agencies willing to be the front runner.

Ultimately, politicians will have to choose what vision to adopt and set out a strategy to achieve that goal. So far, this seems to be not happening consistently in Flanders but is happening elsewhere in the world, sometimes on a city level (e.g., Dubai, Singapore), sometimes on a country level (e.g., Australia, UK). Can the politicians in Flanders afford to wait much longer?

3.3 ARE CITIZENS WAITING FOR THE PLATFORM MODEL? (CONSUMERS)

First, it was examined what channels citizens were still using to interact with governments. Recent research indicated that citizens in The Netherlands still used very traditional methods to get in contact with the government (Ebbers & Pieterson, 2017):

Figure 17. How citizens communicate with the Dutch government - Source: Ebbers et al.

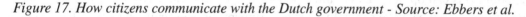

The report concluded that there was a shift towards websites, but telephone usage remained surprisingly high. When we take a look at the Digital Government Infographic 2019 for Belgium (EU, 2019), we see a different picture emerging, whereby citizens used the internet as a prime channel to interact with the Belgium government:

Figure 18. How citizens use internet with public authorities - Source: EU

However, the websites are still being used for basic e-governance applications whereby forms are being downloaded and filled in. In a way that is still the model prevailing from the early stages of e-government.

To understand what channels to use, exploration was done what citizens expect. A recent study from Accenture (Accenture, 2019) revealed the following expectations from citizens in Singapore:

- 80% + greater security and reliability,
- 77% easier interaction,
- 67% single portal for multiple services,
- 62% personalized information,
- 65% payments through preferred apps,
- 75% believe AI would be as fair as humans in handling eligibility and case management

Back in 2014, a more global Accenture study (Accenture, 2014) revealed that citizens were looking for the right service to be available at the right time, and the government must be committed to delivering it. However, citizens are not that happy with the way governments currently interact with citizens. Across the board, almost 60 percent of respondents say they are neutral or not satisfied with current government digital services. Roughly half expect the quality of their government's digital services to meet the standards quality of commercial services such as a website or portal enabling multiple transactions, electronic renewal alerts and emergency broadcasts, and digital post.

The reasons for not engaging with governments and submitting service requests were made clear in another Accenture study(Accenture Governing Institute, 2017), which revealed the following top 5:

- 18% did not know which agency to contact
- 8% knew someone else would take care of it eventually
- 7% said filling out a form takes too long
- 6% did not want to submit private information
- 5% called the helpdesk line, but reported it took too long/person was unhelpful

In summary, a mixed picture was emerging. Citizens seemed to be reluctant to engage with government agencies in a digital way, and if they do, some very traditional means were used. More citizens

Figure 19. Desire to interact digitally with governments - Source: Accenture

would engage with the government if it were more accessible. However, not all citizens agree on what "easy" means(Accenture Governing Institute, 2017).

The desire to use digital means to interact with the government also seemed to differ per country. Accenture reported a mixed picture across the investigated countries (Figure 19).

Caution is needed with these figures as the UK and US were countries having a high number of digital interactions, maybe some digital fatigue was already creeping in. But citizens of certain countries (e.g., UAE) reported a high desire to go digital. Interestingly enough, these are also the countries that have a clear vision, e.g., Dubai, with its Smart Nation vision(Smart Dubai Office, 2018).

On the other hand, expectations are rising rapidly, whereby citizens - driven by the experience offered by established platform companies - want more convenience and self-service than ever before. In the original study of 2014, nearly three-quarters (73 percent) of US citizens indicated that they hold government to the same or higher standard as their commercial providers. In the follow-up survey—just two years later (2016) — the percentage surged to 85 percent (Accenture, 2016a). However, telephone, in-person, and regular mail were still the dominant channels for citizen-government interactions, although citizens who regularly interact with the government reported more digital interaction.

In summary, there is room for improvement. Citizens seem to be reluctant to find their way to the digital channels of the government, so governments, before all, must do a better job in communicating the availability and advantages of the digital channels.

However, the pressure is increasing, as citizens become increasingly mobile and tech-savvy, and they expect government websites and digital capabilities to provide functionality and benefits comparable to those available from the private sector. In other words, citizens are getting used to the convenience of platform models in their private life and expect nothing less from governments.

3.4 Are Companies Waiting to Join the Ecosystem? (As Producers)

Based on interviewing several company representatives in Flanders, there was a "wait and see" attitude emerging when asked if they were interested in joining a government business platform ("Opening Services"). The representatives stated that they first and foremost look for a stable environment where the legislation is clear and where business rules are transparent.

Most also wondered what the "Return on Investment" (ROI) model would be. How will the government ensure a fair-trade approach to all providers? What revenues will be possible and will governments influence price setting, for example, by putting an upper limit on service fees or by standardizing the service fee?

All agreed that companies first want reassurance of business potential. This plays back to the chicken and the egg debate introduced earlier. Companies will only join the platform if there are enough potential "clients" out there, and citizens will only join if there are enough providers that offer services.

As Flanders has a lot of small-sized companies, the interviewees saw the potential to embrace the government platform to get more exposure. McKinsey (McKinsey, 2019) advised companies to take position fast, or they may find it increasingly difficult to catch up as competition is already present on the platform.

Extra guidance was found in the UK(UK Digital Services, 2016) when the government surveyed what companies were looking for when exposing their services on a government platform. Together with the results of the interviews, the following self-compiled list is suggested:

- The platform and its components must be modular, so companies can build and operate services that meet their user's needs
- The platform and its components must be easy to integrate into their service model, to minimize costs and speed up transaction closing
- The platform and its components must be actively supported and maintained by the platform owner (i.e., government) so that companies can be confident that their service will continue to run
- The platform and its components must be supported and maintained in a consistent way to minimize cost and complexity
- The platform and its components must provide access to (enough) citizen data so companies can personalize the value.

In summary, the representatives of companies interviewed were not yet thrilled to become partners in a government platform model ("Opening Services") in Flanders. Most realized they could not ignore this new way of doing business either. They expected governments to take the first step and make the rules of engagement clear.

3.5 Some Critical Notes on Platform Models

During the interviews and literature research, some unexpected social side effects emerged with the rise of platforms in the industry.

Let us take the case of Platform Work, whereby individuals offer their services to others. These platforms are sprawling. Fear was expressed in the researched literature that platforms will result in an erosion of labour laws as individuals will do anything to land an assignment. This would trigger a "race to the bottom" when the price is not regulated, and people start undercutting regulations just to land the job. Some of the platform providers were operating on the edge of labour laws. Take Deliveroo that demanded its bikers not to be employees but independent contractors. This allowed maximum flexibility but tended to bypass social securities (Vandoorne Legal, 2018). This even raised questions in Flemish parliament.

There are also many examples of platforms that are approved (and subsidized) by governments that advocate a sharing economy. These consist of services that individuals offer via a sharing platform like ListMinut, Your private butler, and many more. Hoang et al. (Hoang et al., 2020) investigated who are the winners and losers of advocating your profile on such platforms. They conclude that these platforms tend to favour specific higher-skilled profiles. That means that disadvantaged groups may not benefit from the opportunities offered. The authors expressed the fear that by working in the platform economy instead of traditional occupations, the employees would have fewer protections and workers' rights.

Belgium has regulated this kind of platform model quite well. It has set laws and boundaries for individuals that want to offer their services apart from their professional careers, for example, when retired (FOD Economy, 2020b). FOD Economy even provides a list of platform companies that are approved to operate this model (FOD Economy, 2020a). It is to be remarked that Deliveroo was also on that list.

Another critical note is on the current trust relationship between citizens and governments. Eaves (Eaves, 2018) made the observation that today, many citizens were comfortable sharing data with the government because these organizations were too legally restricted, organizationally siloed, or inefficient to link data together to learn about them. In other words, maximum potential was not used in the shared data. This trust can evaporate quickly when agencies start linking data across organizations. For example, if a person is sick and without a job, should a government warn, facilitate, or prevent that his family might get evicted from their rental house as well?

With new benefits, also come new risks. If the platform fails, it is a single point of failure for the whole ecosystem and for all kinds of services that are available. Politicians will be called to their responsibility. For example, the early failure of the HealthCare.gov platform in the United States caused great distrust into government-run programs. In the official report (US Dep. of Health and Human Services, 2016), it was stated that the most critical issue was the absence of clear leadership. This caused delays in decision making, lack of clarity in project tasks, and the inability to recognize the magnitude of problems as the project deteriorated. Governments must get it right first-time and cannot afford a hit-and-miss approach on establishing a platform model.

3.6 Summary of Stakeholder Analysis

Based on interviews and research, it emerged most governments were not the most progressive when it comes to building out government business platform models, give or take a couple of counties that do have a forward-looking vision for their citizens, including the usage of a platform model.

Politicians seemed reluctant to take a stand; citizens seemed to be still using traditional ways of communicating with the government, and companies were taking a wait-and-see attitude. Nevertheless, there are already some excellent examples where the platform model was successful by design (e.g., VDAB) or coincidental (e.g., Theatre Finder application in Antwerp).

Trust and value were continuously popping up as attention point. All stakeholders were looking for governments to take the first stand, to position the platform and to position how each can play in the redesigned citizen services approach. These insights were used to define or prioritize the practices.

3.7 What About Citizen Collaboration Platforms?

A lot of references to citizen collaboration platforms were found during the literature research. During the interviews, the experts pointed out that for politicians, this is an important model to gather feedback from citizens and work on new legislation.

So, does a citizen collaboration platform fit the defined model of "Opening Services", and can we apply the same set of practices if this feature is pursued?

First, we needed to look back. In history, governments have used different tools to gather feedback on policy issues (e.g., opinion polls and surveys) or draft policies and laws (e.g., comment and notice periods) from citizens. Sometimes tools for consultation provided higher levels of participation (e.g. public hearings, focus groups, citizen panels, workshops) with smaller groups of citizens (OECD, 2001).

The advancement in technology has transformed the relations between governments and citizens. Leaving the "digital divide" or e-inclusion aside, for all citizens, whether on-line or not, there is an increased set of options to participate available. Most governments have embraced new means like on-line discussion groups, e-mail, feedback forms, crowdsourcing, even interactive simulation, or games). Often, there are no clear boundaries as the usage of these technologies may overlap or may be used in combination with each other; however, most were launched in isolation and next to each other.

O'Reilly's (O'Reilly, 2011) definition was followed, whereby government "provides [through ICT) resources, sets rules, and mediates disputes, but allows citizens, non-profits, and the private sector to do most of the heavy lifting," thereby empowering the people, unleashing social innovation, and reinvigorating (...) democracy." Recently, the term e-participation became more prominent as it suggests the combination of technology and participation.

Figure 20. Spectrum of Public Participation - Source: IAP2

IAP2's Spectrum of Public Participation (IAP2, 2019) was designed to assist with the selection of the level of participation that defines the public's role in any public participation process:

The use of platforms is, therefore, a logical step to increase active participation, or, as mapped to the IAP2 model: to involve, collaborate, and even empower citizens.

Evidence was found in the study of Falco (Falco & Kleinhans, 2018) that such platforms should at least share technological features that are the basis for citizen engagement and collaboration, amongst others:

- Collection and sharing of ideas, solutions, local knowledge.
- Discussion and collaboration through opinion maps, surveys, commenting, forums.
- Simulation tools such as budget allocation and 3-D design.
- Voting and ranking of ideas.
- Analytics features of comments, votes, and general user activity on the platform.

Falco et al. warned that despite the abundance of functionalities, their use and actual take-up was not as widespread as it could seem. Despite many examples around, Falco stated that in general local governments seemed reluctant to engage and use such platforms in their public policy and service delivery effort. Despite this warning, there are good - although isolated and locally inspired - examples to highlight. Below are a sample from in Flanders or close by:

- In the Netherlands, a pilot project around **Digital Democracy** ("Proeftuin Digitale Democratie") was launched to create more significant involvement of residents in local policies with the use of open-source digital participation tools.
- In the city of Gent, the citizens provided input on how the city was to spend a part of their **budget** (Burgerbudget Gent, 2019). Cities of Kortrijk or Mechelen also experimented with the so-called "Budget Games" to let citizens decide on budget issues (note: reference only through news articles).
- **"De Wakkere Burger"** (De Wakkere Burger, 2019) is a 'socio-cultural movement' - recognized and subsidized by the Flemish Government that organizes events to stimulate the democratic, transparent and participatory mentality among policymakers.

Internationally, a lot of EU examples can be viewed via eparticipation.eu (EU, 2020) or through the e-participation index of the UN (United Nations, 2020).

These kinds of platforms need to be well regulated and moderated. Otherwise, certain pressure groups might impose too strong a single view. Some specific design criteria around governance are around moderating the platform and guaranteeing an inclusive process, whereby all voices and opinions can be shared despite the social background of participants.

There is value for governments in seeking to build out participation in a standard and coordinated approach. Access to information, consultation, and active participation contribute to good governance by increasing transparency in policymaking, more accountability through direct public scrutiny and oversight, enhanced legitimacy of government decision making, better quality policy decisions, and higher levels of compliance, given public awareness.

In summary, using the platform business model for governments to engage in a more systematic, technology, all-encompassing citizen participation is only logical. However, it requires a decent overall strategy, leadership, and coordination. Otherwise, local initiatives will - even if set up with the best intentions - continue to spread like mushrooms in a forest, thus making it hard for the citizen to know where and to engage.

The assumption – based on limited research but stressed by the experts – was that the same practices for "Opening Services" would apply to citizen participation platforms as well. It fits the model of "Opening Services" with the extra feature for governments of moderating the platform.

REFERENCES

Accenture. (2014). *The Digital Government Divide: Research Shows Citizens Want More*. https://www.accenture.com/us-en/~/media/Accenture/Conversion-Assets/DotCom/Documents/Global/PDF/Dual-pub_11/Accenture-Digital-Government-Divide-Research-Shows-Citizens-Want-More.pdf

Accenture. (2016a). *Digital government: Great expectations, untapped potential Expectations are rising-fast.* https://www.accenture.com/_acnmedia/PDF-30/Accenture-Digital-Citizen-Experience-Pulse-Survey-Highlight.pdf

Accenture. (2016b). *Government as a Platform : Coming soon to a government near you.* https://www.accenture.com/t20160831t013223__w__/us-en/_acnmedia/pdf-29/accenture-government-platform-pov.pdf

Accenture. (2019). *Citizen Survey 2019.* https://www.accenture.com/sg-en/insights/public-service/citizen-survey-2019

Accenture. (2018). *GaaP 2018 Readiness Index.* https://www.accenture.com/_acnmedia/pdf-83/accenture-gaap-2018-readiness-index.pdf

Accenture Governing Institute. (2017). *Getting Citizens Involved: How to encourage greater use of citizen engagement platforms.* https://www.accenture.com/t20171101t154201z__w__/us-en/_acnmedia/pdf-64/accenture-gov17-report-accenture_v.pdf

Andersson Schwarz, J. (2017). Platform Logic: An Interdisciplinary Approach to the Platform-Based Economy. *Policy and Internet, 9*(4), 374–394. doi:10.1002/poi3.159

Andrews, P. (2019). *Government as a Platform: the foundation for Digital Government and Gov 2.0.* The Mandarin. https://www.themandarin.com.au/118672-government-as-a-platform-the-foundation-for-digital-government-and-gov-2-0/

Aspers, P., & Corte, U. (2019). What is Qualitative in Qualitative Research. *Qualitative Sociology, 42*(2), 139–160. doi:10.100711133-019-9413-7

Atos. (2018). Look Out 2020+ Industry Trends Government Thought Leadership Realizing the promise of Government-as-a-platform. *Atos,* 1–8. https://atos.net/content/mini-sites/look-out-2020/assets/pdf/ATOS_LOOK OUT_GOVERNMENT.pdf

Australian Digital Transformation Agency. (2019). *Digital Service Platforms Strategy.* https://www.dta.gov.au/our-projects/digital-service-platforms-strategy/six-keys-success/4-strengthen-digital-leadership-governance-and-accountabilities

Belleflamme, P. (2016). *Platforms and network effects. October.* https://doi.org/ doi:10.13140/RG.2.2.35908.83848

Budgen, D., & Brereton, P. (2006). Performing systematic literature reviews in software engineering. *Proceedings - International Conference on Software Engineering, 2006,* 1051–1052. https://doi.org/10.1145/1134285.1134500

Burgerbudget Gent. (2019). *Een burgerbudget voor Gent 2016-2018.* https://ookmijn.stad.gent/burgerbudget/voorstellen

Burns, T. J. (2008). *Success Factors for Implementing Shared Services in Government.* http://www.businessofgovernment.org/report/success-factors-implementing-shared-services-government

Cheung, M. D. (2014). *IT Governance in IT Shared Services Environments.* https://www.researchgate.net/publication/275331202_IT_Governance_in_Shared_Services_in_Public_Sector

ChoudaryS. P. (2015). *Platform Scale*. https://www.amazon.com/Platform-Scale-emerging-business-investment-ebook/dp/B015FAOKJ6

Christopher, H. Baum | Andrea Di Maio. (2000). *Gartner's Four Phases of E-Government Model*. https://www.gartner.com/en/documents/317292

de Groen, W. P., Lenaerts, K., Bosc, R., & Paquier, F. (2017). *Impact of digitalisation and the on-demand economy on labour markets and the consequences for employment and industrial relations STUDY*. https://www.eesc.europa.eu/en/our-work/publications-other-work/publications/impact-digitalization-and-demand-economy-labour-markets-and-consequences-employment-and-industrial-relations

De Haes, S., Van Grembergen, W., Anant, J., & Huygh, T. (2020). Enterprise Governance of Information Technology. In Enterprise Governance of Information Technology. https://doi.org/ doi:10.1007/978-0-387-84882-2

De Wakkere Burger. (2019). *De Wakkere Burger - Participatie & Democratie*. https://www.dewakkereburger.be/

Dept. U. S. of Health and Human Services. (2016). *HealthCare.gov: Case Study of CMS Management of the Federal Marketplace*. http://www.oig.hhs.gov/reports-and-

Desdemoustier, J., Crutzen, N., & Giffinger, R. (2019). Municipalities' understanding of the Smart City concept: An exploratory analysis in Belgium. *Technological Forecasting and Social Change, 142,* 129–141. doi:10.1016/j.techfore.2018.10.029

DIGIT. (2019). *eGovernment factsheets anniversary report*. https://ec.europa.eu/isa2/sites/isa/files/docs/news/10egov_anniv_report.pdf

Eaves, D. (2018). *The Future of Governance for Digital Platforms*. https://medium.com/digitalhks/part-7-the-future-of-governance-for-digital-platforms-7a876c9ce5e3

Ebbers, W., & Pieterson, W. J. (2017). *Daar gaat een blauwe envelop. 5e Deelrapportage online enquête. Meting 4 mei 2017*. https://cfes.bms.utwente.nl/wp-content/uploads/2018/04/Onderzoek-EBV-Belastingdienst-Rapport4.pdf

Economy, F. O. D. (2020a). *Approved Platform Deeleconomie*. https://financien.belgium.be/sites/default/files/downloads/127-deeleconomie-lijst-erkende-platformen-20190911.pdf

Economy, F. O. D. (2020b). *Deeleconomie - Bijklussen*. https://www.bijklussen.be/nl/deeleconomie.html

Enzo, B. U., Fevre, M. Le, Petrucci, E., Marchionni, P., Biancalana, C., Hiltunen, N., & Yang, C. (2019). *State of the art in the use of emerging technologies in the public sector*. https://doi.org/ doi:10.1787/932780bc-en

EU. (2019). *Digital Government Infographic 2019 Belgium*. https://joinup.ec.europa.eu/sites/default/files/inline-files/Digital_Government_Infographic_Belgium_2019_1.pdf

EU. (2019). *European legislation on open data and the re-use of public sector information*. https://ec.europa.eu/digital-single-market/en/european-legislation-reuse-public-sector-information

EU. (2020). *eparticipation.eu*. https://eparticipation.eu/

Falco, E., & Kleinhans, R. (2018). Beyond technology: Identifying local government challenges for using digital platforms for citizen engagement. *International Journal of Information Management, 40*(February), 17–20. doi:10.1016/j.ijinfomgt.2018.01.007

Fenwick, M., McCahery, J. A., & Vermeulen, E. P. M. (2019). The End of 'Corporate' Governance: Hello 'Platform' Governance. *European Business Organization Law Review, 20*(1), 171–199. doi:10.100740804-019-00137-z

Harms, J. R. (2016). Critical Success Factors for a Smart City Strategy. *25th Twente Student Conference on IT*, 1–8. https://www.semanticscholar.org/paper/Critical-Success-Factors-for-a-Smart-City-Strategy-Harms/0a891c1b3e9642d730e717629032071e6b4997de

HBR. (2019). How Should We Measure the Digital Economy? *Harvard Business Review,* 140–149. https://hbr.org/2019/11/how-should-we-measure-the-digital-economy

HFB. (2019). *Exploitatiegebonden ICT-diensten.* https://overheid.vlaanderen.be/exploitatiegebonden-ict-diensten

Hoang, L., Blank, G., & Quan-Haase, A. (2020). The winners and the losers of the platform economy: Who participates? *Information Communication and Society, 4462*(5), 681–700. Advance online publication. doi:10.1080/1369118X.2020.1720771

IAP2. (2019). *Spectrum of Public Participation.* https://www.ghbook.ir/index.php?name=فرهنگ و رسانه های نوین&option=com_dbook&task=readonline&book_id=13650&page=73&chkhashk=ED9C9491B4&Itemid=218&lang=fa&tmpl=component

IBI. (2017). *Smart City Strategy Success Factors.* https://organicity.eu/smart-city-dont-cities-feel-smart/

IBM. (2018). *Implementing Governance of an API Initiative.* https://developer.ibm.com/apiconnect/2018/02/09/governance-api-initiative/

IBM Government Industry. (2017). *How to implement Tim O'Reilly's vision for Government-as-a-Platform.* https://www.ibm.com/blogs/insights-on-business/government/implement-tim-oreillys-vision-government-platform/

IDC. (2018). *Digital Transformation Means Government Must Think Like a Business, Part 3 : Executing the Vision. September.* www.idc.com

IDC. (2019). *New Procurement and Contracting Strategies for Government Digital Transformation. January,* 1–9. www.idc.com

ISS, & PWC. (2019). *Government transition: from service provider to broker or commisioner.* https://www.servicefutures.com/government-transition-provider-service-broker-commissioner

Jacobides, M. G., Sundararajan, A., & Van Alstyne, M. (2019). *Platforms and Ecosystems: Enabling the Digital Economy. February.* www.weforum.org

Johanesson, P., & Perjons, E. (2014). Design Science. In *An introduction to Design Science.* Springer., doi:10.1007/978-1-4899-6331-4_35

Jungwoo, L. (2010). 10 year retrospect on stage models of e-Government: A qualitative meta-synthesis. *Government Information Quarterly, 27*(3), 220–230. https://www.sciencedirect.com/science/article/abs/pii/S0740624X10000249?via%3Dihub

Kansu, M., & Parker, G. (2018). Transitioning from Services to Platforms : The Financial Services Industry. *MIT Initiative on the Digital Economy.* https://www.db.com/newsroom_news/Whitepaper_MIT_financial_services_platform.pdf

Karippacheril, T. G., Kim, S., Beschel, R. P., & Choi, C. (2016). Bringing Government into the 21st Century: The Korean Digital Governance Experience. In Bringing Government into the 21st Century: The Korean Digital Governance Experience. https://doi.org/ doi:10.1596/978-1-4648-0881-4

Kutlu, O., & Sevinc, I. (2010). AN OVERVIEW OF THE E-GOVERNMENT INITIATIVES IN TURKEY IN RESPECT TO THE EU ACCESSION PROCESS. *International Journal of EBusiness and EGovernment Studies, 2*(2), 1–12. http://www.e-devlet.com

Lacheca, D. (2018). APIs need an identity boost to drive government value. Telecom Asia. 2018 Supplement, P10-10. 1p.

Lenaerts, K., Beblavy, M., & Kilhoffer, Z. (2017). Government Responses to the Platform Economy: Where do we stand? *CPES.* www.ceps.eu

McKinsey. (2018). Harnessing the power of digital in government agencies. *McKinsey on Goverment., 4,* 44. https://www.mckinsey.com/industries/public-sector/our-insights/harnessing-the-power-of-digital-in-us-government-agencies#

McKinsey. (2019). *The right digital-platform strategy.* https://www.mckinsey.com/business-functions/mckinsey-digital/our-insights/the-right-digital-platform-strategy

Müller, S. D., & Skau, S. A. (2015). Success factors influencing implementation of e-government at different stages of maturity: A literature review. *International Journal of Electronic Governance, 7*(2), 136–170. doi:10.1504/IJEG.2015.069495

Nam, T., & Pardo, T. A. (2011). Conceptualizing smart city with dimensions of technology, people, and institutions. *ACM International Conference Proceeding Series,* 282–291. https://doi.org/10.1145/2037556.2037602

Nambisan, S., Zahra, S. A., & Luo, Y. (2019). Global platforms and ecosystems: Implications for international business theories. *Journal of International Business Studies, 50*(9), 1464–1486. doi:10.105741267-019-00262-4

Nielsen, A. P., Lassen, J., & Sandøe, P.OECD. (2001). Citizens as Partners: Information, Consultation and Public Participation in Policy-Making. *OECD Handbook, 20*(424473), 163–178. doi:10.1177/0963662509336713

O'Reilly, T. (2011). *Government as a platform* (Vol. 66, pp. 37–44). https://doi.org/ doi:10.22459/og.04.2018.03

OECD. (2014). *Recommendation of the Council on Digital Government Strategies Public Governance and Territorial Development Directorate*. https://www.oecd.org/gov/digital-government/Recommendation-digital-government-strategies.pdf

OECD. (2017). *Going Digital: Making the Transformation Work For Growth and Well-Being*. https://www.oecd.org/mcm/documents/C-MIN-2017-4 EN.pdf

OECD. (2019). *OECD Better Life Index*. http://www.oecdbetterlifeindex.org/#/11111111111

OECD. (2019). *Strengthening Digital Government*. Issue March., doi:10.1787/9789264307636-

Parker, G. G., Van Alstyne, M. W., & Choudary, S. P. (2016). *Platform Revolution*. W.W. Norton & Company., doi:10.1017/CBO9781107415324.004

Pope, R. (2019a). *Government as a Platform, the hard problems: part 3 – shared components, APIs and the machinery of government*. https://medium.com/@richardjpope/government-as-a-platform-the-hard-problems-part-3-shared-components-and-apis-9c87dba83e66

Pope, R. (2019b). *Playbook: Government as a Platform*. https://ash.harvard.edu/files/ash/files/293091_hvd_ash_gvmnt_as_platform_v2.pdf

Rantanen, M. M., Koskinen, J., & Hyrynsalmi, S. (2019). E-government ecosystem: A new view to explain complex phenomenon. *2019 42nd International Convention on Information and Communication Technology, Electronics and Microelectronics, MIPRO 2019 - Proceedings, May*, 1408–1413. https://doi.org/10.23919/MIPRO.2019.8756909

Reponen, S. (2017). *Government As a Platform: Enabling Participation in a Government Service Innovation Ecosystem*. https://aaltodoc.aalto.fi/bitstream/handle/123456789/24802/master_Reponen_Sara_2017.pdf?sequence=1

RIKS. (2019). *Estonian Government Cloud*. https://riigipilv.ee/en

Santana, E. F. Z., Chaves, A. P., Gerosa, M. A., Kon, F., & Milojicic, D. S. (2017). Software platforms for smart cities: Concepts, requirements, challenges, and a unified reference architecture. *ACM Computing Surveys*, *50*(6). Advance online publication. doi:10.1145/3124391

Santoro, M., Vaccari, L., & Smith, D. (2019). *Web Application Programming Interfaces (APIs): general-purpose standards, terms and European Commission initiatives APIs4DGov study-digital government APIs: the road to value-added open API-driven services*. https://doi.org/ doi:10.2760/675

Sauders, M., Lewis, P., & Thornhill, A. (2009). *Research Methods for business needs*. Pearson Education., doi:10.1080/09523367.2012.743996

SMALLS. (2019). *BE G-Cloud*. https://www.gcloud.belgium.be/nl/home

Smart Dubai Office. (2018). *Smart Dubai 2021*. https://2021.smartdubai.ae/

Smart Dubai Office. (2018). *Smart Dubai 2021*. 0–21. https://2021.smartdubai.ae/

Smedlund, A., Lindblom, A., & Mitronen, L. (2018). *Collaborative Value Co-creation in the Platform Economy* (Vol. 11). Springer., doi:10.1007/978-981-10-8956-5

Smedlund, A., Lindblom, A., & Mitronen, L. (2018). *Collaborative Value Co-creation in the Platform Economy* (Vol. 11). Springer., doi:10.1007/978-981-10-8956-5

Tiwana, A. A. (2014). *Platform Ecosystems, aligning Architecture, Governance and Strategy*. https://www.amazon.com/Platform-Ecosystems-Aligning-Architecture-Governance/dp/0124080669

UK Digital Services. (2016). *How suppliers can benefit from Government as a Platform*. https://governmentasaplatform.blog.gov.uk/2017/08/03/how-suppliers-can-benefit/

UK.GOV. (2019). *UK Gov Digital Marketplace*. https://www.digitalmarketplace.service.gov.uk/

UNDP - Global Centre for Public Service Excellence. (2016). *Citizen Engagement in Public Service Delivery The Critical Role of Public Officials*. https://www.undp.org/content/dam/undp/library/capacity-development/English/Singapore Centre/GCPSE_CitizenEngagement_2016.pdf

United Nations. (2018). *UNITED NATIONS E-GOVERNMENT SURVEY 2018*. https://publicadministration.un.org/egovkb/Portals/egovkb/Documents/un/2018-Survey/E-Government Survey 2018_FINAL for web.pdf

United Nations. (2019). *Human Development Index (HDI)*. http://hdr.undp.org/en/content/human-development-index-hdi

United Nations. (2020). *E-Participation Index*. https://publicadministration.un.org/egovkb/en-us/About/Overview/E-Participation-Index

Universiteit Antwerpen. (2020). *Overzicht databanken*. https://anet.be/submit.phtml?UDses=104716395%3A503538&UDstate=1&UDmode=&UDaccess=&UDrou=%25Start:bopwexe&UDopac=opacuadbobj&UDextra=pattern%3Dobjsys:db

U.S. Department of Veterans Affairs. (2018). *API governance in the public and private sectors*. Github. https://github.com/department-of-veterans-affairs/VA-Micropurchase-Repo/blob/master/2018-06-06_Governance_Models_in_Public_and_Private_Sector/Deliverables/Skylight/skylight_api_governance_research_report.md

Van Alstyne, M. W., Parker, G. G., & Paul Choudary, S. (2016). Pipelines, platforms, and the new rules of strategy. In *Harvard Business Review* (Vol. 2016, Issue April). https://pdfs.semanticscholar.org/7f54/dce38e04cd72c66511ec82b5f590b0e5cecf.pdf?_ga=2.74482194.1243927445.1567858352-69105146.1567858352

Van Veenstra, A. F., & Janssen, M. (2012). Investigating outcomes of t-government using a public value management approach. Lecture Notes in Computer Science (Including Subseries Lecture Notes in Artificial Intelligence and Lecture Notes in Bioinformatics), 7443 LNCS, 187–197. https://doi.org/doi:10.1007/978-3-642-33489-4_16

Vandoorne Legal. (2018). *Een vreemde eend in de bijt*. https://www.vandoorne.com/Kennisbank/2018_q1/platformarbeid-een-vreemde-eend-in-de-bijt/

Vlaamse Overheid. (2015). *Vlaanderen Radicaal Digitaal*. Vlaamse Overheid. https://overheid.vlaanderen.be/vlaanderen-radicaal-digitaal-0#vlaanderen-radicaal-digitaal-ii-2019-2024

Weill, P., & Ross, J. W. (2011). *IT Governance on One Page*. SSRN Electronic Journal., doi:10.2139srn.664612

Welby, B. (2019). The impact of digital government on citizen well-being. In *OECD Working Papers on Public Governance* (Issue 32). https://dx.doi.org/ doi:10.1787/24bac82f-en

Wohlin, C. (2014). *Guidelines for Snowballing in Systematic Literature Studies and a Replication in Software Engineering*. https://doi.org/ doi:10.1145/2601248.2601268

World Economic Forum. (2019). *Platforms*. https://www.weforum.org/platforms

Wu, D., Hesketh, T., Shu, H., Lian, W., Tang, W., & Tian, J. (2019). Description of an online hospital platform, China. In Bulletin of the World Health Organization (Vol. 97, Issue 8, pp. 578–579). https:// doi.org/ doi:10.2471/BLT.18.226936

Yesner, R., & Ozdemir, F. (2017). Understanding Smart City Transformation with Best Practices. In *IDC* (Issue November). https://www.isbak.istanbul/wp-content/uploads/2018/05/IDC-Smart-City-White-Paper-sponsored-by-ISBAK.pdf

Zuiderwijk, A., Susha, I., Charalabidis, Y., Parycek, P., & Janssen, M. (2015). (2015). Open data disclosure and use: critical factors from a case study. *CeDEM 2015: Proceedings of the International Conference for E-Democracy and Open Government 2015, 18*, 197–208. https://www.w3.org/2013/share-psi/workshop/krems/papers/OpenDataFactors_CeDEM

Chapter 2
Research Findings in the Domain of Business Platform Models:
Defining the Practices to Design a Perfect Government Business Platform Model

Yves Vanderbeken
DXC, Belgium

ABSTRACT

First, the author defined the right practices to strategize, design, build, and operate a government business platform model. Then they design a governance mechanism (processes, structures, relational mechanisms) that will provide maximum value. These practices are then converted into a consistent set of validation methods to determine where an organization is today and where it wants to be. Next is a deep dive of three case organizations, using the validation methods, to determine where they want to be and what the best practices are to improving their current operations. Advice is defined at the end of the chapter to guide any government organization on their journey.

1 DEFINITION OF PRACTICES FOR GOVERNMENT BUSINESS PLATFORM MODELS

1.1 Introduction

In this section an answer is provided to research question 1: What are some of the best practices for defining a Government Business Platform model for citizen services?

Based on research, practices were defined that guide governments to the establishment of a successful platform model for "Opening Services". Next to the practices, the associated governance mechanisms were also defined to make sure the platform model is well management and delivers upon the defined vision.

DOI: 10.4018/978-1-7998-7367-9.ch002

1.2 Requirements for defining practices

Before we defined the practices, a better understanding was needed what the practices and governance mechanisms need to conform to. In other words, practices to achieve what purpose?

The example found in the Australian Government (Australian Digital Transformation Agency, 2019) served as guidance. Australia is a pioneer in embracing the business platform model for its citizen services. For a government platform model to be successful, they recommended:

- To determine upfront what value the platform model will bring to all stakeholders in optimizing citizen services – tangible or intangible.
- To evaluate the existence of digital leadership, governance, and accountabilities to set the vision and turn this into an executable strategy.
- To measure how trust and confidence was enabled by the government to allow all stakeholders to function safely within an "Opening Services" oriented model.
- To measure how the government organization transformed their culture, skills, and capabilities to apply a platform model.
- To measure how new technology and data were used to build the platform model, connect stakeholders, and allow the unification of government services towards the platform.
- To address risk and legislative barriers in dealing with the ecosystem and allowing the government organization to transfer the responsibility of the service execution to the ecosystem.
- To foster collaboration and innovation across government and beyond by redesigning the government services to become self-service, proactive, and personalized on the platform.
- To measure how the governance mechanisms were put in place to ensure the platform model is managed and controlled in the best possible way.

Next to these recommendations, an own list of characteristics was added. The practices must be useful, effective, repeatable, and deliver consistent experience when applied to case organizations. This implies:

- **Useful** – the defined practices must be able to represent a 360° view of the approach the organization has taken to design, build and/or roll out their respective platform model. The practices must help identifying where actions or corrections were needed.
- **Effective** – the defined practices and governance mechanisms suggested must allow to verify if an effective platform organization was established.
- **Repeatable** – the practices must be useable in the same way to evaluate different government organizations, i.e., they do not require tweaking or special preparation.
- **Consistent** – the practices must remain stable when used in different government organizations, the results must be comparable without effort.

These requirements were stated to make sure the practices are not theoretical but can be used in case organisations to make a proper judgement where the organisation stands and / or wants to evolve to.

1.3 Approach used

An indirect approach was used, as there was little to no literature available on practices for government business platforms. To define the practices, research was done on what were critical success factors for similar IT projects in literature. This provided the input to link to the specific characteristics of a business platform model for governments and derive the practices. During documentation, the practices were cross-checked with the experts and their guidance was taken into consideration for prioritizing them.

To make sure the right literature was searched, it was necessary to establish a correct working definition for "Critical Success Factor" (CSF) that we could later uplift towards a practice.

To do that, let us go back to the original definition of CSF, which was introduced by Rockart (Rockart, 1978) as *"areas of activity which should receive constant and careful attention from management. The status of performance in each area should be continually measured, and status information should be made available"*. Boynton and Zmud (Boynton & Zmud, 1984) refined this to *"those things that must go well to ensure success for an organization"*. Dickinson (Dickinson et al., 1984) went one step further: *"Critical Success Factors (CSFs) are those events, circumstances, conditions, or activities that require special attention of management because of their significance"*.

The definition of Dickinson was applied as a basis for this research project with the additions that it is *about identifying the factors that have the greatest impact on the project* (Johnson, 2018).

So, the definition that was used in this research as the baseline for defining the practices was:

Critical Success Factors *(CSFs) are those events, circumstances, conditions, or activities that require special attention of management because of their significance. It is about identifying the factors that have the greatest impact on the project.*

With this definition in mind, literature was searched for matching CSFs to a government platform project. As there were only a few articles available that addressed the scope of "Opening Services", a more holistic and rigorous approach was used to analyse the articles and come to a set of CSFs:

Research in literature was done in separate steps to find:

- CSF for IT projects in general - Industry and government focused
- CSF for platforms for industry
- CSF for e-government projects

Together they formed the basis for drafting the practices for government business platform model ("Opening Services").

Figure 1. Approach used to determine CSFs - Own work

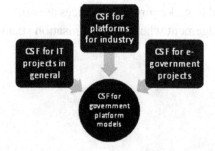

This research project focused primarily on the business platform model for government organizations. As a government has different (business) drivers than industry, the design and exploitation of a government platform will consequently be based on different characteristics. It was therefore interesting and relevant to first understand these differences.

Based on their own research, Ottlewski and Gollnhofer (Ottlewski & Gollnhofer, 2019) came to the following comparison of characteristics between private and public sector platforms (Table 1).

Table 1. Comparison of drivers to apply a platform model (Industry vs. Governments) – Source: Ottlewski & Gollnhofer

Characteristics	Private Sector Platforms	Public Sector Platforms
Platform owner	■ Corporations ■ Start-ups	■ Governmental organizations and institutions - Central, regional, and/or local government ■ Non-governmental institutions
Objectives	■ Pricing, matching, and exchange ■ Profit maximization ■ Monetizing searches, social networks, professional networks, assets, human activities ■ Generating competition, dynamism, and scale ■ Monetizing network effects	■ Connecting and engaging citizens in current societal, environmental, governmental issues ■ Enhancing citizen well-being ■ Efficiency, inclusion, and institutional change ■ Improving the collaboration between citizens and government agencies ■ Synergizing rather than competing ■ Integrating inner and outer resources to create decentralized innovation ■ Reaching socio-economic objectives
Business models – Target Customers	■ Consumers ■ Participating parties on the platform ■ Shareholders ■ Service/product providers on the platform	■ Overall public ■ Citizens ■ Policymakers ■ Governmental institutions ■ Communities, interest groups
Business models – Value proposition	■ Services ■ Stationary offers ■ Digital marketplaces ■ Forums ■ Matchmakers ■ Enablers ■ Hubs ■ Combinations of the above	■ Connecting citizens with providers of services ■ Providing information ■ Receiving help/assistance via the platform ■ Supporting the matchmaking function of the governmental institution versus the ecosystem of providers
Business models – Value Chain	■ Engaging different parties on the platform ■ Connecting actors ■ Creating a market ■ Equipping actors ■ Helping them to provide services to other actors ■ Acting as an intermediary ■ Pairing actors ■ Mediating the service flow ■ Providing a platform for centralized exchange	■ Offerings produced by forming alliances ■ Cooperation between the central, regional, and local governments ■ Integration of private actors
Business models - Profit mechanism	■ Commission for mediating services ■ Revenue as a retailer and service provider ■ Financed by advertising ■ Monetization of platform services (membership fees, user charges)	■ Value not defined in monetary terms, but in citizen well-being ■ Savings for the community by connecting citizens and right partners directly ■ Financed by the public sector or by private-public partnerships, sponsoring to cover the costs involved
Challenges	■ Scaling ■ Chicken/egg problem (consumer and producer acquisition) ■ Participation of platform users	■ Citizens' perception and actual usage of the provided platform solution ■ Tackling the causes, not the symptoms ■ Active engagement of citizens and providers

Whenever a CSF was identified, the above characteristics were taken into consideration to make sure the CSF was relevant to the context of government services and drivers.

1.4 CSFs for IT projects in general

We started by researching what the CSFs were for IT projects in general and for governments specifically. The Standish Group(Johnson, 2018) reported in 2018 per industry what the success rate was and what the contributing factors of success were.

The first conclusion here was that government projects tend to be not so successful with the second-lowest score in the left table. However, Johnson reported that "industry" is not the most accurate or important metric of comparison. He suggested looking for "factors of success". In the right table we can see that that regulated industries like government (and others) tend to suffer from long decision latency:

Figure 2. Factors of Success – Source: CHAOS Report 2018

Industry	Successful	Challenged	Failed	Factors of Success	Points	Investment
Banking	32%	54%	14%	Decision Latency	25	25%
Financial	30%	55%	15%	Minimum Scope	14	15%
Government	26%	52%	22%	Project Sponsors	14	15%
Healthcare	31%	52%	17%	Agile Process	9	12%
Manufacturing	31%	51%	18%	Talented Staff	9	12%
Retail	34%	51%	15%	Team Maturity	9	12%
Services	31%	50%	19%	User Involvement	5	3%
Telecom	24%	53%	23%	SAME	5	1%
				Optimization	5	1%
Other	29%	51%	20%	PM/Execution	5	1%
				Total Points & Yearly Investment	100	100%

Decision latency is described by The Standish Group as: *"The value of the interval is greater than the quality of the decision." Therefore, to improve project performance, organizations need to consider ways to speed-up their decisions"*.

Research from Gaikema, together with the Standish Group (Gaikema et al., 2019) revealed five things you need to do to create a "winning hand" for project success in governments:

- First, a project needs to be small. This means six team members (maximum) with a time box of six months or less.
- Second, the process must be agile, applying, for example, the Scrum methodology.
- Third, the agile team must be highly skilled in both the agile process and the technology.
- Fourth, the product owner or sponsor must be highly skilled.
- And fifth or last, the organization must be highly skilled at emotional maturity.

The study claims that if these five things were done well, there was an 81% chance that a project would come in on time and in budget, with satisfied customers.

Diir and Santos (Diirr & Santos, 2014) studied models for the improvement of IT service processes. They concluded the following CSF to be most mentioned in their research:

- Processes
- Support, commitment, and involvement
- Internal and external resources
- Skills of the people involved in the project
- Structure and culture within the organization
- Implementation strategy of the improvement project
- Collaboration, communication, and conciliation of the people involved
- Strategies for the advertisement of the project and the publication of its results

Coming back to the essence, Poon & Wagner (Poon & Wagner, 2000) suggested that organizations may get it right simply by managing three factors:

- Championing at the executive and operational levels
- Resources
- Linking the system to business objectives

These overall CSFs were kept in mind to come to a general picture.

1.5 CSFs for e-government projects

As stated earlier, e-government is a term broadly used for using new technology to improve citizen services. As such, searching for CSFs on successful e-government projects was a next step in this phase of research.

Based on research, articles (Müller & Skau, 2015) were found that define specifically CSF's for e-government, and also others (Abdelghaffar et al., 2005) defined a specific set for governments. The insights of both articles were merged and slightly adapted based on expert insights in Table 2.

Napitupulu (Napitupulu, 2014) made the first table based on extensive literature research on success factors for e-government projects and later (Napitupulu et al., 2017) refined it to 67 CSF. Muller et al. (Müller & Skau, 2015) went one step further in their research and distinguished between success factors at different stages of e-government maturity. The results of all this research are captured in Table 3:

Othman and Razali (Othman & Razali, 2018) continued on this and defined critical success factors towards integrating e-government services, meaning e-government services that work across silos (of government organisations). As the implementation of e-government typically influences the whole political-administrative system, requiring processes to be streamlined through technology is a must (Müller & Skau, 2015).

Daneels (Danneels, 2017) highlighted that a government organization must be able to understand how the outside-in flow of delivering a service fits or contradicts with the current organization's business processes. Governments need to contribute to the co-creation with inside-out service flows actively. In

Table 2. Analysis of CSF based on articles that define CSF for e-government projects – Own Work

CSF	Rationale
Political Support	Politicians act as CEO of government agencies and are accountable for enhancing the application of democracy by allowing "transparency and accountability". We must acknowledge that in government agencies, resources are allocated according to political priorities and not business needs.
(Culture of) Public Administration	Allowing a new method of delivering services to the public requires a different cultural approach as well. A new operating model is needed
Stakeholders Culture (citizens, business):	The implementation of new service models may require a change in public culture versus the classical approach. Instead of requesting services, citizens can self-service. Usually, citizens are ahead in using technology, but in the platform model, there is a new cultural shift whereby citizens trust external providers to execute the government service.
Financial transactions	Governments are not in the market to make money. Nevertheless, there is a cost associated if governments launch a new technology like a platform. So, the ROI needs to be determined versus the benefits and value this approach brings to citizens.
Usage of latest technology (and associated skills)	Governments need to purchase new technology as well, but usually lack the latest skills and capabilities to use this technology effectively. This puts a brake on the innovation of service delivery.
Collaboration (across existing silos)	One of the barriers to information and knowledge sharing is an organization divided into functional silos. This is specifically true in government organizations, where sometimes even legal restrictions do not allow sharing data. Muller summarized his research stating that one of the barriers to information and knowledge sharing is an organization divided into functional silos. As such, best practices are not shared (Müller & Skau, 2015). Muller advised organisations to improve their information-sharing practices when starting a platform model.

Table 3. CSF at different stages of e-gov maturity – Source: Muller et al.

Category	General CSFs	Low-level CSFs	High-level CSFs
External Environment	Legislation		
	Political and administrative reform		
	Socioeconomic factors		
	Culture		
Organization	Characteristics	Expectations	Result orientation
	Financial resources	Prioritization	
	Infrastructure		
	Collaboration		
	Stakeholders		
Management	Characteristics		Business Process Management
	Commitment		
	Strategy		
	Managing the projects		
Employees	Human resources		
	Fear of change		
	Training and education		
Citizens	Digital divide		
	Training and education		
	Citizen's needs and trust		
Technology	Infrastructure	Costs	Citizen Centricity
	Design and access		
	Security		

other words, governance is not only the process to control the ecosystem, but also the process to drive culture to come to a completely different approach to designing and delivering citizen services.

As a summary, several specific CSFs for the government were taking centre stage here. For example, without political support, the idea of getting funding for an e-government project was limited. Then the culture and collaboration were two repeating CSF that popped up in various studies. These were all internally oriented CSF, meaning factors that the organisation could directly handle and impact.

Building the ecosystem is a priority, meaning governments need to engage actively in building partnerships with non-governmental partners to make the platform work.

For the rest, the list of CSFs was comparable with IT projects in general: without vision and leadership commitment, nothing much will happen. Moreover, people will have to be trained and coached throughout the change as well, their job content might evolve, and new capabilities will have to be obtained.

1.6 From CSFs to the right practices

When the critical success factors were related to business platform models for governments, a slightly different order of importance was deducted. Based on expert interviews and discussions with case owners in Flanders, the following order was established:

1.6.1 Top Priority Practice: Getting the value right

During the literature research, it became apparent that value was the number one driver or practice for designing a platform model, be it in government or industry. It was stressed so often that a chapter was devoted to detailing out the value of a government business platform model.

For example, Danneels (Danneels, 2017) stated in her Ph.D. that value was the number one practice for the citizen - there must be well-articulated value for all stakeholders to become part of the platform either as consumer or contributor. The two-sided network effect must play based on the value. Happy citizens attract interested partners that give more options to the services the government offers.

So, what is the value of applying the platform model to the stakeholders? According to OECD (OECD, 2014), public value refers to various benefits for society that may vary according to the perspective of the actors, including the following:

1) Goods or services that satisfy the desires of citizens and clients.
2) Production choices that meet citizen expectations of justice, fairness, efficiency, and effectiveness.
3) Properly ordered and productive public institutions that reflect citizens' desires and preferences.
4) Fairness and efficiency of distribution.
5) Legitimate use of resource to accomplish public purposes.
6) Innovation and adaptability to changing preferences and demands.

Choudary stated that platform design should start with defining the value that is created or consumed(Choudary, 2019). Cordella & Paletti investigated the government business platform model in relation to the creation of public value in Italy (Cordella & Paletti, 2019). They stated that the creation of public value was not only related to satisfying a single collective need at a time (e.g., more efficient public transportation system) but also with the ability to adapt to changes in value expectations. They concluded that to provide maximum public value, the operational capabilities of public administrations must be flexible and adaptive to the changing needs of the citizen or the society. Value is thus a moving target. They also confirmed that administrations should work together and align their services portfolio, or otherwise, they might create services that serve some values and negatively impact the ones served by other administrations.

This aligns well with the position of Hein et al., that suggest the value creation process to shift from a single company or government perspective into a mutual value co-creation process with an ecosystem of actors (Hein et al., 2019). Following Hein et al., the value is best defined by viewing it from 3 perspectives: Citizen Perspective, Ecosystem Perspective, and Government Perspective. In the next sections, a closer look at these perspectives is provided.

1.6.1.1 Value from the Citizen Perspective

For citizens, time and convenience were stated as important benefits (New Zealand. Department of Internal Affairs, 2017). Certain life events require lots of government contact at once, which can be a negative experience if services are disjointed. This time impact can be large for those running a business or managing amongst disadvantages.

Tiwana (Tiwana, 2014) added to the above that the primary value proposition of platforms for end-users / citizens was that they can more uniquely customize their service request by mixing-and-matching from a diverse pool of service providers. End-users /citizens also benefit from the accelerated pace of innovation of services offered on the platform. Finally, platforms lower search costs and transaction costs associated with finding and acquiring services relative to doing the same without a platform in the middle.

Accenture (Accenture Governing Institute, 2017) stated that a government platform needs to provide the following benefits for citizens:

- Faster access to proactive and tailored services according to unique citizen's needs.
- Get a consistently delightful experience instead of being the "errand boy" of the government agency.
- Involve citizens in the evaluation of existing and/or design of new services.
- More transparency to citizens and business as governments open their data and applications.
- Citizens get faster access to new services and are in total self-control when they request or activate the service request.
- Guaranteed authenticity and privacy as governments control and govern the ecosystem.
- More innovation as industry partners can create new value on top of the government provided data and functionality.

1.6.1.2 Value from the Government Perspective

Governments must be committed to improving the quality of public services and make use of the latest technology to regulate how a service can be delivered by an ecosystem of trusted partners. This will necessitate speed to take strategic decisions adaptability to challenges of the new technological paradigm and changing economic conditions. These are preconditions that will result in citizens' and businesses' satisfaction with public services offered via the platform model (PETROV et al., 2018).

Reponen summarized the factors that encourage governments to follow through with a platform model are shown in Table 4 (Reponen, 2017).

Table 4. List of encouraging factors for governments to use a platform model – Source: Reponen

Encouraging Factor	Mechanism
Operational Efficiency	• Harmonized, integrated systems, and processes • New and modern solutions • Cost Savings
Convenience	• Flexible and agile ways to get new functionalities • Offering that matches individual organization's needs • Centralized organization acting as platform coordinator
Openness	• Open interfaces, collaboration of different actors • Avoiding vendor traps
Learning	• Learning and knowledge sharing

Tiwani (Tiwana, 2014) stated in her book "Platform Ecosystems" that platforms enabled platform owners to innovate on a scale that they could not achieve by themselves. They did this by massively distributing innovation activities from inhouse to outsiders with strong market-based incentives, drive, and deeper expertise in narrow domains and market segments. This simultaneously allowed the platform owner to do more of what it does best and sharpening its focus around what it perceived as its core competence.

1.6.1.3 Value from the Ecosystem Perspective

Why would a producer offer its services on a government issued platform, which is also regulated and controlled by the same government? Here again, Reponen created an overview of all factors that were

Table 5. List of encouraging factors for suppliers to use a platform model – Source: Reponen

Encouraging Factor	Mechanism
Convenience	• Ease of use, simplicity, convenience • Lack of bureaucracy • Short, fast, agile, and uncomplicated process • Decreased ambiguity and risks • Fast experiments
Transparency	• Communication • Openness • Trustworthiness and robustness
Information	• Documentation • Clear "rules of the game" • Clear earnings logic • Access to data • Access to industry experts
Financing	• Getting financing or sponsoring
Business Opportunities	• Growth opportunities • Financial Benefits
Reputation and credibility	• References • Validation, credibility • Publicity, recognition • Ability to demonstrate the viability of a solution
Learning and reciprocity	• Learning • Other's examples • Knowledge sharing • Contacts, networks, community
Social Contribution	• Doing good

appealing to private parties to use a government platform for advertising and positioning their services (Reponen, 2017), which resulted in Table 5.

Tiwani (Tiwana, 2014) stated that platforms enable app developers to use the baseline capabilities of the platform as the foundation for their work. Instead of replicating the functionality that their apps share with other apps, their upfront investment was therefore limited to functionality that their apps did not share with others. Platforms also provided access to an existing pool of customers, who could more easily find the app developer's work and more efficiently transact with them. Therefore, they reduced both search costs and transaction costs between app developers and their prospective customers.

1.6.2 Top Priority Practice: Getting (Political) Leadership

Launching a digital transformation program towards a platform model represents, at first sight, a significant risk for public managers. Public managers tend not to be as entrepreneurial as their industry peers and are more discreet to accept risk.

There must be political will to allow parts of the service delivery to be transferred to an ecosystem of trusted partners. A recent study (Nicholson-Crotty et al., 2019) demonstrated that under certain conditions, public managers might be more risk-seeking, certainly when it comes down to improving efficiency.

According to Brown, governments seemed unlikely to make meaningful progress on deciding what changes they need to make - and why - on adopting the platform model until they also resolved the highly emotive issue of what 'rightfully' belongs in public versus private and voluntary sectors (Brown, Fishenden, & Thompson, 2014).

When we relate this to some of the defined platform models (i.e., "Opening Data" and "Opening Functionality"), we already saw quite some adoption and political support. This was also driven by laws and regulations. For example, the EU mandated all its member states to adopt an "Open Data First" policy (European Commission, 2018).

We must understand that a political decision process operates on a different and longer cycle than in industry. Barcevičius et al. (Barcevičius et al., 2019) defined the following typical policy cycle:

Figure 3. Typical political decision cycle – Source: Barcevičius et al.

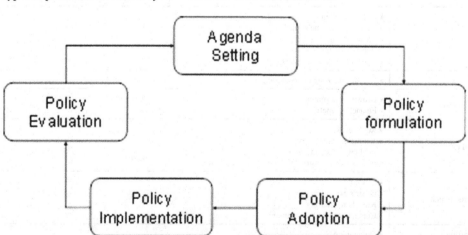

This cycle implies a decent preparation before an action is implemented. Interestingly, the EU proposed a "Policy-Making 2.0" approach to define how digital innovations could impact - and transform - governance processes and policy-making mechanisms. This concept denoted a combination of technologies applied throughout all the phases of the policy cycle to develop better, more participatory, and evidence-based policies (Mureddu et al., 2012).

Even when a new approach to making decisions is taken, political leadership then (and still) needs to be translated into digital leadership. **Digital leadership** is a key dimension of digital government maturity, and the organization's digital dexterity, which is defined is the desire and ability of employees to embrace existing and emerging technologies to achieve better business outcomes(Rosin, 2019).

To be truly successful in rolling out the platform model for "Opening Services", it is not only a political decision. Ownership for digital government transformation towards a platform model must reside with the business (EU - DG Joint Research Centre, 2018). It requires a leadership style and approach that recognizes the transformational impact of digital business and is unafraid to drive the organization in new directions toward reshaping its role, purpose, and operating model (Vitalari & Shaughnessy, 2012).

So, given the right focus, there was evidence in the literature that politicians would favour the adoption of new technologies like the platform model to improve efficiencies in the delivery of services. However, governments do change slowly. Nesta (McFadden, 2016) stated that governments typically found it very difficult to stop doing things. This can lead to the circular argument that we assume the processes must be correct because this it was always done this way.

In conclusion, getting the right political leadership to support the digital transformation of the services delivery model from in-house to an ecosystem led model seems not an easy fight, but not impossible. Once the politicians and the leadership team settle for using a business platform model ("Opening Services"), the strategy can be defined and rolled out.

1.6.3 Top Priority Practice: Establishing the Ecosystem

A lot of literature exists around the importance of building an ecosystem to make a platform work. To understand what an ecosystem is and why it is a critical success factor, several key topics from literature research are summarized below.

Rantanen et al. (Rantanen et al., 2019) defined a government ecosystem as a "*complex socio-technical system incorporating citizens, organizations, companies, as well as governmental agencies, which uses electronic platforms to create and distribute value to its participants*". Complex socio-technical systems typically include the usage of several data sources, the combination of stakeholders, interfaces, and users. Guaranteeing privacy and security are crucial for the technical system that supports the platform.

Although Rantanen et al. projected this on an e-government level, it was generalized here to map to the platform model of "Opening Services".

So, given the ecosystem, what are "Ecosystem Services" then exactly? According to Ref. et al., ecosystem services align multiple players to common objectives and enhance their interactions and capabilities to deliver outcomes, extended value propositions, and superior customer experiences(Ref et al., 2017).

Brown (Brown, Fishenden, & Thompson, 2014) made a case for governments to transition to "a new, diverse ecosystem of state, private and third sector activity, organized around the citizen in the form of services".

Harrison et al. (Harrison et al., 2012) proposed policymakers to engage in strategic ecosystems thinking. The authors stated that governments must evolve from outdated industrial bureaucratic forms to

Figure 4. Ecosystem Players – Source: Rantanen et al.

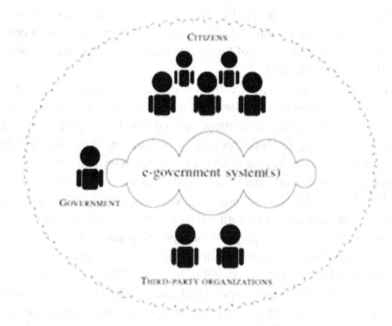

information age networked and interdependent systems. Their perspective on an ecosystem envisioned government organizations as central actors, taking the initiative within networked systems organized to achieve specific goals related to innovation and good government.

To be successful, three aspects were to be considered: a) the ecosystem must be built to improve service delivery, b) create value for all stakeholders, and c) be sustainable(Harrison et al., 2012). Berkman (Berkman, 2005) added some more aspects like the collaboration principle - permitting governments, industry, and other stakeholders to create, grow and reform communities of interested parties that could leverage strengths, solve common problems, innovate and build upon existing efforts.

The objective of a platform ecosystem is to make it simpler to design new types of services that are created around the needs of a citizen rather than the organizational structure of government(Pope, 2019).

The term ecosystem is not unique to the definition of a platform model around "Opening Services". There is enough experience acquired in building ecosystems to reuse data (i.e., "Opening Data"), mandated under the EU Open Data directives (European Commission, 2018). However, when applying ecosystems to "Opening Services", it means that governments not only dispatch data but now also allow the execution of parts of the government service to the ecosystem, consisting of commercial players. In essence, the government facilitates the transaction via the platform, but it is a private partner that delivers the service.

Platforms also provide opportunities for governments to lower expenditure on fixed assets and operational costs by focusing on resource orchestration rather than resource ownership (European Commission. Joint Research Centre., 2018). However, research indicated there is also a cost associated with building the platform and the ecosystem. Moreover, it is not a one-time effort, but an ongoing effort to maintain the services and the number of providers on the platform. So, on top of the investment costs, governments must ask if the cost of launching and maintaining the platform model will be absorbable.

BCG (Lang et al., 2019) advised to make high-value partnerships a priority. Governments should focus on selecting partners with the greatest potential and/or strategic importance first to maximize value creation within the ecosystem. This requires the setup of a good governance model, whereby the government act as the orchestrator, i.e., organizes and manages the ecosystem, defines the strategy, identifies potential participants, and clearly defines all roles, responsibilities, contributions, and interactions upfront, so that all partners know what to expect. It must be clear what the partners will earn from executing the service, so governments should not hide from commercial discussions as well. Flexibility is key to maintain the ecosystem as new partners might join, and existing ones might exit. There must be a form of continuous alignment between the providers and the citizens.

In summary, the success of a government business platform ("Opening Services") will largely depend on the creation - and maintenance - of an ecosystem of partners that have the trust to execute (parts of) the government service. Citizens will have to learn to trust both the government and the providers in delivering the service. Without a proper functioning ecosystem, there will not be much value for citizens to use the platform. Next to the importance of setting up an ecosystem, trust is a practice not to be underestimated. This means that the dynamics of ecosystems as seen in industry platforms are also relevant to government platforms, be it that the drivers are different as governments are not focusing on generating money, but on providing quality services.

1.6.4 Platform Related Practices

A platform is foremost built using technologies to create a digital marketplace where the stakeholders find each other, and transactions can be initiated. As such, much of the practices will impact the technical characteristics of the platform.

Without a well-designed platform from a technology point of view, many of the features will not materialize. A poorly designed platform produces little or no value for users and generates weak network effects, or none at all (Parker et al., 2016). Therefore, the design of a platform should begin with its core interaction—any kind of interaction that is at the heart of the platform's value-creation mission.

1.6.4.1 Platform Design Principles

First and foremost, governments will have to decide how far they want to play in the platform model. According to the EU(EU - DG Joint Research Centre, 2018), governments can play different roles in supporting or using the business platform model:

- **Government as a provider to a Digital Platform** - Providing reusable data to Digital Platforms beyond merely providing open data creates a window of opportunity for governments to stimulate new economic activity or develop a self-sustaining business model.
- **Government as an owner of a Digital Platform** - Creating Digital Platforms in governments shifts from a focus on citizen value to a focus on ecosystem value.
- **Government as an owner or funder of a Digital Platform** - Digital Platforms are leveraged in digital transformation initiatives in governments and blur the boundaries between traditional industry segments and government sectors.

An important design criterium is the Platform Openness, which refers to the degree to which the platform allows other companies in the ecosystem to develop applications or services that boost the value of the platform. Benlian et al. (Benlian et al., 2015) stated that there are two criteria to consider when designing a platform:

- **Transparency**: the extent to which it allows third parties to fully understand how to create and distribute applications on a platform, following all platform-related governance mechanisms set in place
- **Accessibility**: the degree to which the platform provides resources that allow developers to contribute to the platform by creating and distributing third-party applications without having to face platform-specific restrictions

It is important to understand what this exactly means as this will drive the design of the platform. The authors provided the guidelines shown in table 6 to decide how far to go in the design of a platform.

Table 6. Design Criteria for a Platform – Source: Benian et al.

	Transparency	Accessibility
Technical Platform	• The platform offers features that allow developers to communicate and exchange with other developers (exchange among developers) • The documentation of the technical platform includes all relevant information for the development of applications (technical documentation) • The platform offers features to receive instant technical support from the platform provider (technical support by provider)	• It must be easy to make oneself familiar with the platform's technical standards (learning curve towards technical standards) • The platform offers helpful tools that make the development of applications easier (availability of development tools) • The platform supports technical interoperability (i.e., compatibility) with other systems or platforms (technical interoperability) • The scope of functionalities that is made available to developers (via APIs) is limited (functional scope) • The technical performance of the platform constrains the functioning of applications (technical performance) • The initial costs for technical requirements (e.g., annual fees for the developer community, hardware requirements) are limiting the access to the platform (cost of required technical equipment) • The costs of selling applications on the platform's marketplace (e.g., revenue share paid to the platform provider, fees for billing system, etc.) constrain developers in distributing their applications (cost of selling) • The terms and conditions of the marketplace (e.g., on pay-out schedules and thresholds) constrain developers in their sales activities (distribution restrictions in terms and conditions) • The application review and marketing guidelines constrain developers in distributing their applications (constraints through app review and marketing guidelines)
	• The platform provider openly communicates the review and marketing guidelines (communication regarding app review and marketing guidelines) • The terms and conditions of the platform's marketplace (i.e., about promoting and selling apps) are transparent (transparency of terms and conditions) • The notification practices of the platform provider (e.g., about planned changes in the terms and conditions) are transparent (notification practices) • The search, filter, and ranking mechanisms of applications (i.e., application discoverability) on the marketplace are clear to developers (transparency of market mechanisms) • The platforms marketplace allows and supports communication between application developers and end-users (communication with end-users	

In the case of "Opening Services" and as part of this research project, the scope was limited to providing matching functionality via a platform model. Nevertheless, one should allow space for third parties to develop new applications or services on top of that as well. For example, a third party can offer an AI-inspired algorithm that provides better matching based on behaviour data (which raises the question of privacy and GDPR, see data related practices). As such, the table above provides good overall guidance to decide on all aspects.

The "Leading Edge Forum" organization (Leading Edge Forum, 2019) provided further guidance to **design the platform for change**. They advised not to build for stability but constant evolution. Business platforms must start small, evolve, fail fast, learn from it, and move forward. In traditional organizations, evolution tends to slow when scaling starts, whereas, in platform organizations, they run in parallel. Platform organizations should be built for constant evolution.

Don Tapscott, author of Grown Up Digital (Tapscott, 2009), stated that younger, digitally native generations now demanded better services, more convenient access to information, and a way to personalize or customize the services they receive from the government. They wanted the public sector to organize service delivery in ways that maximize convenience to the citizen as opposed to the bureaucracy. This puts convenience right at the centre of the design of a platform. The services offered on the platform must be addressing a real need of a citizen and help them in addressing that problem (e.g., getting to the right healthcare or education). It must be practical and execution-oriented, not dumping information on the citizen.

1.6.4.2 Platform (Technology) Architecture

With the above design principles in mind, the practices for architecting the platform from a technology point of view could be defined. This proved to be an art on its own, given the many articles and opinions available. The below was restricted to describing the main technical aspects from a conceptual point-of-view.

The going-in position was that for the successful creation of a government business platform model for "Opening Services", the underlying technology layers must be successfully set up and managed as well. The argument is that one cannot run a successful "Opening Services" platform without a well-functioning technology base layer or a proper data access strategy. On top of that, the functionality must be accessible through a well-architected and secured API strategy. A data insight layer (i.e., advanced analytics) is necessary to learn more about the interactions and the behaviour of the consumers to be able to provide personalized services.

What sets the "Opening Services" layer apart is the Matching algorithms that bring together the ecosystem providers with the demands of the consumer (i.e., citizen).

Figure 5. Matching as core functionality – Own Model

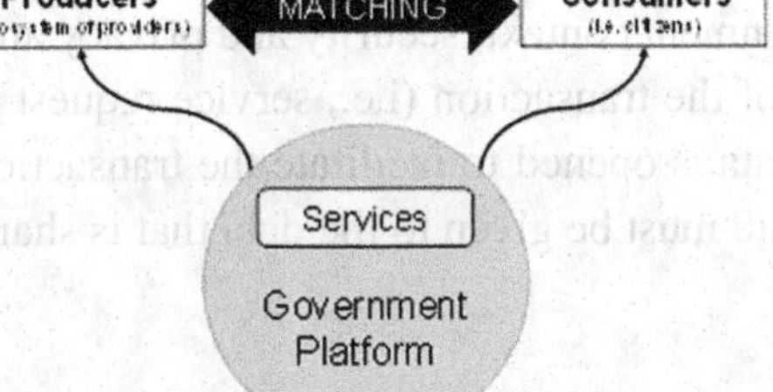

81

Figure 6. Conceptual Technology build-up for a platform – Own Model

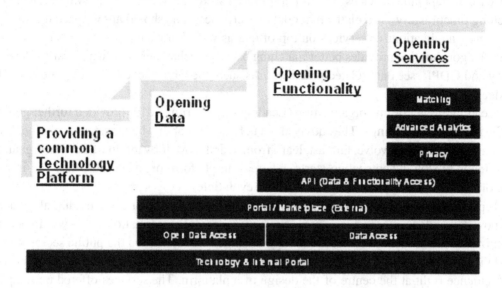

This means that each layer inherits the characteristics of the underlying layer, as depicted below:

On the Technology layer, it is important to standardize the provision or sourcing of infrastructure efficiently and allow scaling according to the needs of the platform. The creation of a portal or catalogue of technical services is also defined at this level.

Next up is to standardize access to data, often originating government applications in the back office, but also from / to commercial applications. The experience from Open Data can be reused here, but we need to ensure the quality of the data is excellent, and privacy is guaranteed when this is linked with providing a personalized citizen service. Most governments have established an Open Data portal than can be used as a source for searching (metadata) information.

To dynamically access information from government applications or data sources, an API standardization approach must be defined and imposed on the platform model. It must be clear how the providers can access and/or integrate functionality into their service delivery model (which can also be in the form of an application).

What sets the "Opening Services" model apart is that a **matching algorithm** is used to connect the requests of the citizen for a service to a possible series of producers that can fulfil this need. The algorithm is key. This algorithm will use data that is collected from the platform (i.e., behavioural analysis) to personalize the services and proactively offer to citizens. The combination of the two makes the platform hyper-relevant to the citizen. Technology like Machine Learning and Artificial Intelligence can help achieve this level of proactiveness, running in the background to continuously provide the right information based on the behaviour of the citizens or previous visits. Governments tend to be careful to let algorithms decide what a citizen will get, so transparency is equally important in using the technologies.

When we operate in a government context, security and privacy are crucial to guarantee that data is only used in the right context of the transaction (i.e., service request and execution). The government must ensure only the correct data is opened to facilitate the transaction, and not resell any transaction data to third parties. Special care must be given to the data that is shared with the ecosystem.

Janssen and Estevez (Janssen & Estevez, 2013) were already referenced for their concept of I-government (see introduction). As part of this work, the authors also created a platform reference architecture, in which public organizations introduce platforms to facilitate innovation and interactions with other public organizations, business, and citizens, and focus on their orchestration role:

Figure 7. How Public and Private Platforms interconnect in I-Government style – Source: Janssen and Estevez

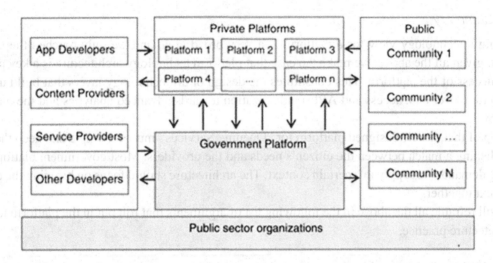

The authors suggested that private platforms can tap into government platforms to reuse functionality or data as offered by the government. This hybrid model is also recognized by Ottlewski(Ottlewski & Gollnhofer, 2019) and defined as *"Online platforms that are sponsored/financed/hosted/initiated by private owners but located in a domain that is traditionally situated in the public sector"*. Most of these platforms tend to have a social function like a neighbourhood watch or a cooperation with charitable/ non-profit partners in a societal context.

This model is consistent with the definitions of "Opening Data" and "Opening Functionality", as presented earlier. Quite some of the examples listed earlier focus on a combination of both government and private platforms merging.

Tiwana (Tiwana, 2014) stated there are four technical properties for establishing a platform architecture:

1. **Simple**. The architecture of a platform should be simple enough to be comprehensible, at least at a high level of abstraction. This means that the platform should be conceptually decomposable into its major subsystems, the platform's functionality reused by many apps should be identifiable, and interactions between the platform and apps should be well defined and explicit.
2. **Resilient**. One defective app should not cause the entire ecosystem to malfunction. The key to resilience is to ensure apps are weakly coupled with the platform through interfaces that do not change over time. This approach of keeping platform–app dependencies to a minimum also makes the entire ecosystem more stable in its performance.

3. **Maintainable**. It should be possible to cost-effectively make any changes within the platform without inadvertently "breaking" apps that depend on it. Conversely, changes in an app should not require parallel tweaking in the platform. Designing for maintainability also increases a platform's composability (i.e., capacity to integrate with new apps).
4. **Evolvable**. Evolvability means the capacity to do things in the future that it was never originally designed to do. For this, the architecture—particularly the interfaces—of a platform must change over time.

1.6.4.3 Summary

Describing a technology view for a platform is a very broad area, and we only touched the surface. However, given all the above, we make the case that platform technology architecture is a key practice for the success of the platform. This constitutes the design of all technology layers that build up from infrastructure, over data access and API standardization towards advanced analytics and the matching algorithm.

On top of that, a well-designed platform for "Opening Services" must be simple and keep the focus on establishing a match between the citizen's needs and the providers. Most government platforms are matching demand and supply in a certain context. The architecture should focus on bringing the content and context together.

We will reframe all the above in the following list of arguments that roll out in the platform technology architecture practice:

- Establish a platform architecture and roadmap, including a standardized approach to infrastructure, data, API, security, application, and business logic technology.
- Integrate platform technology seamlessly with older legacy technologies.
- Foresee technology on the platform to support real-time citizen insights and pro-active service proposals.
- Data is used for environmental sensing, machine learning and/or predictive analytics to optimize the pro-activeness of the platform.
- The Matching algorithm needs specific attention to connect the right providers with the citizens, using behavioural data to offer proactive and highly personalized services.

1.6.5 Governance Related Practices

We already discussed and argued that the ecosystem is a top priority practice for a platform model. However, ultimately, an ecosystem needs to be governed, coached, and steered towards participating in the matching process that is being offered on the platform. Not only the ecosystem needs to be governed, but also the government organization (business and IT) needs to be governed correctly to design, manage, and provide the platform professionally.

Uludag (Uludag et al., 2016) states that we must distinguish between platform governance and platform ecosystem governance. Based on these recommendations, some adjusted features were defined:

- **Platform governance** focuses on setting the right architecture and IT standards (e.g., API & data access) but also defining the services that will be offered on the platform and the associated matching algorithm or criteria whereby a transaction can occur. This is like an inside-out view of establishing a successful platform model.

- **Ecosystem governance** focuses on selecting and/or attracting providers (and keeping them) to the platform, innovate the service delivery model together and allow room for the providers to earn a fair share out of every transaction.

The main difference between platform and ecosystem governance lies in the fact that ecosystem providers or even the citizens cannot be directly controlled by the platform owner via hierarchical power or authority.

Parker et al. (Parker et al., 2016) make a case for self-governance to allow effective platform management. They state that well-run platforms govern their activities following the principles of transparency and participation.

The Leading Edge Forum went one step further and suggested collaborative governance as a practice (Leading Edge Forum, 2019). The authors suggested investing more time and resources into the ecosystem governance than in platform governance. Building/participating in a platform organization means governments need to think of the process more as a multi-party conversation and put much of our effort and creativity into communicating and negotiating. Governance should also be collaborative.

When defining the governance model for the platform model in Australia, PWC (PWC, 2007) described governance as "*the framework for establishing accountability roles and decision-making authority for digital strategy*". We can reuse this definition and apply it to the business platform model as more than ever before ecosystem governance includes the establishment of policies and continuous monitoring of their proper implementation by members of the governing body of an organization.

Uludag (Uludag et al., 2016) compared different models for architecting a platform and applying ecosystem governance. The research of Gawer (Gawer, 2014) was particularly interesting to this research project, as she laid out the characteristics for a platform. When we relate the research from Gawer to the scope of this research project, we can replace her definition of Industry Platforms with government platforms, as all characteristics apply. As such, we can agree with the definition of Uludag on what a Platform Ecosystem is about: "*A platform ecosystem consists of the platform, secondary applications developed for it, the actors providing, extending, and using the platform and applications as well as their interactions and the effects of these interactions*".

As a summary, the importance of a correctly designed and effective ecosystem governance goes on top as a practice above setting up a good internal platform governance. This goes hand in hand with opening the culture of a traditional government organization from doing it all themselves towards promoting an open ecosystem whereby industry partners help deliver the service, and success is determined by the level collaboration of the two.

1.6.6 Data Related Practices

Governments have always collected, analysed, and used information in their legislatively mandated duties; such activities are essential to delivering any service to citizens (Harrison et al., 2012).

For a government business platform to be operating successfully and providing proactive and personalized service advise, literature analysis revealed three aspects of data & information management to be considered:

- **Providing data** - about the services design and the citizens to the ecosystem.
- **Gathering Data** - Learning from the behaviour of the citizens using the platform or from the providers themselves. For example, in the case of the VDAB Matching Platform (see case description), the VDAB asks data from the service providers on how fast the transaction has resulted in a job offer.
- **Using Data** - use of all the data gathered to provide proactive or personalized services advise or make the catalog of services relevant to the citizen's demands and situation.

This requires specific technology to be installed on the platform to provide, gather, or use the data. This is where advanced analytics and artificial intelligence come to play.

Before all, the platform provider needs to decide the types of data, the time horizon for storing data, the formatting of data as well as the terms of use with different parties. The cost reduction, security concerns, and regulation are the three major factors that affect decisions on data management.

The disruption caused by technologies like artificial intelligence, the Internet of Things (IoT), big data, behavioural/predictive analytics, and blockchain technologies—offers the greatest potential in the public sector to transform the way governments engage with citizens (Engin & Treleaven, 2019). These technologies may improve public services in a decentralized, lower cost, more efficient and personalized manner when used on a business platform model ("Opening Services")

As one of the main objectives of a business platform model ("Opening Services") is to provide personalized and proactive services, this can only be triggered by some form of data analytics. A form of analytics that is being to enable that is behaviour analytics. This centres on understanding how and why citizens act, thus enabling predictions about how citizens are likely to act in the future. It enables providers on the platform to make the right responses to the right citizen request at the right time (Engin & Treleaven, 2019). This approach is commonly used on private platforms like Airbnb and booking. com, but still in early deployment phases in government platforms.

Greater aggregation and use of citizen data by the platform can drive improved matching of providers with consumers, benefiting both. This needs to be balanced versus concerns on privacy: citizens need to be properly informed about how the collected data will be used, especially by the industry players, and how governments will protect the citizen's rights in this versus delivering a proactive and/or personalized service.

Ensuring transparency is a key practice in this field. With the rise of Artificial Intelligence technologies, the main argument against their use in the public sector is their dependence on the data that is often biased, incomplete, and/or imperfect, and their lack of transparency of the processes that produce the outcome (Engin & Treleaven, 2019).

In summary, there are 2 practices that emerge from using data and (advanced) technologies on a platform: proper data management and ensuring transparency to all stakeholders what data to use, how it affects the service proposition, and where the red line is.

1.6.7 Challenges to overcome

One can assume that implementing a government business platform model ("Opening Services") does not come without several challenges. Therefore, specific research was done to capture a number of those concerns versus the defined practices. This will give extra guidance to governments on what to expect when defining, designing, and implementing the platform.

Brown et al. (Brown, Fishenden, Thompson, et al., 2014) investigated the digital approach of the UK for several years and defined several challenges the UK government had to overcome (Table 7).

Table 7. Challenges to overcome - UK digital gov. experiences – Source: Brown et al.

Challenge	What is it about?
People	Resistance to change is commonly experienced with any new way of working. Agile, open team practices can deeply affect the culture and values that are in place and result in strong pushback from disoriented individuals.
Tools	Most business processes are not targeted at rapid delivery cycles and extensive experimentation in creating new products and services. Tooling can be a severe inhibitor to agile ways of operating if not aligned with the innovative practices.
Governance	The measures and metrics used to govern the project assume a traditional view of project progress and success. Adjustments are required to provide a balance between governed progress and the need for fast learning cycles.
Citizen - Customer	Any rapid delivery cycle demands more frequent feedback with customers and other stakeholders. Getting the input needed to learn is essential. Yet many consumer–supplier relationships do not readily support such interactions
Financial Control	Product funding cycles are frequently based on progress through various stage-gates such as 'design complete' and 'first customer shipment'. In more agile delivery cycles, flexibility is required to fund activities with different risk profiles, delivering functionality in small slices.
Organizational Structure	Eventually, the whole management structure of the organization becomes stressed when empowered teams interact directly with consumers in rapid iterations of new product features. The command-and-control view of decision making can be directly at odds with the shifting priority-based delivery model of the agile teams

Many of the above challenges related to the usage and institutionalization of the Agile approach to define, design but also release functionality via the platform to the citizens. As civil servants were not used to this approach, it could hamper the realization of the platform related vision.

The Leading Edge Forum (Leading Edge Forum, 2019) researched several reasons for failure to use the platform model. Their analysis - although industry-agnostic - goes alongside with the research of Brown and is summarized here below:

- **Issue 1: Lack of clarity.** There must be clarity from day one what the vision is trying to establish, or confusion will cause misdirection.
- **Issue 2: New, yet unknown rules**. Many of the strategic and economic models of the past (such as Porter's Five Forces) are much less helpful in a platform organization context when one moves from a linear to a platform approach. Traditional techniques are thus not enough to capture the dynamics of a platform model.
- **Issue 3: Measurement challenges.** Most of the platform's value is indirect, certainly for governments. Therefore, new measurements need to be developed to see how governments improve service delivery.
- **Issue 4: Testing challenge.** Testing a platform organization needs to reconcile feedback from all stakeholders. Fail fast and fail forward are more complex in a platform organization and its ecosystem than in a traditional pipeline business and its suppliers.
- **Issue 5: The n-body problem.** One cannot build a successful platform organization without continuously engaging, negotiating, and coming to a consensus with others in the ecosystem.

- **Issue 6: Influence, not control.** Because the platform organization does not directly control the means of production in an ecosystem, there is not always control possible. As governments like to remain in control, this advocates for a good relationship with the ecosystem and influence them to change their behaviour, or they risk losing their trust relationship.
- **Issue 8: Ecosystems are highly contextual.** Each platform organization tends to be unique in terms of the domain it is serving, industry, geography, regulatory, and other considerations. One, therefore cannot copy/paste another platform and enforce one's services onto it. A new platform organization and ecosystem must be significantly adapted to the context.

The Leading Edge Forum also provided advice on how to handle the above challenges, which was adapted to governments:

- Put time and resources into collaboration and communication with all stakeholders, rather than focusing on development.
- Build the platform for constant evolution, not stability. Start small, evolve, fail fast, fail forward.
- Design the organization from the outside in, not the inside out. Citizens must see value and trust is key to access personal data.
- Start building the ecosystem from day one. Focus on business issues that one can using the platform.

Based on experience in dealing with smart city projects over the years, Boorsma also listed a set of pitfalls of government platform projects in his recent book(Boorsma, 2017). These include:

- **Lack of Clear Objectives**: Clear objectives are a must-have. Without them, defining what constitutes success or failure, or the threshold for scaling and replication is impossible.
- **Stuck in Silos**: Many organizations, public and private alike, have traditionally been organized in silos - enclosed environments that harbour their hierarchies, maintain their systems and practices, and gather and retain their data.
- **No Plan to Replicate or Scale**: Many platform projects commence as pilots or a set of pilots without a plan for a scaling of efforts.
- **Digital Divides and the Lack of Community Communications**: digital divides within organizations often slow the initiative down - or worse. If the goals, the means, the risks, and the rewards of the endeavour are not collectively understood early in the project, it can diminish the project and its outcomes at essentially every level.
- **Legacy IT, Sub-optimal Networks**: From a technology perspective, old, existing technology may not be optimal for integration into effective architectures that require to be seamless, secure, and ultimately interoperable to implement high-end platform solutions.
- **Closed Architectures**: digital propositions that involve proprietary or closed architectures have a big risk of failure.

All the listed challenges are in adherence with the compiled list of practices and complemented the research findings.

Table 8. Final list of practices – Own Work

Best Practice	Details
Vision	• There is a clear and documented vision that explains the ultimate goals of embracing the platform model, what it will bring as value and how it will be realized. • The platform model is defined and designed to contribute to optimized government service delivery using the latest technology available (i.e., Cloud, Big Data, API). • The organization is customer-centric and creates platform value through innovation. • The defined platform strategy aligns with the overall business strategy or government objectives. • Using the platform allows the organization to provide better citizen services in a more efficient way. • The vision explicitly mentions trust as a binding factor. Government organizations will ensure a trusted relation between citizens and providers can be established throughout the entire duration of the transaction.
Strategy	• There is a clear top-down driven strategy outlining steps how the platform model and services will be realized. • Services on the platform will be transversal. This necessitates transforming the whole business process and operating model of all organizations involved rather than one or more operational silos. • Digitization of underlying processes and the information around them is a prerequisite.
(Political) Leadership	• (Political) leaders have a compelling long-term vision established around the platform model and how it will add value to all stakeholders, especially citizens. • Political leaders inspire government organizations to embrace the platform model and create a roadmap or strategy to make it real. They strive for efficiency across silos. • There is political support at the highest level and the vision can be linked to the government declaration, where the fundaments for improving citizen services are stated. • A minister is made accountable and responsible for realizing the platform model. • Organizational leaders actively identify and realize new opportunities using the platform model. • Organizational leadership empowers civil servants to collaborate to achieve platform objectives.
Governance	• Establishing accountability roles and decision-making authority for the platform strategy. • Establishment of policies and continuous monitoring of their proper implementation by ecosystem providers to the platform. • Working across organizational boundaries, breaking down silos in favour of the service. • The platform organization is designed from the outside in, not the inside out. • Business and IT become one team with one goal - the platform is the joined product. • Corporate governance to regulate the platform model for citizen services is well described. An independent governance body can audit the organization.
(Innovation) Culture	• Innovation activities are conducted regularly. • Civil Servants are empowered to take calculated risks. • Civil Servants work collaboratively and are supported in cross-skilling and knowledge sharing. • Everyone in the organization has a mandate to think creatively and innovate. • Rigorous and systematic approaches are taken to innovation and change management. • Civil Servants are empowered to work autonomously. • Both small experiments and enterprise initiatives are used to realize innovative ideas.
Value Alignment	• The value is clear to all stakeholders and is continuously monitored by the government. • Citizens and ecosystem providers effectively communicate with the government organization to co-create value. • All staff work in sync to implement a platform vision and know and understand the platform strategy. • The organization works as part of a platform ecosystem.
Business Agility	• The organization quickly senses, creates, and responds to changes in the environment and identifies latent citizen needs. • The organization can pivot based on analysis of customer insight and key performance metrics. • Processes and systems are used to react to rapid business change. • Technology is never a bottleneck for making changes to established ways of working.
Platform Architecture & Infrastructure	• A platform architecture and roadmap are established, including a standardized approach to infrastructure, data, API, security, application, and business logic technology. • The chosen platform technology integrates seamlessly with older legacy technologies. • The Matching algorithm gets specific attention to connect the right providers with the citizens, using behavioural data to offer proactive and highly personalized services. The algorithm is under constant review and adapted.
Risk Management	• Services, systems applications, and tools protect the platform from cyber-attacks and other security risks. • Risk factors are defined, managed, and assessed when it comes to the platform technology investment. • Scalable technology is used to meet demand (e.g., Cloud). • Risk management is an embedded part of the culture and processes of the platform organization. • Risk management encompasses all facets like legal, compliancy, security, operational risk, etc.
Talent and Skills	• Government employees are skilled in platform operating models or can access external expertise with these skills as needed. • High-quality technical staff are attracted to the platform organization. • Government employees can quickly find solutions to business or citizen problems. • A culture of innovation is cultivated. Fail fast, fail early is encouraged.
Customer Experience Design	• User experience research is used to understand citizen pain points to design tailored services on the platform/catalogue. • Customer / Citizen experience is fully integrated across all areas of interaction on the platform. • Customer / Citizen experience is continuously improved on the platform. • Citizens can effectively communicate or provide feedback to address complaints and resolve issues. • Citizens can rate the providers or provide feedback on the received services, at least to the organisation.
Business Ecosystem Design	• It is clear to the ecosystem providers what services they are authorized to execute. • Trust between the partners and the government is established on an ongoing basis. • Openness and transparency for designing the service is ensured. • A clear definition of the value for the ecosystem partners ("what is in it for me") is defined. This should include pricing and renumeration for the services provided. • The ecosystems advices and helps to evolve and innovate the service portfolio over time. • Organisations invest in moderating the continuous evolution of the ecosystem, onboarding new providers, evaluating existing ones.
Technology Ecosystem Design	• Standardize digital connections with the business network (e.g., through API sharing). • Set technological foundation that optimizes collaboration with suppliers and end-users. • Interoperable technology platforms enable the delivery of more efficient outcomes (e.g., through re-use of common services, such as digital identity services).
Data Driven	• Organisations provide the necessary skills and technologies to gather, process and provide the right data on the citizen behaviour. • The rules for partners to collect and use behaviour data are clear. • Technology to support real-time citizen insights and pro-active service proposals is foreseen on the platform. • Privacy is ensured when collecting data. • Data is used for environmental sensing, machine learning and/or predictive analytics to optimize the pro-activeness of the platform. • If possible, organisations should provide quality Open Data to the ecosystem that they can use in their transaction.

1.7 Summary - list of practices

Based on the literature researched, referenced, and documented in the previous sections, Table 8 was created as a summary, representing the necessary practices for managing a business platform model for governments ("Opening Services"):

Next, a tagging was added to indicate where in the organization the responsibility lays for handling the stated practices. Determining the level of responsibility was based on the management triangle model of Anthony (Anthony, 1965). In that model, three levels of management were put forward:

- **Strategic** level where the Board of Directors and Executive Management provide direction and goals for the organization.
- **Tactical** level where Senior and middle management set the framework to reach the goals.
- **Operational** level – where lower management and administration deal with the actual implementation of the tasks required to reach the goals.

This resulted in the mapping of responsibility in an organization versus the practices defined (Table 9).

Table 9. Final List of practices as mapped to organizational responsibility – Own Work

Strategic level	Tactical Level	Operational Level
Vision	Business Agility	Talent and Skills
Strategy	Business Ecosystem Design	Customer Experience Design
(Political) Leadership	Platform Architecture & Infrastructure	Risk Management
Governance	Data Driven	Technology Ecosystem Design
(Innovation) Culture		
Value		

The above table contains the definition of the practices and will be further enhanced in the next chapters.

2 PLATFORM GOVERNANCE MECHANISMS

2.1 Introduction

In this section an answer is provided to Research Question 2: What are the criteria that constitute a well-governed design for a Government Business Platform model for citizen services?

Governments, just like enterprises, are dynamic systems that need a standardized approach to be effective in organizing and executing work. One cannot assume that the earlier defined practices will be enough for individuals to act together in the interest of realizing a business platform model.

Enter governance to make sure the organization effectively and efficiently strives to achieve its stated goals. In this chapter, a governance model to support designing, building, rolling out and sustaining a platform model for governments was researched and documented.

2.2 Approach Used

Earlier management theories for organizing work were based on the separation of thinking (management) and doing (workers). For example, Frederick Taylor (Taylor, 1911) implied in his book Scientific Management that workers were to be controlled by management that set out the rules for executing standardized, routine repetitive tasks. Henri Fayol (Wikipedia, 2020a) believed that management was essential for administration and their focus was on forecasting, planning, organizing, staffing, and commanding. Mary Parker (Wikipedia, 2020b) was far ahead in her thinking in the Theory of Dynamic Administration when she defined organizing as a continuous process, where the focus is on harmonious human relationships in a joint organizational effort.

Modern theories like from Lawrence and Lorsch in their book "Organization and environment" (Paul R. Lawrence, 1967) acknowledge that there is no best way of organizing management behavior; it all depends on the specific situational and environmental situation, such as the phase of the enterprise life cycle, or the availability of resources. As such, there are no universally applicable laws of organization.

Hoogervorst (Hoogervorst, 2017) demonstrated that in post-modern theories, the emphasis is more on the ritualistic nature of (strategic) planning and controlling the execution. This while considering the everchanging complexity, dynamics, and uncertainty of the organization. Hoogervorst acknowledged that change is a constant and governance models are therefore not carved in stone.

In the context of a platform model, Tiwani stated that governance is also how a platform owner influences its ecosystem (Tiwana, 2014). For her, the role of governance is to orchestrate foremost the integration of unique contributions of providers into a platform's ecosystem. Finally, she states that the optimal governance structure for a platform is often the simplest one at the least cost. This view contradicts with the structured approach that many governance frameworks prescribe, claiming that governance models need to be well designed to the business objectives to bring the value.

The research followed the viewpoint that a governance model must be defined for purpose and re-designed or adapted frequently according to the changing needs. A governance model is therefore more than a set of management guidelines to simply document and follow progress, each governance model needs constant monitoring and steering mechanisms build inside.

With the increasing importance of IT systems for realizing the goals of governments and enterprises, the need for specific governance between business and IT organizations has come to the forefront. At its most basic definition, IT governance is the process by which decisions are made around IT investments (Symons et al., 2005). IT governance as defined by Calder and Moir (Calder, 2009) is a matter of optimising the use of IT investments through strong collaboration and communication between the business and IT's leaders and their strategies.

Many frameworks became available that provide guidance on IT governance and how to set up the right processes and controls to give reasonable assurance that the intended goals and/ or value are achieved for the realization of the organization's strategy.

Research was done to find the most appropriate framework to build a governance model for our specific case. We found guidance in the following international ISO standards:

- ISO27001 (Calder, 2013), which details the requirements for establishing and maintaining an Information Security Management System (ISMS), encompassing people, processes, and technology.

- ISO20000 (ISO/IEC, 2005), which sets out a specification for a service management system, including coverage for ongoing maintenance and continual improvement.
- ISO38500 (Calder, 2008), which is an overarching framework of principles and guidance for the governing body of an organisation, specifically IT. This standard is based on the earlier Calder Moir Framework (Calder, 2009).

Next, other frameworks were added to the research, like the IT Service Capability Maturity Model (CMMI Institute, 2018), Six Sigma (Smith, 2002), IT Balanced Scorecard by Robert Kaplan and David Norton (Kaplan & Norton, 1992), ITIL (Axelos, 2016), M_o_R (AXELOS, 2012), Prince2 (Axelos, 2015), MSP (Axelos, 2020), eTOM (TM Forum, 2020) and PMBOK (PMBOK, 2017).

Specific research was done into COBIT2019 (ISACA, 2019b) as an overall control framework that provides best practices, tools and guidance for the effective management and governance of enterprise IT.

Alreemy (Alreemy et al., 2016) defined a set of critical success factors to select the right governance framework. They include attention to stakeholder involvement, management support, financial support, organizational effects like culture, alignment between business and IT, regulatory compliance, IT structure, IT staffing, and managing the implementation in project-controlled way. Based on these success factors, COBIT2019 was selected as the most suitable governance framework for this research. The following picture represents how the COBIT2019 framework links with the other frameworks:

Figure 8. Mapping COBIT vs other IT governance frameworks - Source: www.isaca.org

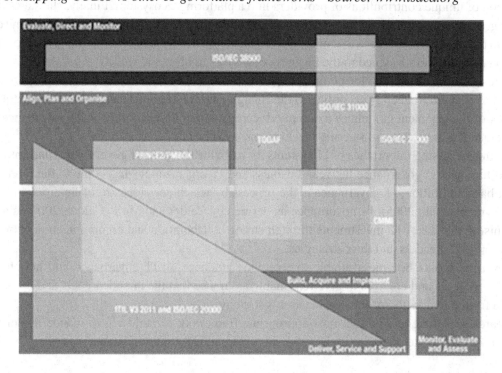

COBIT2019 also provided the right framework for building an IT governance model, being the integrated "Enterprise Governance of IT (EGIT)" approach. This approach is about defining and implementing processes, structures, and relational mechanisms that enable both business and IT stakeholders to execute their responsibilities in support of business/IT alignment and the creation and protection of IT business value (De Haes et al., 2020).

Now that a framework is chosen, further study was needed to define a list of prioritized processes, organizational structures, and relational mechanisms relevant to a business platform model for governments.

2.3 Defining the Scope of the Organization to Design For

As stated before, a governance model needs to be defined according to the specific needs and dynamics of that organization. As such, it was needed to define the characteristics of the government organisation we have in mind for this work in a bit more in detail, so that we all have a clear view on what we are designing a governance model for.

COBIT2019 has introduced a design guide (ISACA, 2019a) for tailoring a governance system. In the examples of the Design Guide and Toolkit, Example 3 refers to a large and high-profile government agency. However, after studying the example and because not all governments around the world are organized and structured the same way, the necessity for defining a more detailed and different type of organisation was needed.

The Government of Flanders, a regional government in Belgium, was chosen as the target for going through the design guide. Over the last 30 years, federal power was gradually and increasingly distributed to the regions (Flanders, Wallonia, Brussels). Each region now has responsibility to define its own policies that are fit for purpose. The regional government is elected and has its own government declaration. The administrations operationalizing the citizen services are independent and each have their own budget. They each report to a minister and have agreed an overall services agreement how the government declaration will be executed. Cities and local communities have their own IT budget to define local initiatives. Local political leadership is elected, and a local coalition is formed. Few mandatory IT standards are enforced from the regional government to all its organisations. There is a central IT Department defining guidelines and horizontal services (e.g., for data exchange), but often not enforced as standard.

These organisations typically range in size from 300- to 5.000+ people. They have their own IT Team, decide on IT sourcing, applications, and infrastructure to use internally and externally. Some do not have a dedicated CIO, often the board of directors decides on IT investments. The IT responsible can tap into several small or large contracting agreements to outsource (parts of) the IT system. Central guidelines (like a central governance model) cannot be enforced.

Our target organization for creating a governance model is thus an autonomous operating organisation that has the mandate and budget to define how technology is used to define and execute citizen services.

2.4 Governance Processes

2.4.1 Approach Used

To determine the list of processes suitable and linked to the defined practices, the following approach was used:

Figure 9. Approach for determining processes - Own work

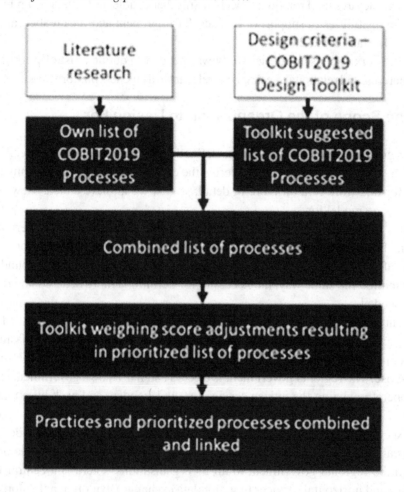

When defining the different levels of "Government as a Platform" earlier, a list of processes was defined based on literature research. This was based on own research and insights.

Next to that, the COBIT2019 Design Toolkit was used to determine a theoretic set of processes, based on defining the right values for the design criteria matching the organization criteria.

The two lists were then combined into one set of weighted processes. Adjustment factors were applied to certain processes to align the research with the results of the Design Toolkit. On top of that, the top priority processes were defined. Criteria for what a top priority process is, were also defined.

Finally, the processes were attributed to the practices, so that there is clarity what processes to focus on for the relevant practices. In the following chapters this approach will be documented in full detail.

2.4.2 Applying the COBIT2019 Design Toolkit

The COBIT2019 Design Toolkit was used to define the list of processes. The toolkit consists of 10 guiding design factors. For each design factor, a set of questions needed to be answered. The practices defined earlier were used as guidance to select the right value or score the questions for each of the design factors. Comments were added on why each score was given.

Based on the provided answers, the Design Toolkit calculates a set of processes to put in place and what their weight or importance is to realize the overall goals of the defined organization. Because the prioritized results did not match the literature research, adjustment factors were used to align the priority with the practices defined. The toolkit allows and even promotes this. This was done by the author and documented further on in this chapter.

2.4.2.1 Design Factor 1: Enterprise Strategy

Organizations can have different strategies, which can be expressed as one or more of the archetypes in the assessment. The goal of this design factor is to determine how the organization rates versus each of the archetypes.

Literature research indicated that the business platform model for governments necessitates a different approach (i.e., vision and strategy) to provide services to citizens. The focus is much more on innovating the citizen services delivery model and using technology. Governments do not need to make a profit, as such cost leadership is ranked low. The market for a government is defined, and therefore growth or acquisition is not that relevant. Governments do recognize that they need to provide a good service to their citizens and remain a loyal and stable partner, that is why Client Services / Stability is ranked high.

This resulted in the input scoring shown in Table 10.

Table 10. Scoring of Design Factor 1 (Enterprise Strategy) – Source: COBIT2019 Design Toolkit

Value (Archetype)	Importance (1-5)	Reason for score
Growth/Acquisition	1	Governments are not seeking growth or profits but use the platform model for improving service delivery to their citizens. Hence the importance here is very low.
Innovation/Differentiation	5	Governments are using the platform model to innovate citizen service delivery and be more efficient using the latest technology, hence the high score.
Cost Leadership	1	Here again, although driving efficiencies is a goal, governments are not seeking cost leadership as such. Therefore, a low score was attributed.
Client Service/Stability	4	The platform business model is all about providing good citizen services in a stable and trusted ecosystem, hence a high score here.

The most suitable archetype for an organisation that applies the platform business model is thus the innovation / differentiation one.

2.4.2.2 Design Factor 2: Enterprise Goals

COBIT2019 states that enterprise strategy is realized by the achievement of (a set of) enterprise goals. These goals are defined in the COBIT framework and are evaluated for prioritizing certain processes. Governments have different enterprise goals than companies. Where in companies, the main driver is revenue and profit, governments are seeking to provide value, comfort, and security to their citizens.

The enterprise goals were thus scored as shown in Table 11.

Table 11. Scoring of Design Factor 1 (Enterprise Goals) – Source: COBIT2019 Design Toolkit

Value	Importance (1-5)	Why?
EG01—Portfolio of competitive products and services	1	This is rated low as government organizations have no commercial objectives.
EG02—Managed business risk	2	A government cannot go "out of business", worst case they stop the platform model and revert to a classical approach.
EG03—Compliance with external laws and regulations	5	Governments will have to make sure the necessary regulation is available to administer a fair transaction on the platform. Governments will have to set specific rules to provide trust to the citizens, even if they are the platform provider themselves. The ecosystem must be well controlled.
EG04—Quality of financial information	4	Transparency of all transactions and how the ecosystem is rewarded is key for success.
EG05—Customer-oriented service culture	4	Even if there are no commercial objectives, providing a customer-oriented culture is key to success. This is necessary to create the network effects. If either the citizen as a consumer or the ecosystem as a provider is not feeling well threated, this will diminish the success of the platform.
EG06—Business-service continuity and availability	3	There needs to be ensured continuity of the services on the platform. Irregular delivery or availability will kill the reputation of the platform.
EG07—Quality of management information	4	Even for governments, it is important to understand the dynamics that are happening on the platform, so decent metrics and management information is necessary. If only to report back to the politicians what the effect of the platform is.
EG08—Optimization of internal business process functionality	3	The goal is not to make a profit or optimize business cost TCO; but using the platform model can provide a shift in workload internally as the transaction will be executed on the platform, between the citizen as a consumer and the ecosystem as a provider.
EG09—Optimization of business process costs	2	Although the platform will create efficiency in the business process, and potentially lower the TCO, it is not a primary goal for governments.
EG10—Staff skills, motivation, and productivity	4	Using the platform model will require new skills and capabilities. If the staff is not trained and motivation is low, then the platform will not be attractive or used.
EG11—Compliance with internal policies	5	This goes together with EG03. Additionally, the government will have to comply with its own rules it has set to all platform providers
EG12—Managed digital transformation programs	4	Using new technologies to deliver services must be well managed. If the platform is not well managed, with new skills and capabilities, the success will be low.
EG13—Product and business innovation	5	Using the platform is a new way to deliver government services, driven by technology. So, innovation and agility are key to the success of the platform.

2.4.2.3 Design Factor 3: Risk Profile

COBIT2019 states that the risk profile identifies the sort of IT related risk to which the enterprise is currently exposed and indicates which areas of risk are exceeding the risk appetite. Next was thus to determine the risk profile for the organisation using the government platform business model ("Opening Services"). Each risk is scored on 2 dimensions: impact to the organisation when the risk occurs and likelihood the risk will hit the organisation.

The risk profile was scored as indicated in Table12.

Table 12. Scoring of Design Factor 3 (Risk Profile) – Source: COBIT2019 Design Toolkit

Risk Scenario Category	Impact (1-5)	Likelihood (1-5)	Why?
IT investment decision making, portfolio definition & maintenance	3	3	Like any other project, the platform model will necessitate investment in both technology and resources to set up, both from the business as from the IT side. Without a proper process for decision making on the budget, the risk is that the platform does not get created or launched.
Program & projects life cycle management	4	3	The design, build, roll-out, and maintenance of the government platform must be well managed, or the risk is that the reputation is negative before it gets launched. This will hamper the network effects.
IT cost & oversight	2	2	The necessary processes must be in place to provide cost oversight, or the risk is that costs for the platform model will go out of control. This is, however, not much different than for other projects at the government level.
IT expertise, skills & behaviour	5	5	Having the right skills and capabilities is key to design and launch a good functioning platform model, also redesigning the services requires new skills. If this is not well provided for, the risk is that the reputation of the platform is impacted negatively.
Enterprise/IT architecture	4	4	This is one of the practices identified. If not done right, the risk is that the services are not well displayed, have the wrong information, or are not transparent to citizens. Interoperability of APIs is a key risk factor if not done correctly.
IT operational infrastructure incidents	3	2	Government platform models are a bit less critical than Amazon or Uber. However, when not managed, the risk is that consumers and providers will not be able to facilitate a transaction.
Unauthorized actions	3	4	Security and guaranteeing that citizen data is well kept is a key item to deal with. If the citizen's data is breached by unauthorized actions, then the whole platform will not be trusted anymore.
Software adoption/usage problems	3	3	Using the platform model will require the adoption of new technology. This needs to be well managed. If the software is not well designed, the risk is that the platform does not function properly, and services cannot be ordered.
Hardware incidents	2	2	It was stated in the literature that platforms should adopt Cloud technology to host the platform. When designed well, the risk of hardware incidents is neglectable, although also Cloud providers can have outages. Good SLAs are needed.
Software failures	2	3	The risk of software failures on a platform is not larger or smaller than for other software. The impact is larger as the citizens will immediately see the error. So, this needs a decent software quality process to be in place.
Logical attacks (hacking, malware, etc.)	4	5	Like unauthorized access, security is a key process to install and maintain on the platform. If the platform is breached, the government will be in the press, something no politician wants to have.
Third-party/supplier incidents	2	2	Incidents by ecosystem providers will damage their reputation. Nevertheless, the government (as platform owners need to remain vigilant and kept in the loop or the risk is that the reputation of the government will be hampered.
Noncompliance	5	2	Governments cannot afford to be non-compliant with their own rules. As there is much focus on being compliant, the likelihood of this occurring is low.
Geopolitical Issues	2	2	Government platforms are usually local, regional, or at the country level because the services are specific to the country. In the future, if for example, the EU decides to centralize services across Europe, this might become more relevant (e.g., Open Data platforms, where a consolidation towards the EU platform is now a given).
Industrial action	1	1	The impact is on the ecosystem provider side here.
Acts of nature	2	2	This is not different than for other IT systems.
Technology-based innovation	5	3	New technologies are being used when platforms are launched, so this needs careful preparation and control. The impact is big for all stakeholders, certainly for the government as a platform provider.
Environmental	1	1	This is not seen as a risk unless global warming puts the data centre below sea level.
Data & information management	4	4	A key practice is to manage information about the citizen behaviour on the platform. Therefore, a good data management process is needed. If not managed properly, the stakeholders will miss essential insights in the transaction of the citizen with the ecosystem.

2.4.2.4 Design Factor 4: I&T Related Issues

In this Design Factor, COBIT2019 zooms in on the risks and nominates several related issues. This idea is to rate the importance of the issues. This will further refine the risk profile of the organisation and the

Table 13. Scoring of Design Factor 4 (I&T Related Issues) – Source: COBIT2019 Design Toolkit

I&T-Related Issue	Importance (1-3)	Why
Frustration between different IT entities across the organization because of a perception of low contribution to business value	1	Even if there is frustration inside the government organisation, this should not impact the performance of the platform. With good leadership, this can be well mitigated.
Frustration between business departments (i.e., the IT customer) and the IT department because of failed initiatives or a perception of low contribution to business value	1	We have seen that one of the practices is agility. If managed in an agile way in which failures are allowed, then frustration is a side effect.
Significant I&T-related incidents, such as data loss, security breaches, project failure and application errors, linked to IT	3	This is one of the defined practices and risks.
Service delivery problems by the IT outsourcer(s)	2	Good delivery SLAs of the suppliers are a defined practice.
Failures to meet IT-related regulatory or contractual requirements	3	This is a key practice and identified risk.
Regular audit findings or other assessment reports about poor IT performance or reported IT quality or service problems	2	Governments take great pride in compliance, so auditing gets done regularly. This is not a real issue for platform models, not any more or different than for other IT projects.
Substantial hidden and rogue IT spending, that is, I&T spending by user departments outside the control of the normal I&T investment decision mechanisms and approved budgets	2	This is not different than for another IT Project.
Duplications or overlaps between various initiatives, or other forms of wasted resources	3	Research indicated that most government agencies create platforms on their own (note: focus on "Opening services"). For Providing Functionality and data, this seems to be more centralized.
Insufficient IT resources, staff with inadequate skills or staff burnout/dissatisfaction	3	This was identified as a key practice.
IT-enabled changes or projects frequently failing to meet business needs and delivered late or over budget	2	This is not different than for another IT Project.
Reluctance by board members, executives, or senior management to engage with IT, or a lack of committed business sponsorship for IT	3	This was identified as a key practice.
Complex IT operating model and/or unclear decision mechanisms for IT-related decisions	2	This is not different than for another IT Project.
Excessively high cost of IT	1	This is not different than for another IT Project.
Obstructed or failed implementation of new initiatives or innovations caused by the current IT architecture and systems	3	This was identified as a key practice.
The gap between business and technical knowledge, which leads to business users and information and/or technology specialists speaking different languages	3	This was identified as a key practice.
Regular issues with data quality and integration of data across various sources	2	This was identified as a key practice.
High level of end-user computing, creating (among other problems) a lack of oversight and quality control over the applications that are being developed and put in operation	1	This is not different than for another IT Project.
Business departments implementing their information solutions with little or no involvement of the enterprise IT department (related to end-user computing, which often stems from dissatisfaction with IT solutions and services)	3	As government agencies tend to choose technology on their own, this is a serious risk and needs to be well mitigated.
Ignorance of and/or noncompliance with privacy regulations	3	This is key to governments as they are both the regulator and user that needs to comply.
Inability to exploit new technologies or innovate using I&T	3	This was identified as a key practice.

need for associated processes to implement.

The assessment (Table 13) was done with regards to the government platform business model ("Opening Services").

2.4.2.5 Design Factor 5: Threat Landscape

The threat landscape determines how an organisation is under threat from internal and external risks. High treat organizations can be due to its geopolitical situation, industry sector or profile.

The following assessment was done on Threat Landscape with regards to the government platform business model ("Opening Services"):

Table 14. Scoring of Design Factor 5 (Threat Landscape) – Source: COBIT2019 Design Toolkit

Value	Importance (100%)	Why?
High	90%	Security, keeping citizen data safe, transparency without being hacked are key drivers to keep the reputation of the platform high. The fact that the platform is operated by the governments even raises the stakes higher. No politician wants to have a press article that the platform is hacked, or services are being abused (or data sold to third parties).
Normal	10%	See above.

Based on this assessment, a government organisation using the platform business model is considered under high threat from external factors like hacking or security breaches.

2.4.2.6 Design Factor 6: Compliance

COBIT2019 checks for compliance requirements to which the organization is subject to. The questions determine if the organization is basically low, normal, or highly exposed to compliance.

The following assessment was done on compliance with regards to the government platform business model ("Opening Services"):

Table 15. Scoring of Design Factor 6 (Compliance) – Source: COBIT2019 Design Toolkit

Value	Importance (100%)	Why?
High	100%	Governments need to comply with their own rules set for platform companies. First, the government needs to define rules to regulate the platform interactions. This is not something governments are used to. Think about the debate of regulating Uber in Brussels / Belgium. When governments now apply the model themselves, they will need to comply with their own rules. Furthermore, ecosystems must be fair, and transparency is key to get the trust of the citizens. Governments are in a dual role here.
Normal	0%	
Low	0%	

Basically, it was determined that a government is high on compliance.

2.4.2.7 Design Factor 7: Role of IT

COBIT2019 now checks the role of IT for the organisation. The design factor determines if IT is strategic or supportive for the defined business goals.

The following assessment was done on Role of IT with regards to the government platform business model ("Opening Services"):

Table 16. Scoring of Design Factor 7 (Role of IT) – Source: COBIT2019 Design Toolkit

Value	Importance (1-5)	Why?
Support	1	See below.
Factory	1	See below.
Turnaround	1	See below.
Strategic	5	IT is strategic for the well-functioning of the platform. Many new technologies will be applied, and architecture is defined as a key practice. Without IT, there is no platform model, and services will not be orchestrated between citizens and the ecosystem.

It was without any doubt determined that the role of IT is strategic for the realisation of a government business platform model. A platform is based on technology and thus needs to get a very central place in the governance model.

2.4.2.8 Design Factor 8: Sourcing Model for IT

In this design factor, COBIT2019 validates how IT can be sourced for achieving the business goals. The model validates three methods of IT Sources and determines how much emphasize on certain processes like supplier management (to name an example) and contract management structures should be.

The following assessment was done on Sourcing Model with regards to the government platform business model ("Opening Services"):

Table 17. Scoring of Design Factor 8 (Sourcing Model for IT) – Source: COBIT2019 Design Toolkit

Value	Importance (100%)	Why?
Outsourcing	20%	Research has indicated that technology-oriented skills and capabilities for the platform model can be sourced from the market. This includes the outsourcing of certain maintenance tasks as well.
Cloud	60%	One of the defined practices is to apply Cloud as a preferred delivery mechanism due to the availability of technology, make sure the platform is resilient and scalable when services are introduced or when network effects start to play.
Insourced	20%	Keeping the architecture, design, and interoperability standards inhouse is key to keeping a neutral view. This is part of the defined practices to put much internal focus on a good architecture design.

In the case of the business platform model for governments, a higher importance to Cloud and Outsourcing was attributed. This is because technology is the cornerstone of a successful platform model and as technology is moving forward very fast, attracting resources internally would slow down the roll out of such a model.

2.4.2.9 Design Factor 9: Implementation Methods

Next, COBIT2019 checks what implementation method the organization is experienced with. This will determine the need for more rigid project management processes or adopting other methods like agile.

The following assessment was done on Implementation Methods with regards to the government platform business model ("Opening Services"):

Table 18. Scoring of Design Factor 9 (Implementation Methods) – Source: COBIT2019 Design Toolkit

Value	Importance (100%)	Why?
Agile	70%	One of the practices is to build the platform iteratively. This will mean that the agency will have to adopt an agile approach to design and deliver new functionality of services to the platform.
DevOps	30%	Inherit to using Cloud technology is the DevOps approach, whereby new services or APIs can go to production without interrupting the platform operations.
Traditional	0%	Working in the traditional waterfall approach is not considered, see practice of Agile

Literature research on platform models strongly advised to use an Agile approach. It is one of the defined practices. This may impose stress on traditional oriented organisations that want to adopt a business platform model. In the governance model, extra means for coaching Agile approaches will be needed.

2.4.2.10 Design Factor 10: Technology Adoption Strategy

The last design factor of COBIT2019 checks if the organisation has a culture of being a first mover on using new technology or rather a slow adopter. This will determine what processes for IT need to be further prioritized.

The following assessment was done on Technology Adoption Strategy with regards to the government platform business model ("Opening Services"):

Table 19. Scoring of Design Factor 10 (Technology Adoption Strategy) – Source: COBIT2019 Design Toolkit

Value	Importance (100%)	Why?
First mover	0%	This might sound contradictory, because one of the overall practices of a platform company in the commercial space, is to be the first mover, be the first to establish the platform and lock the ecosystem to the platform. Being a first movers does not guarantee everlasting success, but the action is where the volume of transactions is, and people are always looking for innovators to challenge the incumbents.
Follower	20%	
Slow adopter	80%	Hence it may sound strange to score the government platform model as a slow adopter. That is because the commercial play is not there for a government model. Research has indicated governments are generally late in adopting new technology, and this is certainly the case for the platform model. However, this does not need to be a drawback, if the government learns from the best practices and applies this consequently, a good services platform is better than being the first mover.

2.4.2.11 Results from the Design Toolkit

After filling in the ten design factors with the above-mentioned values, the Design Toolkit calculated the importance of each process. The Toolkit thus indicated how important each process is, by giving the process a calculated score between -100 and +100. The higher the score, the more important the institutionalization of this process is recommended.

Below is the resulting graph that the toolkit provides:

Figure 10. COBIT2019 Design Toolkit - Suggested processes & weights - Source: COBIT2019 Toolkit

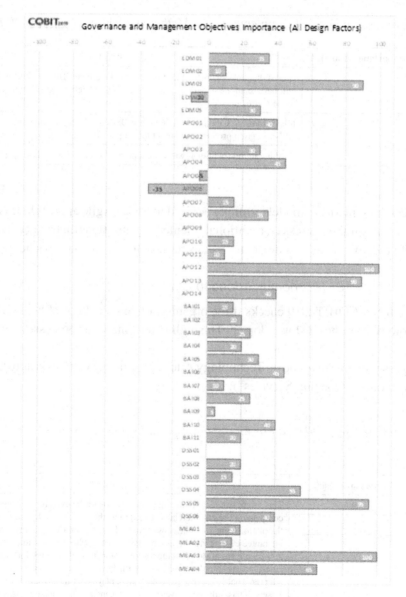

However, discrepancies were noted versus the literature research. For example, the chart gave process APO02 – Managed Strategy a score of zero (0). Given the strategic context of government business platform models ("Opening Services"), this was not reflecting the reality.

The toolkit allowed for re-prioritizing the defined processes, by attributing a different weighing factor to each score. How this was done, is explained in the next paragraphs below.

2.4.3 Combining research and design toolkit results

The Design Toolkit provided a first list of suggested processes to be institutionalized for governing a government business platform model ("Opening Services"). Next, an analysis was done to compare the results of the toolkit versus the literature research. The result is documented in the table below, including an overview of decisions taken about the prioritization of COBIT2019 processes.

To understand the decision taken for each process, it is necessary to understand the build-up of the table below:

- The first column contains a list of all COBIT2019 processes
- The second column "Design Toolkit Value" lists the score that the design toolkit attributed for each process, given the answers provided.
- The column "Research Priority" indicates if this process was also identified in the literature research and its importance. In the literature research, only a High – Low – Medium judgement score was used by the author.

Next, decisions were made what to do with each process, if we are to keep it in the consolidated list, or if we need to add it.

- **Keeping** the process, means that the process was important in either the literature research or the Design Toolkit.
- **Adding** the process, means that either a high score was attributed in the Design Toolkit or in the literature research. Processes that got either a low score in the Design Toolkit or a low judgement in the literature are marked as not a priority.
- Lastly, there were also processed that got a medium score in the Design Toolkit but were not captured in the literature reviewed. For these ones, the decision was to **integrate** them into the list, meaning either as a separate process or link it to another one.

Given the above, the following columns in de table below can now be explained:

- The column "Action" describes what decision is taken. Either the process is kept ("Keep"), or added to the list ("Add"), or given a low priority ("Not a top priority") or can be integrated with other processes ("Integrate")
- The column "Remarks" provides an explication to justify each action and provides traceability to the consolidated resulting list.

The table is thus:

Table 20. COBIT2019 variation analysis versus functional platform models - Own Work

COBIT2019 Process	Design Toolkit Value	Research Priority	Action	Remarks
EDM01 - Ensured Governance Framework Setting and Maintenance	35	High	Keep	The toolkit provides a much lower value than expected versus the priority given based on literature research.
EDM02 - Managed Benefits delivery	10	High	Keep	The toolkit provides a much lower value than expected versus the priority given based on literature research.
EDM03 - Ensured Risk Optimization	90	Not selected	Add	Process "APO12 - Managed Risk" was put forward, so this is a correct add on for the validation.
EDM04 - Ensure Resource Optimization	-10	Not selected	Keep	The focus was put on process "APO07 - Managed Human Resources", but the value of the toolkit is surprisingly low.
EDM05 - Ensured stakeholder engagement	30	High	Keep	The toolkit provides a much lower value than expected versus the priority given based on literature research.
APO01 - Managed I&T Management Framework	40	Not selected	Add	This process was not withheld in the research, but it is correct to have this defined as a process.
APO02 - Managed Strategy	0	High	Keep	The Design toolkit attributes a much lower score than research indicates. Note that for this process, contact was made with the author of the toolkit, but no indication of an error in the toolkit was discovered. As such, the conclusion remains valid.
APO03 - Managed Enterprise Architecture	30	High	Keep	The Design toolkit attributes a much lower score than research indicates.
APO04 - Managed Innovation	45	Not Selected	Add	Innovation is key to platform design and was identified as a practice, hence this process will have to be in place for managing innovation on the platform specifically.
APO05 - Managed Portfolio	-5	High	Keep	The Design toolkit attributes a much lower score than research indicates. Portfolio is defined in the literature as both the technology options as the business services that are offered on the platform.
APO06 - Managed Budgets and Costs	-35	High	Keep	The score of the toolkit is much lower than expected.
APO07 - Managed Human Resources	15	High	Keep	The score of the toolkit is much lower than expected.
APO08 - Managed relations	35	High	Keep	The Design toolkit confirms the importance of managing relationships in the ecosystem but attributes a lower-than-expected score.
APO09 - Managed Service Level Agreements	0	High	Keep	The Design toolkit attributes a much lower score than research indicates. Note that in this case, contact was made with the author of the toolkit, but there is no indication of an error in the toolkit, so the conclusion remains valid.
APO10 - Managed Vendors	15	High	Keep	Managing a platform requires a good relationship with the ecosystem; this can be technology or business inspired. Literature suggests spending quite some time on this. We will keep this process.
APO11 - Managed Quality	10	Not Selected	Add	Was not selected in the literature, does not get a high score in the design toolkit, but cannot be ignored either.
APO12 - Managed Risk	100	High	Keep	The score of the Design Toolkit is in line with the found evidence in the literature, and managing risk is crucial to the success of launching a platform model.
APO13 - Managed Security	90	High	Keep	The score of the Design Toolkit is in line with the found evidence in the literature, ensure a secure approach to facilitating transactions or processing citizen data is crucial for the success of a platform model.
APO14 - Managed Data	40	High	Keep	Seen the importance of collecting and analysing citizen behaviour data on the platform and how to convert these insights into proactive and personalized propositions, the score of the design toolkit is low versus what the literature states.
BAI01 - Managed Programs	15	Not Selected	Not a top priority	The Design Toolkit scores this lower than expected; but all programs need to be managed correctly to be able to design, launch and operate the platform model.
BAI02 - Managed Requirements Definition	20	Not Selected	Not a top priority	Literature proposes the agile approach, where the backlog is used to determine requirements. This does not hide away that requirement definition must be well managed, regardless of the method used.
BAI03 - Managed Solutions Identification and Build	25	Not Selected	Not a top priority	Via literature research, we stressed the importance of the architecture process, but eventually, all solutions need to be in line with standards (if any exist for the platform model) - the score reflects the literature.
BAI04 - Managed Availability and Capacity	20	Not Selected	Not a top priority	Due to preference in the literature to use Cloud as an infrastructure environment, the need for availability and capacity is more a monitoring than a planning process. However, we must manage the workload as more usage will increase the run time costs as well.
BAI05 - Managed Organizational Change	30	High	Keep	Rolling out a platform model requires organizational change; this is one of the defined practices. It is therefore surprising that the score of the design toolkit is not so high.
BAI06 - Managed IT Changes	45	Not Selected	Add	Using the platform model will impose the introduction of new technologies. The literature suggests an agile approach to testing what these technologies bring to the table. The Design Toolkit gives a reasonable high score to this process; hence it will be added to the list.
BAI07—Managed IT Change Acceptance and Transitioning	10	Not Selected	Not a top priority	Both the research and the toolkit are aligned stating this is not the highest priority, although change is an important element in defining a platform model.
BAI08 - Managed Knowledge	25	High	Keep	Research shows this to be an important element as a new business operating model is being introduced, requiring many people in the organization to get the necessary knowledge (which is not only about technical aspects) to drive the change.
BAI09—Managed Assets	5	Not Selected	Not a top priority	Both the research and the toolkit are aligned, stating this is not the highest priority. As platforms will run in the Cloud, asset management reverts to service management as the assets will be owned by the Cloud provider.
BAI10—Managed Configuration	40	Not Selected	Not a top priority	Both the research and the toolkit are aligned, stating this is not the highest priority. This is because this process is assumed to be part of the service that the Cloud provider will deliver.
BAI11—Managed Projects	20	Not Selected	Not a top priority	Both the research and the toolkit are aligned, stating this is not the highest priority. Nevertheless, being agile and going for many little steps forward is no excuse for not having a decent project management structure, approach, and standards

continued on following page

Table 20. Continued

COBIT2019 Process	Design Toolkit Value	Research Priority	Action	Remarks
DSS01—Managed Operations	0	Not Selected	Not a top priority	As most platform models will operate using a Cloud provider, the IT-related function is assumed to be the responsibility of the Cloud provider and part of the Service SLA to the government. On the business front, the organization will have to make sure the service descriptions, ecosystem agreements, and KPI are well defined.
DSS02—Managed Service Requests & Incidents	20	Not Selected	Not a top priority	For the IT side, we can assume this to be the responsibility of the Cloud provider. On the business side, any incidents with transactions failing to establish on the platform needs to be investigated by the business team as well.
DSS03—Managed Problems	15	Not Selected	Not a top priority	Same assumptions as for DSS02 apply.
DSS04	55	Not Selected	Not a top priority	Same assumptions as for DSS02 apply.
DSS05—Managed Security Services	95	Not Selected	Integrate	As security is a key practice for operating a platform model, this deserves a place in the process model, even if the platform is hosted at a Cloud provider, the government will still have to manage the overall security on the platform.
DSS06—Managed Business Process Controls	40	Not Selected	Integrate	Even if most of the technical infrastructure is delivered and operated by a Cloud Providers, the government as a business will still have to set up its own business (transactions) controls and follow up on what is happening on the platform.
MEA01 - Manage Performance and Conformance Reporting	20	High	Keep	Research indicates this is to be carefully controlled, see the specific situation that governments are both the regulator and provider of a platform. The Design Toolkit scores this very low. This process will be kept as it is a key practice to deal with.
MEA02—Managed System of Internal Control	15	Not Selected	Not a top priority	Both the research and the Design Toolkit score this low, but this does not mean that no system for internal control is needed.
MEA03 - Managed compliance with external requirements	100	High	Keep	The design toolkit score is aligned with research. Governments need to comply with legal rules on operating the platform in a secure, transparent, and value add way to all stakeholders.
MEA04—Managed Assurance	65	Not Selected	Integrate	Research indicates that compliance for government organizations is important, so a form of assurance process is needed. The Design Toolkit provides a big enough score to justify this process is in place.

The above exercise was needed to apply step 4 of the toolkit ("*Conclude the scope of the governance system*"). In this step 4, the toolkit allows to adjust the priorities of each process, by adding a weighing factor, based on own insights or – in this case – literature research.

The following logic was applied in adjusting the scores:

- Where literature indicated it was high priority, the weighing factor was adjusted upward with adding +50 to the original score of the design toolkit.
- The exception was process APO02 - Managed Strategy, where +100 was added to the score. As described, this was because literature indicated this to be of high importance, but the Design Toolkit originally calculated a score of zero (0).
- For process APO13 - Managed Security and process DS05 - Managed Security Services, the score was upgraded to +100. This was because neither the literature nor the original score gave a high score to security, but the subject of security was frequently mentioned in the list of practices, be it under the practice of Risk. This way, the author decided to give more attention to security.

Figure 11 is a screen shot from the Design Toolkit that illustrates how the application of the weighing factor adjusted the priority. This gave the following readjusted scoring the Design Toolkit.

The next step was to get a prioritized list of processes extracted from the literature review and the Design Toolkit. For convenience, the numeric scores of the Design Toolkit were converted into a Low – Medium – High priority scale. To do that, the author applied the following arbitrary filter based on the score of the design toolkit:

Figure 11a. Adjusted weight factors per process - Source: COBIT2019 Design Toolkit

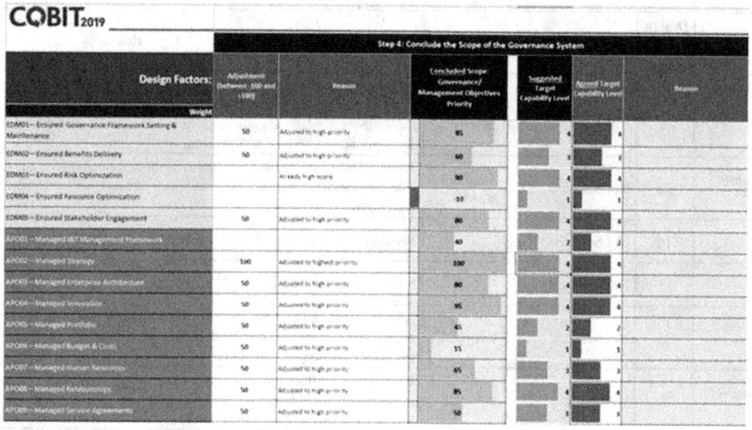

Figure 11b. continued

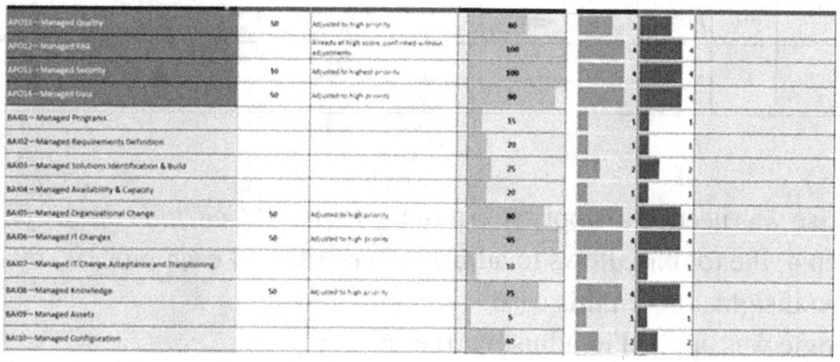

Figure 11c. continued
Note: the capability level that the design toolkit also delivers was not further used in this research project.

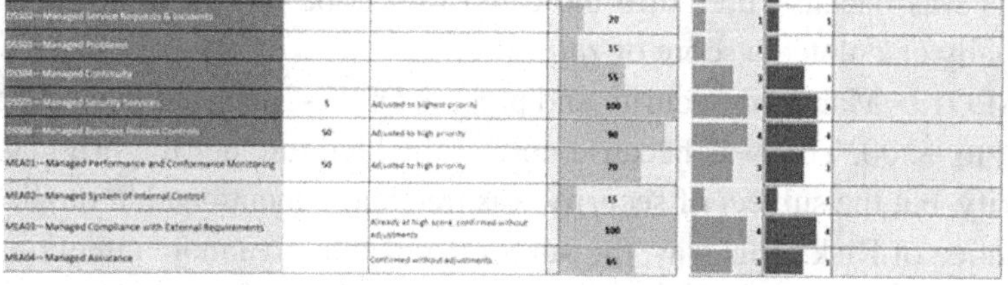

- If the value of the Design Toolkit is less than 40 (i.e., < 40), the process is tagged as "Low Priority"
- If the value of the Design Toolkit is minimal 40 (i.e., >= 40) but less than 80 (i.e., < 80), the process is tagged as "Medium Priority"
- If the value of the Design Toolkit is minimal 80 (i.e., >= 40) but less than 100 (i.e., < 100), the process is tagged as "High Priority"

- If the value of the Design Toolkit is 100, which is the maximum score the Design Toolkit can attribute, then the process is tagged as "Top Priority"

Table 21 indicates the assigned priority of each of the selected COBIT2019 processes. For convenience, the original and adjusted score of the Design Toolkit is displayed and the last column depicts the assigned priority, applying the filter as described above.

Table 21. Adjusted scores of COBIT2019 Design Toolkit - Own work

COBIT2019 Process	Original Score	Adjusted Score	Assigned Priority
EDM01 - Ensured Governance Framework Setting and Maintenance	35	85	High
EDM02 - Managed Benefits delivery	10	60	Medium
EDM03 - Ensured Risk Optimization	90	90	High
EDM04 - Ensure Resource Optimization	-10	-10	Low
EDM05 - Ensured stakeholder engagement	30	80	High
APO01 - Managed I&T Management Framework	40	40	Medium
APO02 - Managed Strategy	0	100	Top
APO03 - Managed Enterprise Architecture	30	80	High
APO04 - Managed Innovation	45	95	High
APO05 - Managed Portfolio	-5	45	Medium
APO06 - Managed Budgets and Costs	-35	15	Low
APO07 - Managed Human Resources	15	65	Medium
APO08 - Managed relations	35	85	High
APO09 - Managed Service Level Agreements	0	50	Medium
APO10 - Managed Vendors	15	65	Medium
APO11 - Managed Quality	10	60	Medium
APO12 - Managed Risk	100	100	Top
APO13 - Managed Security	90	100	Top
APO14 - Managed Data	40	90	High
BAI05 - Managed Organizational Change	30	80	High
BAI06 - Managed IT Changes	45	95	High
BAI08 - Managed Knowledge	25	75	Medium
DSS05—Managed Security Services	95	100	Top
DSS06—Managed Business Process Controls	40	90	High
MEA01 - Manage Performance and Conformance Reporting	20	70	Medium
MEA03 - Managed compliance with external requirements	100	100	Top
MEA04—Managed Assurance	65	65	Medium

To provide extra clarity to reader, the above table was sorted on priority and the scores were left out. This results in Table 22.

Table 22. Adjusted scores of COBIT2019 Design Toolkit - ranked per assigned priority - Own work

COBIT2019 Process	Assigned Priority
APO02 - Managed Strategy	Top
APO12 - Managed Risk	Top
APO13 - Managed Security	Top
DSS05 - Managed Security Services	Top
MEA03 - Managed compliance with external requirements	Top
EDM01 - Ensured Governance Framework Setting and Maintenance	High
EDM03 - Ensured Risk Optimization	High
EDM05 - Ensured stakeholder engagement	High
APO03 - Managed Enterprise Architecture	High
APO04 - Managed Innovation	High
APO08 - Managed relations	High
APO14 - Managed Data	High
BAI05 - Managed Organizational Change	High
BAI06 - Managed IT Changes	High
DSS06—Managed Business Process Controls	High
EDM02 - Managed Benefits delivery	Medium
APO01 - Managed I&T Management Framework	Medium
APO05 - Managed Portfolio	Medium
APO07 - Managed Human Resources	Medium
APO09 - Managed Service Level Agreements	Medium
APO10 - Managed Vendors	Medium
APO11 - Managed Quality	Medium
BAI08 - Managed Knowledge	Medium
MEA01 - Manage Performance and Conformance Reporting	Medium
MEA04—Managed Assurance	Medium
EDM04 - Ensure Resource Optimization	Low
APO06 - Managed Budgets and Costs	Low

This means that for designing and managing a business platform for governments ("Opening Services"), following the defined characterises of the organisation, the relevant COBIT2019 processes are hereby identified and prioritized.

2.4.4 Final list of processes for the defined practices

Although the above table stands on its own as an artifact, the real value from this artifact comes when we combine this with the list of best practices, that was designed as artifact from research question 1.

In the below table, this merger is visualized. The following steps were executed to come to this table:

- First, we took the table with the final list of best practices, where the list of practices was also mapped to the levels of responsibility in an organization. This table was the canvas to map the process upon.
- Next, the processes were linked to the right practices and added to the canvas. This was based on insights obtained in the literature research. As an example, to make sure the practice of Strategy is taken care of, the COBIT2019 processes of APO02 - Managed Strategy and APO05 - Managed Portfolio were linked and suggested.
- The processes that were marked Top Priority as extra marked in the table.

Table 23 is the merger of prioritized processes with the best practices and organizational responsibility.

Table 25. Linking COBIT2019 prioritized processes to organization - Own Work

Strategic Level	Tactical Level	Operational Level
Vision **See Strategy**	**Business Agility** No prioritized processes were allocated here, as this is more referring to applying a methodology like Agile & Scrum	**Talent and Skills** >> **APO07** - Managed Human Resources >> **EDM04** - Ensure Resource Optimization
Strategy >> **APO02** - **Managed Strategy** (Top Priority) >> **APO05** - **Managed Portfolio**	**Business Ecosystem Design** >> **EDM05** - Ensured stakeholder engagement >> **APO08** - Managed relations >> **APO09** - Managed Service Level Agreements >> **APO10** - Managed Vendors	**Customer Experience Design** No prioritized processes were allocated here, as this is more referring to applying a methodology like Design Thinking
Leadership >> **BAI05** - **Managed Organizational Change** >> **BAI08** - **Managed Knowledge** >> **MEA01** - **Manage Performance and Conformance Reporting** >> **MEA04—Managed Assurance** >> **APO06** - **Managed Budgets and Costs** >> **MEA03** - **Managed compliance with external requirements** (Top Priority) >> **APO13** - **Managed Security** >> **DSS05** - **Managed Security Services**	**Platform Architecture & Infrastructure** >> **APO03** - Managed Enterprise Architecture	**Risk Management** >> **DSS06**—Managed Business Process Controls >> **APO12** - Managed Risk (**Top Priority**) >> **EDM03** - Ensured Risk Optimization (**Top Priority**)
Governance >> **EDM01** - **Ensured Governance Framework Setting and Maintenance** >> **APO01** - **Managed I&T Management Framework** >> **APO11** - **Managed Quality**	**Data Driven** >> **APO14** - Managed Data	**Technology Ecosystem Design** >> **BAI06** - Managed IT Changes
(Innovation) Culture >> **APO04** - **Managed Innovation**		
Value >> **EDM02** - **Managed Benefits delivery**		

The above table is the artifact that provides part of the answer to research question 2. It is now clear – after all the literature research and design work – what processes to put in place to realize and manage the defined practices successfully. As such, it is now clear how to successfully manage and control a government business platform model ("Opening Services"). This is by no means a generic model; it is specific to the characteristics of a government organization, as set in the beginning of this chapter.

The exercise on itself is repeatable if other criteria where to be chosen, given new insights. The process is described in a transparent way and the reader can follow how each decision contributed to the final list. Therefore, it acts as guidance for refining the processes as the situation changes.

However, this is only a part of a governance model. Next is to complement this table with defining the organizational structures and relational mechanisms.

2.5 Organizational Structures

In this section, the organizational structures were determined to govern a government business platform model ("Opening Services").

Organizational structures represent the enterprise's key decision-making roles or structures like Board, Executive Committee, CEO, CIO, etc.(De Haes et al., 2020). Although governments have different stakeholders, the assumption taken here was that most industry structures can be found in a government organization as well, as depicted in Figure 12.

Figure 12. Illustrative organizational structures - company versus government - own work

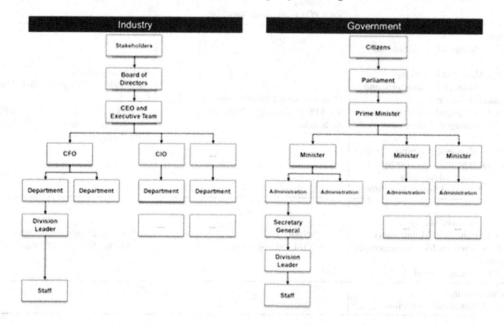

We therefore used the guidance of De Haes et al. as a starting point in this chapter and applied it in the context of a government organization.

As Gartner stated, *"Digital Leadership is a Team Sport"* (Gartner, 2016), and because a platform model brings technology and business together, the first suggestion is that leadership for a platform organization must not come from either the business or the IT, it must come from both.

Therefore, a **joined Platform Leadership Team** is put forward as the structure where business and IT share the same goals and budget to realize the platform model in governments. This seems illogic as the traditional organizational structures typically focus on aligning the two but keeping them separate. In a Platform Leadership team, one does not make that difference anymore, business and IT are one team.

Figure 13. Joined Platform Leadership Team visualized - Own Work

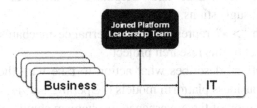

The above suggestion is only underpinned by limited literature examples (see below) and experience from working with the case organization of VDAB, where this suggestion was applied. Further research will be needed to confirm if this suggestion can be become a generic suggestion.

This suggestion is following the advice of a blog by De Haes (De Haes, 2017) when he states that *"established organizations will have to redefine their approach to governance to unlock the potential value of digital transformation"*.

Finally, there is evidence that this suggestion is already applied. For example, at the Australian Government, where the one-team approach was put in place to take care of the platform strategy (Australian Digital Transformation Agency, 2019):

Figure 14. Platform Governance Model – Relational Structures. Source: Australian Government & PWC

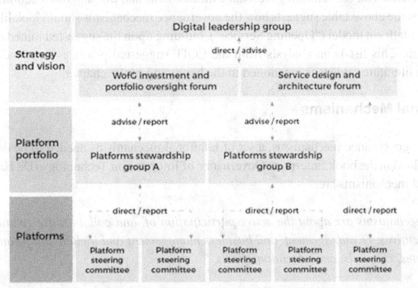

Inspired by the advice from De Haes et al. (De Haes et al., 2020) that *"each organization has to select its own set of Enterprise Governance of IT (i.e., EGIT) mechanisms, suitable for its specific context (i.e., sector, size, culture)"*, the organizational structures were further defined.

The following table started from the defined structures in the EGIT book and adapted them to fit the organizational structures specifically for a government business platform model ("Opening Services"). One may wonder why we refer here to EGIT, but the book mentions specifically that the EGIT organizational structures are aligned with COBIT2019.

The table is built up as follows:

- Column "IT Governance Mechanism" and "Original Definition" are taken from the EGIT, referenced above, with added suggestions.
  - Items marked with ">>" represent new governance mechanisms to be added, which are unique to the subject of this research project
- Column "Suggested Action" describes what action to take with the governance mechanism in relation to government business platform models ("Opening Services"). The actions are
  - "Extend": the definition of the governance mechanism should be extended.
  - "Keep": this governance mechanism can be kept as per definition in EGIT (see reference above)
  - "New": this governance mechanism is not in the EGIT table, but is suggested to be added as it is relevant to the subject of this research project
  - "No impact, but…": this governance mechanism can be kept as per definition in EGIT (see reference above), but an additional remark is made about the scope or validity of the mechanism.
  - "NA" refers to "not applicable", indicating that the governance mechanism is not applicable to the subject of this research project.
- Column "Extended definition for Platform" provides a reason for each of the suggested actions why this is necessary in the context of this research project.

Table 26 indicates all the relevant governance mechanisms and the suggested actions.

In summary, the above table suggests how the governance mechanisms should look like for a government business platform model ("Opening Service"), building upon the suggested joined digital platform leadership team. This list is an analysis from the EGIT suggested practices combined with insights obtained from literature review as mentioned in the beginning of the chapter.

2.6 Relational Mechanisms

To finalize the governance mechanisms, a set of relational mechanisms needed to be documented. De Haes et al. defined in the book Enterprise Governance of Information Technology (De Haes et al., 2020) what relational mechanisms are:

"Relational mechanisms are about the active participation of, and collaborative relationship among, corporate executives, IT management, and business management and include job rotation, announcements, advocates, channels, and education efforts."

Table 26. IT Governance Mechanisms for Platform model, based on De Haes & Van Grembergen- Own Work

IT Governance Mechanism	Original Definition	Suggested Action	Extended definition for Platform
IT strategy committee at the level of board of directors	Committee at the level of board of directors to ensure IT is regular agenda item and reporting issue for the board of directors	Extend	Extend to have a Platform strategy committee reporting to the board of directors. If not possible, the platform should be on the agenda of both the business and IT strategies.
IT expertise at the level of board of directors	Members of the board of directors have expertise and experience regarding the value and risk of IT	Extend	Extend to include platform expertise at board level, especially about the value and the ecosystem approach, agreeing to embrace private industry to deliver (part of) the services of that organization.
(IT) Audit committee at the level of board of directors	Independent committee at the level of board of directors overviewing (IT) assurance activities	Extend	Extend to Platform audit committee assuring (at least) risk, compliance, fair trade, and value for all stakeholders.
CIO on executive committee	CIO is a full member of the executive committee	Extend	The CIO should indeed continue to be a full member of the executive committee but also sit the in the Digital Transformation Team and assist the Digital Transformation Leader or be nominated in that position.
CIO reporting to CEO and / or COO	CIO has a direct reporting line to the CEO and / or COO	Keep, but...	In this role, the CIO has a double role. This person must make sure all IT & technology projects are managed. Additionally, the CIO must ensure that a large set of requirements from the Digital Transformation Team, representing the platform model, are covered and managed. This might result in conflicting interests, so the CIO must closely align IT projects to the overall (platform) strategy.
>> Digital Transformation Leader	Not existing	**New**	This person represents the digital transformation initiatives (e.g., the platform model) and reports directly to the CEO.
IT Steering Committee at the executive / senior management level	Steering committee at the executive or senior management level responsible for determining business priorities in IT investments	No impact, but...	A new Platform Steering Committee is suggested to coexist and focus on digital transformation of which the platform model is one of the key initiatives.
IT governance function / officer	Function in the organization responsible for promoting, driving, and managing IT governance processes	NA	See the role of Platform Governance Officer.
>> Platform Governance Officer	Non-Existing	**New**	Governance was defined as a key practice, consisting of both the ecosystem governance as the platform governance. This requires a different focus, and thus it is suggested to separate this from the IT Governance function / officer.
Security / compliance / risk officer	Function responsible for security, compliance and/or risk, which possibly impacts IT	Extend	Extend scope to Platform model.
IT Project steering committee	A steering committee composed of business and IT people focusing on prioritizing and managing IT Projects	Extend	Extend to include platform technology to manage and steer, IT investment in infrastructure or the introduction of new technologies that will be used in the platform architecture.
IT Security steering committee	A steering committee composed of business and IT people focusing on IT-related risks and security issues	Extend	Include Platform security as one of the fixed agenda items - but not only IT risk, also Business and Ecosystem risk.
Architecture steering committee	A committee composed of business and IT people providing architecture guidelines and advise on their applications	Extend	Include Platform architecture as one of the fixed agenda items, including platform infrastructure, API standardization, and technical ecosystem design.
>> Platform Steering Committee	Non existing	**New**	Create a new steering committee that focuses solely on designing and maintaining the platform. This brings together business and IT people to discuss future projects, standardization of IT, but also applying new capabilities like Agile and Customer Experience to build an efficient platform marketplace.
Integration of governance / alignment tasks in roles and responsibilities	Documented roles & responsibilities include governance/alignment tasks for business and IT people	Extend	A one-team approach for designing and managing platform models is advocated, therefore, the roles and responsibilities need to be defined and clear in the context of the platform model as well.

This definition was applied to determine the organizational structures the relational mechanisms that are relevant to a government business platform ("Opening Services") organization.

The analysis is compiled in the below table. The following fields exist:

- "Relational Mechanism": this is the relational mechanism that is proposed by De Haes et al. (De Haes et al., 2020) and was evaluated in the context of the scope of this research project
  - Items marked with ">>" represent new relational mechanisms to be added, which are unique to the subject of this research project
- "Original description": this field is copied from the EGIT book for completeness
- The Column "Suggested Action" describes what action to take with the relational mechanism in relation to government business platform models ("Opening Services"). The actions are
  - "Keep": this relational mechanism can be kept as per definition in EGIT (see reference above)
  - "Extend": the definition of the relational mechanism is kept but the scope and impact should be extended.
  - "Redefine": the definition of the relational mechanism is kept but must be redefined to align with the scope
  - "New": this relational mechanism is not in the EGIT table, but is suggested to be added as it is relevant to the subject of this research project
- Column "Platform Model aligned description" provides a reason for each of the suggested actions why this is necessary in the context of this research project

Table 27 indicates all relevant relational mechanisms and the suggested actions:

In summary, the above table follows the same analysis approach as for the organizational structures. It suggests how the relational mechanisms should look like for a government business platform model ("Opening Service"), building upon the suggested joined digital platform leadership team. This list is an analysis from the EGIT suggested practices combined with insights obtained from literature review as mentioned in the beginning of the chapter.

2.7 Summary – defined governance mechanisms

This section determined the processes, organizational structures, and relational mechanisms to manage a government platform organisation. Together they form the designed for purpose governance model for the defined practices.

COBIT2019 was chosen as the framework to use to create the right list of processes. Using the CO-BIT2019 Design Toolkit, the list of relevant and weighted processes was derived and adjusted to the findings of the literature research. As COBIT2019 now also incorporates EGIT, also the organizational structures and relational mechanisms were defined.

With the above defined, an answer was provided to research question 2 and a set of governance mechanisms are defined. This concluded the theoretical research. Next, we will turn our attention to using this information in the field. The next chapter will explain how this information was used to prepare for field research.

Table 27. Mapping of relational mechanisms to platform model based on De Haes & Van Grembergen – Own Work

Relational Mechanism	Original Description	Suggested Action	Platform Model aligned description
Job Rotation	IT Staff working the business units and business staff working in IT	Keep	For the platform model, we suggest business and IT to become one team, bringing together expertise that is focused on establishing all defined practices (and processes) together. This requires the necessary job rotation from the business and/or IT team to the newly formed platform team.
Co-location	Physically locating business and IT people together	Keep	Fits with the principle that business and IT should be one team, thus also sitting together
Cross-training	Training businesspeople about IT and / or training IT people about business	Extend	Extend to train all stakeholders (also external ones) about the platform model and the skills needed to fully embrace this model, innovating the current approach to service design and - most of all - delivery through a set of industry partners.
Knowledge management (on IT Governance)	Systems (e.g., on the intranet) to share and distribute knowledge about IT governance framework, responsibilities, tasks, etc.	Extend	Extend to also include the ecosystem and platform governance.
Business / IT Account Manager	Bridging the gap between business and IT through account managers who acts as in-between	Redefine	Replace account manager by the role of Product Owner as defined by SAFE / Agile, as business and IT are one team, working together. This also fits with the agile approach as a suggested key practice.
>> Platform Account Manager	Non existing	New	This role focuses on defining, maintaining, and growing the ecosystem that will provide services on the platform.
Executive / Senior management giving the good example	Senior business and IT management acting as "partners"	Keep	Although a one-team approach is advocating, there will be other business and IT departments remaining in the organization that operates outside of the platform team. Therefore, management must allow the platform team to function as an autonomous entity while not eating away all resources and budget.
Informal meetings between business and IT executive / senior management	Informal meetings, with no agenda, where business and IT senior management talk about general activities, directions, etc.	Keep	Extend to include platform strategy, how to extend the ecosystem, how to redesign service design further, how to break silos, etc.
IT Leadership	Ability of CIO or similar role to articulate a vision for the role of IT in the organization and ensure that this vision is clearly understood by managers throughout the organization	Keep	Leadership is defined as a Top Best Practice and supporting the platform team must be well articulated in all layers of the organization. IT and business leaders should do this together.
Corporate internal communication addressing IT on a regular basis	Internal corporate communication regularly addresses general IT issues	Extend	Extend to specific communication about the platform model performance, not only in technical terms, but also in ecosystem growth, transactions facilitated, and value delivered to citizens.
IT Governance awareness campaigns	Campaigns to explain to business and IT people the need for IT governance	Extend	Extend to include platform and ecosystem governance, which will require different skills and talent to emerge inside the organization.

3 VALIDATION METHODS FOR PLATFORM ORGANIZATIONS

3.1 Introduction

The practices and governance model defined earlier can now be tested in one or more situations or cases. However, first an extra step was added. The artifacts are considered theoretical and not yet in such a state that they could be applied in a repeatable way, gathering the same information in a consistent way, and coming to objective conclusions across case organisations. Therefore, this section defined several standardized validation methods to apply in every case organization. This also facilitated consolidation and comparison of case information collected.

3.2 Evaluation Methods

To understand where an organization is versus the practices, several evaluation methods were created:

- **Method 1: Mapping the practices to an organizational Capability & Impact quadrant** In this method, we converted the practices in a structured questionnaire that converts the result into a magic quadrant model. This method was partially inspired by the digital maturity index model, as defined by MIT Sloan School, together with Capgemini (Westerman et al., 2012). This was mapped to the specific context of governments applying the platform model. This method will provide insights to the organization how well they are implementing the practices and in what quadrant they are positioned.
- **Method 2: Governance Design Compliance Checklist** This method goes one step further and defined several criteria to verify if the organization has the correct processes, structures, and relational mechanisms in place for successfully defining or maintaining the government business platform model ("Opening Services").
- **Method 3: The Government Business Platform Maturity Model** The practices were also converted into a maturity model. This way, a government organization can immediately see where they are and how they must act to improve and grow in maturity.
- **Method 4: The Balanced Score Card** Finally, the practices were converted into a Balanced Score Card that demonstrates how all elements interact and can be measured as one program.

Table 28 summarized how the artifacts are linked to the validation methods, in other words which validation method addresses which artifact.

Table 28. Relationship Artifacts and Validation Methods - Own work

Artifact	Validation Method
Practices	Method 1: Organizational Capability
Governance Mechanisms	Method 2: Governance Compliance Checklist
Practices and Governance Mechanisms	Method 3: Platform maturity model Method 4: Platform Balanced Score Card

In the next paragraphs, these methods will be introduced individually and in detail.

3.3 Method 1: Mapping the practices to an organizational Capability & Impact quadrant

The inspiration for this method came from two existing models:

- Reference to the digital maturity model defined by MIT Sloan School, together with Capgemini (Westerman et al., 2012).
- Reference to a quadrant model that PWC made with the Department of Science, Information Technology and Innovation of the Australian Government (Shahiduzzaman et al., 2017) for measuring the digital readiness of a government organization.

Based on these two, the information and assessment approach were converted in a similar approach for measuring how ready an organization is to engage with the platform business model ("Opening Services").

3.3.1 Step 1: Definition of Indicators

Two indicators were defined, representing the X- and Y-axis of a resulting quadrant chart:

- (X-Axis): The Platform Capability Indicators measure the strength of the organization's foundation in applying the platform model.
  This goes beyond physical technology infrastructure and encompasses the aspects of the organization that allow it to derive value from technology. This includes strategy, talent and skills, risk management, and customer experience.
  Questions indicate there is a strategy defined to execute the vision, the necessary skills and talent are present, the ecosystem approach is defined, the supporting technologies are in place, and the redesign of the experience is all worked out.
  The higher the score, the more capabilities there are available in the organization to engage.
- (Y-Axis): The Platform Impact Indicators measure how the platform model and associated characteristics are being accepted by the organization to respond to citizen demand for improved service offerings.
  In other words, this is about is how well the organization understands the citizen (including the business processes of the citizen's lives and the ecosystem) and is using the platform model to respond to the citizen's needs.
  Questions indicate there is a vision agreed, leadership is fully supporting this vision, the value is clear, the service platform is defined, and the governance model is aligned to support the vision.
  The higher the score, the more the organization understands what to achieve and a plan on how to engage.

3.3.2 Step 2: Definition of the Quadrant Map

Now, we can map out and visualize the different meanings of the X- and Y-Axis on a quadrant map (Figure 15).

Figure 15. Platform Maturity Magic Quadrant - Own work

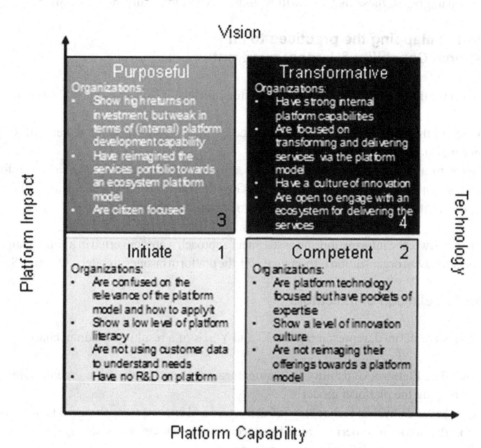

The idea behind this quadrant model is that organizations reach the highest level of maturity when they have both a strong platform foundation (in terms of platform capability indicators) and a good understanding of how to leverage this foundation for a strategic business advantage (platform impact indicators).

Organizations typically follow four distinct stages to navigate the digital maturity journey. These stages are: Initiate, Competent, Purposeful, and Transformative. Each of these stages deals with the organizational level of digital maturity and criteria to be met to move to the next stage.

3.3.3 Step 3: Mapping of the practices to the indicators

Next, the assembled list of platform practices was mapped to the capability and impact indicators, maintaining the Strategic, Tactical, and Operation level view that was defined earlier.

The mapping shows that the capabilities are more on the Tactical and Operational level, while the impact is more on the Strategic level.

Table 29. Mapping of practices to Capability & Impact Dimensions - Own Work

Level	Capability	Impact
Strategic	Strategy	Vision
		(Political) Leadership
		Governance
		Innovation Culture
		Value
Tactical	Business ecosystem design	
	Platform Architecture & Infrastructure	
	Technology Ecosystem Design	
	Data Driven	
Operational	Talent and skills	Business agility
	Risk management	
	Customer experience design	

3.3.4 Step 4: Questions per Indicator

The following list of questions was created to evaluate the platform model and put in a table with the following columns:

- Indicator: refers to the practice that is being tested
- #: provides a sequence number to the question, this has no further meaning
- Question: contains the question to be answered
- Score: contains space where the reply can be noted. The numerical score can be one of the below values:
  - 1 = Strongly disagree
  - 2 = Disagree
  - 3 = Neutral
  - 4 = Agree
  - 5 = Strongly Agree

There are two tables (Table 30 and Table 31) of questions defined that each represent an axis.

3.3.5 Step 5: Calculating the score of the indicators

The next step is to calculate the overall score by adding up the entries in the column "Score" and position them in a resulting graph. We add up the numbers per questionnaire and apply the following scoring approach to determine in which quadrant the answer lies (Table 32).

This scoring does not involve any other calculations than adding the individual scores. The consolidated number that maps to a quadrant, based on an axis between 0 and 150 points. The combination of the two scores determines in which quadrant the indicators map.

Table 30. Platform Quadrant - Capability Indicators Questions - Own Work

Indicator	#	Question	Score (1...5)
STRATEGY	1	My organization has a clear, coherent, and actionable strategy that shows the path and steps towards the adoption of the business platform model to optimize service delivery.	
	2	There is a strategy defined in my organization that focuses on transforming the whole business (end-to-end) rather than transforming one or more operations or silos.	
	3	There is a business strategy defined in my organization that strives for maximum reconfiguration and availability of the services in our organization on the platform maximizing the usage of an ecosystem of providers that deliver (parts of) the service.	
BUSINESS ECOSYSTEM DESIGN	4	My organization has an active collaboration and scouting process with (potential) ecosystem providers, and we discuss jointly how the providers can deliver (part of) the service.	
	5	My organization has a clear business roadmap so that the ecosystem providers know what services will be launched on the platform when and how they can help execute them. There is a joined process and governance to prioritize and / or decide.	
	6	My organization orchestrates the ecosystem providers in an open and co-creative approach. Rules will be set for setting up a trust relationship, but providers can be flexible and innovative in the way they commit to delivering the services.	
PLATFORM ARCHITECTURE & INFRASTRUCTURE	7	My organization funds and resources the platform services and technology transformation adequately.	
	8	Platform related investment considers an organization-wide approach (people and culture), rather than investing only in technology and/or developers.	
	9	My organization can effectively integrate new technologies with older 'legacy technologies' on the platform.	
	10	My organization has the technological platform infrastructure and corresponding solutions in place to support real-time citizen insights.	
	11	My organization has systems, applications, and tools in place that enable the execution of efficient business processes on the platform (e.g., API approach).	
	12	My organization has the technological infrastructure and corresponding solutions in place to support service execution on the platform.	
TECHNOLOGY ECOSYSTEM DESIGN	13	We connect with our partners digitally. For example, we use modern business system integration platforms, such as API-enabled cloud-based services, to enable efficient business interactions that otherwise would not be possible.	
	14	We leverage functional reusable components from our government (e.g., ID verification) to avoid ambiguity in functionality.	
	15	We have a roadmap established so that the ecosystem providers know what functionality or components will be made available when.	
DATA DRIVEN	16	My organization has established an appropriate business intelligence system to help us make timely decisions based on (patterns of) interactions on the platform.	
	17	My organization has the process and technology in place to understand citizen behaviour and adapt the service provisioning process accordingly to become more proactive and / or tailored.	
	18	My organization is data-focused and uses data for environmental sensing/machine learning/predictive analysis to provide proactive service advice to a citizen.	
TALENT AND SKILLS	19	Employees have the skills and competencies to facilitate the design and build of services on the platform or can access these skills from partners or suppliers as needed.	
	20	We find it easy to attract high-quality technical staff to our organization because of our vision to be a leader in platform service design and ways of working.	
	21	My organization continuously invests in developing digital skills of employees, especially those that relate to the execution of our platform strategy.	
RISK MANAGEMENT	22	My organization has services, systems, applications, and tools in place to appropriately protect the organization from cyber-attacks and other security risks.	
	23	My organization actively and regularly assesses technical, business, and social risk factors when it comes to platform technology investment.	
	24	My organization considers the scalability of platform infrastructure to meet the demand driven by changing number of services available on the platform, the growing number of consumers and providers, growing number of interactions, or legislation requirements.	
	25	My organization has embedded a proactive risk management approach within the culture and processes of the platform organization.	
CUSTOMER EXPERIENCE DESIGN	26	User Experience research is conducted by my organization to better understand citizen pain points as part of designing better services made available through the platform.	
	27	My organization can design and deliver a tailored product to fulfil specific citizen's needs.	
	28	My organization provides citizens with a fully integrated experience in all areas of interaction including government and ecosystem providers.	
	29	My organization continuously improves its digital and physical experiences on the platform to deliver genuine value to the citizen.	
	30	Citizens can effectively communicate with my organization to address complaints and help resolve issues that originate the platform.	

Table 31. Platform Quadrant - Impact Indicators Questions - Own Work

Indicator	#	Question	Score (1...5)
VISION	1	My organization has a long-term (e.g., 5 years and beyond) goal that reflects its ultimate point of success using the platform model.	
	2	My organization is customer centric and creates digital value by addressing citizen's problem in a new and innovative way.	
	3	Using the platform business model is an essential element of realizing the vision of my organization.	
	4	The platform strategy in my organization is aligned with the overall business strategy.	
(POLITICAL) LEADERSHIP	5	Leaders in my organization have a compelling long-term goal and associated objectives for my organization how to use the platform model.	
	6	Leaders in my organization can communicate their future foresight throughout the organization.	
	7	Leaders in my organization actively identify and realize opportunities for realizing platform enabled efficiency.	
	8	Leaders in my organization have empowered employees to work in cross-functional teams and collaborative environments.	
GOVERNANCE	9	Efficient and agile processes and systems are used to react to rapid business change.	
	10	Employees can quickly identify the core of a business or customer problem and self-organize to address the solution timely manner.	
	11	My organization empowers staff to work autonomously as required, when providing with an appropriate level of vision, guidance, and coordination to maintains focus.	
INNOVATION CULTURE	12	In my organization everyone has a mandate to think creatively, innovate and provide input to the platform team.	
	13	My organization takes a rigorous and systematic approach to innovation or change management with regards to service design using the platform model.	
	14	My organization has a process for introducing new technologies to improve efficiency of the business operations.	
	15	My organization conducts both small iterative experiments, and enterprise-wide initiatives to realize innovation that has business impact.	
	16	My organization conducts innovation activities as a regular task.	
	17	Employees feel empowered and take calculated risks to be successful.	
	18	Employees regularly work in interdisciplinary teams and are supported in cross-skilling and knowledge sharing.	
	19	Teams work collaboratively on projects and share developments all the way through, factoring in feedback and new insights to improve as they go.	
VALUE	20	Citizens can effectively communicate with my organization to co-create value.	
	21	My ecosystem providers can effectively communicate with my organization to co-create value.	
	22	All staff (e.g., technology and management) in my organization work in sync towards implementing our (platform) vision.	
	23	My organization fosters an integrated digital ecosystem. For example, we share data and/or provide integration points so that ecosystem providers can create value-add services that complement our own, increasing impact, and citizen satisfaction.	
	24	Everyone in the organization knows of, understands, and can act on our platform strategy.	
	25	There are very few technical issues in the delivery of our platform services.	
	26	When technical issues do occur in service delivery, we are able to resolve them within an acceptable period (i.e., within customer expectations of our industry).	
	27	My organization's platform initiatives are currently generating efficiencies versus the defined metrics (e.g., faster service provision, government cost reductions), and the impacts are increasing over time.	
BUSINESS AGILITY	28	My organization has a proven ability to identify citizen's latent needs.	
	29	My organization has demonstrated ability to pivot its purpose, products, and service based on analysis of citizen insight and key performance metrics.	
	30	Technology is no longer a bottleneck in our organization. For example, our technical delivery teams can implement services faster than we can generate new service delivery ideas.	

Table 32. Platform Quadrant - Scoring - Own Work

Quadrant	Capability Score	Impact Score
Initiate	0 - 75	0 - 75
Competent	76 - 150	0 - 75
Purposeful	0 - 75	76 - 150
Transformative	76 - 150	76 - 150

3.3.6 Step 6: Mapping the scores on the quadrant model

Finally, we can map the result of the questionnaire towards to the following template:

Figure 16. Platform Quadrant - Template - Own Work

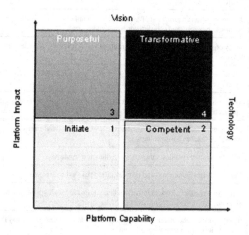

3.4 Method 2: Governance Compliance Checklist

This method was based on the Enterprise Governance of IT (i.e., EGIT) model (De Haes et al., 2020) that stipulated that a governance model is the combination of the rightly designed processes, structures and relational mechanisms to provide the value that the government organization is seeking when designing and maintaining a government business platform model ("Opening Services").

3.4.1 Process Checklist

For the design, rollout, and maintenance of the business platform model ("Opening Services") to be successful, COBIT2019 was used as a reference framework and determined the following processes to be minimally in place. The list was turned into a checklist, which can be used to validate the status inside an organization (Table 33).

Table 33. Process Checklist - Own Work

COBIT2019 Process	In place Y/N?	Comments
APO02 - Managed Strategy		
APO12 - Managed Risk		
APO13 - Managed Security		
DSS05 - Managed Security Services		
MEA03 - Managed compliance with external requirements		
EDM01 - Ensured Governance Framework Setting and Maintenance		
EDM03 - Ensured Risk Optimization		
EDM05 - Ensured stakeholder engagement		
APO03 - Managed Enterprise Architecture		
APO04 - Managed Innovation		
APO08 - Managed relations		
APO14 - Managed Data		
BAI05 - Managed Organizational Change		
BAI06 - Managed IT Changes		
DSS06 - Managed Business Process Controls		
EDM02 - Managed Benefits delivery		
APO01 - Managed I&T Management Framework		
APO05 - Managed Portfolio		
APO07 - Managed Human Resources		
APO09 - Managed Service Level Agreements		
APO10 - Managed Vendors		
APO11 - Managed Quality		
BAI08 - Managed Knowledge		
MEA01 - Manage Performance and Conformance Reporting		
MEA04 - Managed Assurance		
EDM04 - Ensure Resource Optimization		
APO06 - Managed Budgets and Costs		

The following columns are in the table:

- "COBIT2019 Process" contains the name of the COBIT2019 process that is being considered
- "In Place (Y/N)": Here it is noted if the organization has this process in place or not. Only a Y for Yes or N for No can be used, to make it easier.
- "Comments": It is advised to add comments to support the Y/N indication. This will explain why a Yes or No is the correct answer.

3.4.2 Organizational Structures Checklist

A similar checklist was created to validate if the following structures are in place for managing the business platform model ("Opening Services") (Table 24).

Table 34. Organizational Structures Checklist - Own Work

Platform related organizational structures	In Place (Y/N)?	Comments
Platform Strategy committee at the level of board of directors		
Digital Platform Transformation Leader, reporting to CEO		
Platform Steering Committee		
Platform Governance Officer		
Platform Security / compliance / risk officer		
Platform Project steering committee		
Platform Security steering committee		
Platform Architecture steering committee		

The following columns are in the table:

- Platform related organizational structures: Refers to the organizational structure that is being validated.
- In Place (Y/N): Here it is noted if the organization has this organizational structure in place or not. Only a Y for Yes or N for No can be used, to make it easier.
- Comments: It is advised to add comments to support the Y/N indication. This will explain why a Yes or No is the correct answer.

3.4.3 Relational Mechanisms Checklist

Next to processes and structures, we can also check if there are some relational mechanisms in place to raise the awareness of the business platform model ("Opening Services") in the organization. Table 35 created checklist can be used to validate the existence of the relevant relational mechanisms.

The following columns are in the table:

- Are the following relational mechanisms in place for Platform Model: Refers to the relational mechanism that is being validated.
- In Place (Y/N): Here it is noted if the organization has this relational mechanism in place or not. Only a Y for Yes or N for No can be used, to make it easier.
- Comments: It is mandatory to add comments to support the Y/N indication. This will explain why a Yes or No is the correct answer.

Table 35. Relational Mechanisms Checklist - Own Work

Are the following relational mechanisms in place for Platform Model?	In Place (Y/N)?	Comments?
1. Co-location: The people working on the platform are all sitting together and operate as one team, regardless of the organizational link		
2. Knowledge management There is an active Knowledge management process in place for platform business model for business and IT people inside the organization, ecosystem partners and consumers of the services		
3. Ongoing awareness training on Platform model strategy Training business and IT people about the value, risks, inhibitors, and usage of Business platform model inside the organization Training of ecosystem partners what the value is of engaging on the platform		
4. Articulate the vision on Platform Model The ability of digital transformation leader or similar senior leadership role to articulate a vision for Business platform model and ensure that this vision is clearly understood by managers throughout the organization		
5. Internal Communication Internal communication on new services made available on the platform		
6. Network outside the organization Tap into a network of expertise on Business platform model outside of the own organization		
7. Platform Account Manager There is active engagement on defining, maintaining, and growing the ecosystem on the platform		

3.5 Method 3: Introducing the Government Platform Maturity Model

During interviews with case owners, many asked where they stand today versus other government organizations and what the next steps would be. Now that the Practices and Governance Design Criteria are defined, these were turned into a compass that government organizations can use to determine what level of maturity they have reached and what the next level of maturity is and implies.

Established in the 1970s (Nolan & Gibson, 1974), the notion of a maturity model got popular with the establishment of the Capability Maturity Model (CMM) in late 1980. Since then, many models were published like Six Sigma or the EFQM.

The concept of maturity models is not without criticism (Mettler, 2009). Mettler states that there is still a significant lack of knowledge on how to design theoretically sound and widely accepted maturity models. Most of the models seem to be based on "good practice" or "success factors" derived from projects that have demonstrated favourable results to an organization or an industry sector.

Mettler defined two approaches to build a maturity artifact: either it is positioned as a method that focuses on the specification activities (i.e., the how), or it is positioned as a model where state descriptions (i.e., what) are described versus a desired to be state.

Given the criticism of research and the limited theoretical underpinning in this study, the below maturity model is positioned as a first - and theoretically inspired - attempt to bring together the practices based on own research and validation with a selected set of case owners. It is positioned as a "model", wherein we describe the state.

Elaborating on this and based on the decision parameters defined by Mettler(Mettler, 2009), the following characteristics were applied in the creation and usage of the model:

- The model is practitioner-based as it is the results of own research and discussions with case owners.
- The reliability is verified on a limited scale (i.e., with the case owners).
- The practicality is that the model provides a general recommendation.
- The accessibility is free, meaning there are no cost involved in using the model.
- The application method is self-assessment, anyone can use the model as a compass.
- It is meant as an informal appraisal mechanism.
- Respondents are typically business representatives.
- It is positioned as an informal appraisal tool.
- The frequency of application can be repeated per own insights.

Finally, the definition of Kohlegger et al. (Kohlegger et al., 2009) further inspired the creation of the below model: "*A maturity model conceptually represents phases of increasing quantitative or qualitative capability changes of a maturing element in order to assess its advances with respect to defined focus areas.*"

First, we needed to determine the maturity levels and associated values. We started from the classical approach of defining 5 maturity levels that build upon each other. The CMM model (Jayaram, 2003) defines the following levels (Table 36):

Table 36. CMM Maturity Levels – Source: Jayaram

Capability Maturity Levels	Level 1	Level 2	Level 3	Level 4	Level 5
	Initial	Managed	Defined	Predictable	Optimizing
	The implemented processes are chaotic and disorganized, mainly because the processes are not defined and documented. Therefore, the processes could not be executed repeatedly, and individual group efforts are mostly present	This is the foundation level since this level enables that processes are executed repeatedly. Processes are established, defined, and documented through basic management practices. This allows that success can be recurring	Best practices are identified through documentation and are integrated into standard or common processes.	The organization can control and manage its processes through data collection and analysis	The processes are being constantly optimized based on their quantitative knowledge.

Next, we made the definition of the maturity levels more specific towards using a business platform model. For that, we introduced the ambition level as the lead indicator of the maturity level:

Table 37. Ambition levels for a government business platform model ("Opening Services") – Own Work

Specific	Level 1: Exploring	Level 2: Developing	Level 3: Embracing	Level 4: Using	Level 5: Leading
Ambition	**Exploring** the business platform Model	**Developing** a business platform model	**Embracing** it for first experimental cases or well-defined scope	**Using** a business platform model for certain services	**Leading** the business platform model, actively putting services out to the ecosystem

The ambition is often translated into a vision. As such, this can be a leading indicator to a state where one is today and where one wants to aspire or become. To fully understand each level, the following guidance applies:

- **Level 1: Exploring** the business platform Model: The organization is looking at others what they are doing. They are not a platform provider today, their services are still being handled through the classical channels (which might be digital, but not from a platform model). They are considering options, developing a strategy, or are already preparing a business case. They might already use data or functionality from existing platforms. Sometimes they even offer data and/or functionality to platforms (e.g., Open Data), but none involve services from the portfolio of the organization

- **Level 2: Developing** a business platform model: the organization has decided to adopt the business platform model, but it is still under development, i.e., not rolled out yet. This means, there are no services yet offered via a platform model, except maybe those that are on another existing platform that is not owned or managed by this organization (e.g., Open Data). The ecosystem is under development. Designing the services is ongoing. An implementation plan is being drafted; business and IT have acknowledged this to be a project.

- **Level 3: Embracing** it for first experimental cases or well-defined scope. The organization has defined which of its services can be put on a business platform model and how the ecosystem can execute them. The IT department has rolled out a first version of the platform or has decided with another (government) stakeholder to incorporate the services on that platform. The ecosystem is aware of the strategy and is engaged in facilitating a transaction on the platform. Certain services might already be live, be it in experimental faze or with a very well-defined scope.

- **Level 4: Using** a business platform model for certain services. The organization has rolled out a business platform model (or integrates with another existing one), and services are defined. The ecosystem is defined, and transactions are happening. Monitoring is done. A plan exists to increase the number of services soon.

- **Level 5: Leading** the business platform model, actively putting services out to the ecosystem. The organization is mature in the usage of a business platform model, i.e., it is now a core part of the way the services of that organization are delivered to citizens. The business and IT departments of the organizations have embraced the model and fully support the rollout. The platform represents most of the service request, which does not mean other channels are stopped entirely. The ecosystem is up and running, and transactions happen frequently. Reviews are happening to refine the model and continuously verify the trustworthiness of the providers. New services are being added, while behaviour data is collected to tweak existing services. Feedback is collected from citizens.

Table 38. Government Platform Maturity Model in detail - Own work

Level	Dimension	Level 1: Exploring	Level 2: Developing	Level 3: Embracing	Level 4: Using	Level 5: Leading
Standard Capability Level		**Initial**	**Managed**	**Defined**	**Predictable**	**Optimizing**
Strategic	**Ambition / Vision**	**Exploring** the Business Platform Model	**Developing** a business Platform model	**Embracing** it for first experimental cases or well-defined scope	**Using** a platform model for certain services	**Leading** the model, actively putting services out to the established ecosystem
	Strategy & (Political) Leadership	At best, the **leadership has expressed the desire to investigate** if the platform model is suitable for the delivery of services via an ecosystem. If so, discussions have started or are ongoing with politicians if this model fits the government declaration (which is to be the overarching vision). A team has been tasked to identify potential use cases for optimizing service delivery	The leadership has a **strategy defined** and has an agreement with the political level to start developing the business platform model. An investment has been approved. The business case for selected service delivery is approved (Benefits, Cost & Value are clear). A roll-out plan is defined, and a dedicated team is involved	**Politicians and leadership are aligned** with pursuing the business platform model. A roll-out plan exists, and services are being opened on the platform (either own or another one) Investments are approved, expected benefits are clear.	**Politicians and leadership are monitoring** the realization versus the defined strategy. Clear definition of the platform service delivery model, selected services are integrated on another government platform model.	**Politicians fully support the (further) rollout** of the platform model and agree to increase the number of services gradually. A strategy is defined by the leadership team. It describes how to achieve the goals and investments for the coming period. Budgets have been forecasted and approved.
	Governance	There is **no specific governance model** designed yet to facilitate the strategy of becoming a platform provider. The existing governance processes and structures still apply	A **specific governance model is being designed** to bring the value associated with the decision to start developing a platform model	A **specific governance model is finalized** and gradually introduced	A **specific governance model is defined** and in place	The **governance model is fully rolled out** and refined to improve the value of the platform model continuously
	Innovation Culture	**Lack of systemic innovation activities**, very low efficiency of the innovation process. There is no specific process for platform innovation. Culture is focused on execution and efficiency within the existing process	**Innovation activities** are carried out at least as **fragmentary.** The organization can process improvements. The platform model is recognized as an innovative way to deliver services. Culture is open to changing the existing processes.	**Everyone** in the organization **is supporting** the program to build out a platform model for the selected services. The organization carries out most innovation activities. There is an increased release of control of the process towards the ecosystem. Culture is open to allow and support the execution of services to the ecosystem	The organization can **adapt** to changing conditions and still attain its goals. The execution is now delegated to the ecosystem that can innovate their delivery model as well. The culture of co-creation and sharing results is established.	The organization and the ecosystem **share the same culture of innovating and improving** service delivery via the platform model. Innovation is agreed upon and executed together with the ecosystem. The culture of fully supporting the ecosystem and operate as one to the benefit of the citizen is accepted by all stakeholders and participants
	Value	**Value is being investigated**, but not clear yet.	**Value to the organization is defined and clear**, but not or inconsistently measured.	**Value is defined and consistently measured,** but no targets have been set yet.	**Value is achieved** versus the expectations and targets.	**Value to all stakeholders is clear, measured, and reported** (in a transparent way). It can be calculated that the value is optimized throughout the platform
Tactical Level	**Business Agility**	**Traditional structures and ways of working** with resource optimization and command and control. Everyone focuses on optimizing their part of the process. When complexity grows, and speed is required the organization suffer badly from decision latency and can no longer meet the customers' needs	**Experimenting and breaking silos** to establish platform thinking Teams become well-functioning and high performing. The team has a mandate to make quick business decisions. A Service Design team is reworking existing processes to allow the ecosystem to execute (part of) them.	**Platform thinking prevails.** Every service is now optimized for delivery on the platform by the ecosystems. Teams are shifting skills and starting to collaborate across departments, and/or silo is inside the organization. Business value is increasing rapidly alongside quality and innovation.	**Service Design based on the platform model prevails.** Most services now run on the platform and innovation prevails over internal optimization There is transparency on what is going on with the ecosystem. Strategic goals and cost of employees are separated, and budgets are set on a team-level	**"Platform First" thinking is normal.** Services are organized around customers value instead of internal processes. All stakeholders feel empowered and enjoy collaborating on innovating and improving service delivery. Citizens feel awesome using the products and services via the platform.

Table 38. Continued

Level	Dimension	Level 1: Exploring	Level 2: Developing	Level 3: Embracing	Level 4: Using	Level 5: Leading
	Business Ecosystem Design	**Partner Cooperation.** Providers are platform aware and ready to discuss value creation	**Task Collaboration.** Providers are collaborating in an opportunistic or per initiative way, where there is value, partners are ready to step in.	**Business Model Collaboration** Co-creation of value propositions. It is clear what services will be offered on the platform, and providers know how they can realize them.	**Ecosystem Partnerships.** Ecosystem thinking and behaviour is established. The government now orchestrates the providers and service design. Open innovation is established. The next generation of service delivery is established.	**Service Delivery Integration.** Purpose oriented - no matter who delivers (part of) the service Continual innovation Evaluation of the performance of all providers by citizens
	Platform Architecture & Infrastructure	**No specific or dedicated Platform architecture exists inside the** organization. Case by case considerations on what infrastructure to use (internal or Cloud)	**Defining the Platform architecture** Preference for Cloud infrastructure provisioning	**Platform Architecture defined** and experiments are being rolled out Cloud infrastructure as standard, exceptions possible	**Platform Architecture standardized** and being increasingly rolled out Cloud infrastructure as default, exceptions unlikely	**Platform Architecture rolled out and optimized** Extra platform features like billing, metrics, etc. are standardized and endorsed by the ecosystem Containers based infrastructure Architecture open to other sources of data (e.g., IoT, sensor)
	Data Driven	**Lack of Data for Analytics Projects** that focus on platform behaviour or providing proactive proposals of eligible services to citizens	**Isolated Data Projects** that focus on platform behaviour or providing proactive proposals of eligible services to citizens	**Secure, Reliable Data Repository:** Data warehouse or lake systems with well-defined management and governance are utilized to provide a foundational system for reporting, data science and key operational users originating platform transaction and citizen behaviour data	**Governed Self-Service Access:** Stakeholders from the ecosystem have access to expanded data for exploration with data access granted based on levels of expertise. Reporting teams focus on operational analytics while business users run queries and extract data as needed	**Insights Driven:** Data-driven insights are integrated in processes and accessible across the ecosystems to measure and drive action, resulting in the ability to seamlessly integrate data and insights into new business policies and processes on the platform - with respect of privacy and GDPR
Operational Level	Talent and Skills	**Platform capabilities undefined,** using classical roles inside Business & IT for the moment	**Platform capabilities defined** but not yet established as separate team. Selected people are being trained, or specific expertise (business and/ or IT) is bought from the market	**Platform capabilities defined and agreed,** first team established, but still linked to existing Business and / or IT organization	**Platform capabilities defined and building** out separate teams. The capabilities are predictable, meaning the organization knows where to invest where to get the necessary skills and expertise	**Platform capabilities fully established** and rolled out in the organization.
	Customer Experience Design	**Isolated attempts** within the organization to exploring service design as a new methodology and unite with other service design enthusiasts to start a first initiative.	**Pioneering** to get service design established in the organization, with service design projects and the creation of evidence of its value	**Scaling** - Service design expands throughout the organization through unifying tools and methodologies and teaching of its capabilities	**Integrated** - Siloed organizational structures are torn down and transformed into a design-led foundation, including the ecosystem partners. Service design is embedded in the daily way of working through integrated systems and metrics	**Thrive** - Service design now thrives in the organization through leadership and experimentation, and service design is integrated in the company culture. Methodologies are being evolved as the organization is pushing the service design towards the platform
	Risk Management	**Unsure** how risk management is to be established for a platform business model. No or isolated risk procedures that focus on the platform.	**Awareness** of need to establish a separate risk management process for the platform business model. Some first risk procedures emerge.	**Common understanding** of risk management process for platform models across the organization (both business as IT side)	**A formal risk management process** for platform models is in place and communicated to all stakeholders (also the ecosystem partners)	The risk management process is **embedded** in the ecosystem via formal risk management processes, including decision making

continued on following page

Table 38. Continued

Level	Dimension	Level 1: Exploring	Level 2: Developing	Level 3: Embracing	Level 4: Using	Level 5: Leading
Design Criteria	**Technology Ecosystem Design**	**Ad-hoc** and Sporadic usage of APIs that may be externally exposed - governed via IT standards	API standardization approach **developing** across the organization.	API standardization approach **defined and established.** APIs externally exposed and integrated in third party (platform) systems in a standard and repeatable approach. Certain data or functionality is exposed via APIs and integrated in another platform (e.g., data is available on the Open Data platform)	**API standards are managed** by stakeholders of the ecosystem All externally exposed data & functionality under constant monitoring and adjustments	**API standards are optimized** Open and Ecosystem APIs The platform system and integration architecture is fully based on APIs.
	Processes	The designed processes for the business platform model are handled in **an ad-hoc way.** Few activities are explicitly defined for the platform model and success depends on individual efforts and heroics	The designed processes for the business platform model are **developing.** Basic management of the defined processes is established to track progress. The necessary discipline is in place to repeat earlier success	The designed processes for business platform model are **defined and established** The processes are documented, standardized, and integrated with an organizational-wide method	The designed processes for business platform model are **managed** All processes and controls are implemented, measured, and adjusted according to the usage of the platform. Detailed measures of the processes and output quality are collected.	The designed processes for the business platform model are **optimized.** Processes are continuously reviewed and optimized following the evolution of the platform, its services, and the value generation. Improvements are enabled by quantitative feedback from all stakeholders and from plotting innovative new ideas and technologies to the platform
	Structures	**Platform PMO is ad-hoc** Informal, basic roles, responsibilities are decentralized The Platform model is an irregular item on the existing agenda of business and IT fora.	**Platform PMO is developing** Platform Strategy is an item on the IT Steering Group - Business Case being formed, the budget is allocated for investment	**Platform PMO is Defined and Established** Platform Steering Group as part of overall IT Steering group - separate item on the agenda	**Platform PMO is Managed** Dedicated Steering groups are formed for management and roll-out of the platform model.	**Platform PMO is Optimized** Optimal integration of Platform PMO with the rest of the business and the ecosystem. Skills are optimized to optimize the operating model and continuously work with the ecosystem partners
	Relational Mechanisms	**Platform Leadership Team Structure is ad-hoc** Business and IT people are working together in an unstructured way to explore the platform model. Existing business and IT leadership are handling the platform as an extra item on their workload	**Platform Leadership Team Structure is developing** A Platform Team is identified but still operates as a virtual team inside the organization. Team members get dedicated time to work on the platform model. Existing business and IT leadership are mandated to cover the platform strategy	**Platform Leadership Team Structure is Defined and Established** A dedicated Platform Team with clearly defined R&R, scope, and metrics is installed in the organization. Active participation of principle stakeholders both inside the organization as in the ecosystem	**Platform Leadership Team Structure is Managed** The Platform Team operates as a separate unit inside the organization. The Platform Leader frequently reports to the CEO on progress and improvements.	**Platform Leadership Team Structure is Optimized** Business, IT, and the ecosystem operate as one integrated team with dedicated leadership, maintaining shared goals and objectives, SLAs, and one budget. Leadership interacts with the Board on optimizing strategy of the organization towards (more) platform services.

Next, we added the defined practices as extra rows and defined the value of each practice per level. These results are consolidated in the following Government business Platform Model maturity model, including a basic description of what it means to perform on the level (Table 38).

3.6 Method 4: The Platform Balanced Score Card

Defining practices as a long list was necessary but becomes more relevant if we can combine these into a set of objectives and metrics that can be used in the organization to define or measure the progress of using the platform business model ("Opening Services").

This means we needed to find a mechanism to convert the practices into a logical and coherent sequence combining the strategic, tactical, and operational levels into one integrated and linked dashboard.

To achieve that, the inspiration came from the original balanced scorecard concept, popularized by Robert Kaplan and David Norton(Kaplan & Norton, 1996), which is based on four fundamental perspectives: financial, customer, internal business process, and learning and growth.

However, some of these perspectives are not that relevant to governments. For example, the financial perspective in a government is not that critical but usually replaced by service delivery satisfaction. Therefore, we turned to the adapted model introduced by De Haes and Van Grembergen (De Haes et al., 2020) as a compass.

Figure 17. BSC from the original model to the De Haes & Van Grembergen model

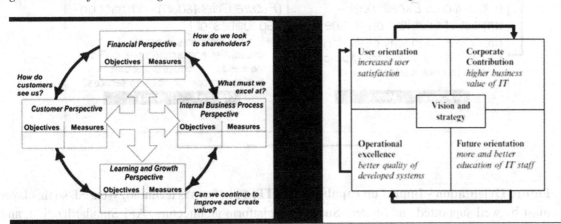

One of the reasons to prefer this model is the "cause and effect" implication that goes with a business platform model. A platform is not just an IT artifact, nor is it just a set of processes that is operated by the business (using technology). No, a platform requires business and IT to operate as one team, and goals need to be interlocked. The platform is the product that the business brings to the market, and technology is the main contributor to that success. In effect, the platform is both the product and technology. So, for every euro spent on the technology side, there is to be an effect on the business side.

Some state that the BSC is an outdated model and does not serve modern IT purposes anymore (Cram, 2007), but to link the strategy and execution in a "cause and effect" approach, we found this model still to be relevant. Fortunately, Cram concludes in his research article that "*Its resilience in remaining part of present-day control frameworks indicates that the IT BSC is not an outdated management fad, but an important component of modern organizations*".

We can now plot the defined practices onto a BSC. This will make sure the platform model is well defined both from business as IT side and recognized as a strategic project inside the organization that gets the right political support and contributes to the success of the government organization.

The following perspectives for a BSC for a government platform business model ("Opening Services") were defined (Figure 18).

Figure 18. Platform perspectives mapped to the BSC model – Own Work

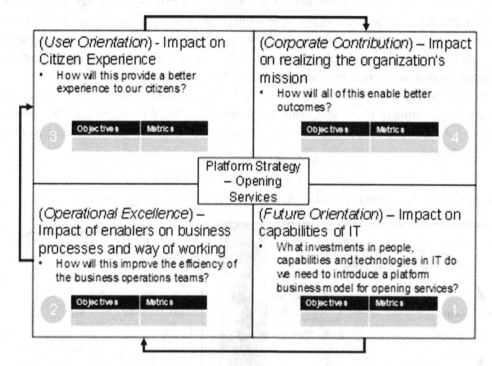

- **Future Orientation - Impact on capabilities of IT**: a platform is technology-based, so this layer must be well supported and funded. Success in platform architecture, API Standardization, and data-driven only materialize if the organization has the necessary capabilities sourced or available in-house. Even more than with other strategies, investing in IT is critical to the success of the business. Not reaching the objectives stated here will be an inhibitor to realizing the business goals.
- **Operational Excellence - Impact of enablers on business processes and way of working**: the big difference in a platform model is that the platform is the product, and thus the role of business changes dramatically from sourcing functionality to providing access to data and functionality for the ecosystem to perform correctly. This requires a different approach to organizing the business towards being a director of services and building out the ecosystem. For that, different objectives are required.
- **User Orientation - Impact on Citizen Experience**: the reason a platform is used is to create added value in executing services faster and more efficiently by using the ecosystem as a provider. However, ultimately, the citizen is the judge of this. It must be measured if the services offered on the platform are meeting the demand and fulfilling the increasing request for personalized and proactive services. The success of the platform is thus measured by the feedback of the citizen, not only based on the number of transactions.
- **Corporate contribution - Impact on realizing the organization's mission**: every government organization has a set of services they deliver to citizens. By choosing to use the platform model, the organization has chosen to optimize delivery and become a director instead of the executor, using an ecosystem of providers. This requires good leadership commitment and a budget to turn this into reality. Practices as (political) leadership, vision, and strategy play here.

Table 39. Linking the BSC perspectives to practices, suggested objectives, and metrics – Own Work

BSC Perspective	Defined Critical Success Factor	Suggested Objectives	Suggested Metrics
Future Orientation - Impact on capabilities of IT	Platform Architecture & Infrastructure	• Allocate the necessary budget to provide all technologies that comprise the platform • Validate platform architecture with the ecosystem (e.g., API Standards) • Adopt a Cloud-First strategy	• Percentage of IT Spend on platform related technologies versus overall IT Spend • Number of non-compliance issues reported • Number of servers in use in-house versus in the Cloud
	Talent and Skills	• Build out new skills for business and IT people • Attract the right platform expertise • Define job roles for the platform (both leadership, business as IT)	• Number of trainings planned, executed • Number of new hires per skill type • Number of new technologies introduced and in production state • Number of new hires per skill type • Number of platform roles defined • Number of talents per platform role
	Technology Ecosystem Design	• Drive automated and proactive decision-making using AI	• Number of processes that can make autonomous decisions • Percentage of Service Response time improvement vs. manual decisions
Operational Excellence - Impact of enablers on business processes and way of working	Business Agility	• Redesign current services to the platform model	• Number of services that are offered on the platform versus traditionally • Trend report of services converted to the platform
	Business Ecosystem Design	• Facilitate discussions with partners that want to join the ecosystem	• Number of new partners added to the ecosystem last month and trend showing the evolution
User Orientation - Impact on Citizen Experience	Customer Experience Design	• Involve citizens in the evaluation of the platform experience	• Percentage of positive feedback from consumers versus total feedback
	Data-Driven	• Define ethical handling of data on the platform • Educate staff and ecosystem partners on the risks of improper data handling • Be transparent on what data is used and how	• Number of compliance violations • Executive Involvement • Number of info sessions, trainings, etc. on the importance of data in a platform organization • Number of transparencies reports available to consumers of the platform services
Corporate contribution - Impact on realizing the organization's mission	Vision	• Vision is defined and endorsed by the board	• Communication of the vision to the organization
	Strategy	• The vision is converted into a strategy and budgets are approved • Managed platform transformation programs	• The strategy and budget are approved at board level (portfolio management) • Number of platform programs on time and within budget • Percent of stakeholders satisfied with platform delivery • Percent of platform transformation programs with regular reported status updates
	Leadership	• Politicians express their approval for pursuing the platform model	• Communication to organizational leadership
	Governance	• Implement governance mechanisms that is conformant to the design criteria created in Work Product 2	• Percentage of design criteria implemented versus the defined design criteria in Work Product 2
	(Innovation) Culture	• Knowledge, expertise, and initiatives for business platform innovation	• Level of business executive awareness and understanding of I&T innovation possibilities • Number of approved initiatives resulting from innovative I&T ideas • Number of innovation champions recognized/ awarded
	Value	• Redesign government services to a platform-based service, whereby (parts of) the service is executed by an ecosystem	• Number of services that are transferred to the platform • Percentage of services still traditionally oriented versus offered on the platform • Determine cost of quality for government (what did it cost to set up the service, what did it bring in return for the ecosystem and the citizens) • Number of initiated transactions that result in an actual service order and delivery • Most value metrics like ROI, TCO, number of services requested, time to market, efficiency gains, citizen satisfaction, etc., remain valid as well
	Risk Management	• Keep data safe and relevant to the service transaction only – number of data breaches is zero • Develop and keep actual a risk procedure for both technical as business risks	• Number of reported attempts to data breaches • Effectiveness of risk response strategies in mitigating platform related risks • Classical metrics like number of risks reported, number of risks open, number of risks closed, etc., remain valid

Now that we have a model and a canvas, we needed to convert the practices into a set of objectives and metrics. The idea behind the metrics is to get a better understanding of reality so that the organization can improve decision-making.

However, Tiwani stated that control via metrics is rarely required in platform ecosystems (Tiwana, 2014) as the system is self-regulating via the feedback loop. Consumers that provide negative feedback on selected providers will impact other potential consumers. Providers must, therefore, do their utmost best to provide the best possible service via a transaction on the platform. According to Tiwani, a form of process control by the platform owner (in this case, the government), provides an alternative for a metrics-driven control mechanism.

Keeping Tiwani's remark in the back of our head, we still defined the following set of objectives and metrics as a first cut (Table 39).

It is assumed that every organization will have to define and prioritize its set objectives and metrics. Therefore, the above table is a suggested list, based on literature review and interviews with case owners.

An example of how this is applied can be found with the government of Australia (Australian Digital Transformation Agency, 2019), where they have defined a balanced set of performance measures for their digital program including:

- Strategic measures that demonstrate progress against strategic goals, for example, the percentage of transactions completed using digital platforms compared to non-digital alternatives
- Project performance measures that relate to the implementation of digital initiatives, for example, the percentage of user stories completed
- User experience measures that address how satisfied a user is with a digital platform, for example, user satisfaction and engagement scores over time
- Financial measures that are related to the costs and benefits of running a digital platform, for example, the cost per transaction for a platform
- Operational measures relating to the general operation of digital platforms, for example, system availability of the platform
- Data and security measures that provide visibility of security and data controls, for example, the effectiveness of access controls and the number of cybersecurity incidents detected

To measure the success of a government business platform model and link it to the BSC model, innovators and evaluators must know what impact was intended; they must know who they were seeking to provide value to (Harrison et al., 2012). Fundamental questions like:

- What was the outcome sought or expected from launching the government business platform initiative, and how is it connected to using an ecosystem?
- What was the intent, and has it been realized?
- What change in the government's way of working was sought?
  - More accountable public officials,
  - Improved policy outcomes,
  - More engaged citizenry,
  - Reductions in political and economic inequity,
  - Better services,
  - etc.

…must be addressed when evaluating the platform business model. The question of what was expected in terms of impact must be identified and considered in developing measurement models upfront.

4 EVALUATING GOVERNMENT PLATFORM ORGANIZATIONS

4.1 Introduction

As the defined practices and governance mechanisms were wrapped in a series of validation methods, it was time to validate if these proved valuable in real organisations. We ventured out to case organisations in Flanders with two goals:

- To understand if the practices were confirmed to be representative to real life platform organisations.
- To determine to what extent the validation methods were deemed useful, effective, repeatable, and deliver consistent data across real case organizations. In other words, to make sure the results are consistent and comparable to these and other organisations.
- To determine what the status is of the use case organisations versus a government business platform model, how far they have come, how they can/should apply the defined practices and governance mechanism and ultimately what the lessons learned are from reality.

This was determined through applying the validation methods in real cases and collect feedback from the case owners.

4.2 Selecting the Right Case Organisations

The artifacts and validation methods were created for a specific context, namely, to help government organizations design, roll out and maintain a business platform model ("Opening Services"). Below criteria were defined to select the right case organisations to maximize the relevance of the artifacts:

- The case organisation cannot be fictitious.
- It must relate to an existing government organization in Flanders, operating on regional or local level or a combination.
- The vision of applying a platform model for improving citizen services in the organization exists. In general, the vision must be documented and traceable to either a section in the current Flemish government declaration or to an expressed political view. In other words, there must be (political) leadership support.
- The organization must have at least expressed the desire to evolve to a platform model that fits the "Opening Services" characteristics as defined before. Organizations that fit the "Opening Functionality" characteristics but want to evolve are welcome as well.
- The organization must already have translated the expressed vision into a strategy, whether this draft or not.
- Preferably, the government organization has already a platform up and running or will do this within the next 12 – 18 months, preferably within the current legislation agreement.

This narrowed the number of potential organisations down, but also made it more relevant. This does not mean that organizations that are not that far yet cannot use the artifacts as guidance. It means they were not considered in the scope of this research project as prime candidates to validate the artifacts.

In applying the above criteria, three organizations and key responsible persons were selected that fit them all:

- **VDAB** – Agency of Employment of Flanders, operating on regional level. Here, the platform model was operational, and the ecosystem was established. Interviews were done with the VDAB Program Manager of the Matching Platform.
- **LB365** – Platform for common local government services, operating on municipality level. Here, two founding members of the platform were interviewed several times, i.e., the Program manager and COO V-ICT-OR together with Policy Manager – Department of Organization Development, City of Brasschaat.
- **ACPAAS** – Antwerp City Platform as A Service, operated by Digipolis, the IT Department of the City of Antwerp – operating on city level. Here, the platform model was operational, ecosystem was established, but business was not yet embracing the "Opening Services" characteristics. The contact people here were the Chief Enterprise Architect and the president of Digipolis Antwerp.

As such, there was a case on city, local and regional level that fitted the profile.

A naturalistic evaluation approach was used. This involved data collections via interviews, discussing the artifacts with real case responsible people, based on a real platform model with the desire to solve real problems. In other words, the case owner was able to speak from real experiences.

In the below sections, one will find both the interview notes and the filled-in validation models. This way, we can present a view of where the organization is today and where it wants to be versus the defined practices and design criteria.

4.3 Questions Addressed in Interview

To be consistent, the following questions were used in interviews with the stakeholders of the case organisations.

- **On (Political) Leadership:** Do you see enough (political) leadership in your region (i.e., Flanders) or organization to move forward with a business platform model ("Opening Services")? How was the interest of the politicians established? Do your goals of using a platform model fit in the agenda of the politicians?
- **On Vision:** Is there a vision established at your organization to improve citizen services using the platform model?
- **On Strategy:** does your organization have a strategy defined and agreed to execute upon the vision, i.e., to define or roll out a platform model to improve citizen service delivery?
- **On Value:** has your organization thought about the value that the platform model will bring for each stakeholder (i.e., government, ecosystem, and citizens)? In general, do you see value in adopting the platform model for optimizing citizen service delivery?

- **On Governance:** has your organization defined a set of specific governance mechanisms to get the most value out of the platform model, from an organizational point of view? In general, do you see the importance of governance to drive the strategy?

- **On Business Ecosystem Design:** What are your experiences in setting up the ecosystem? How was this handled inside your organization? How do you maintain the interlock with the ecosystem players? How do you onboard new providers? What do you offer them as value?

- **On Technology Ecosystem Design:** how did you build out the technical standards for making the data and functionality available to the ecosystem to be able to execute the services? What are your experiences here?

- **On (Innovation) Culture:** does your organization embrace innovation? In general, does the definition and roll-out of a platform model require a different culture inside your organization?

- **On Risk Management:** does your organization apply a risk-based approach to manage and mitigate risks? Is security high on your agenda?

- **On Platform Architecture & Infrastructure:** did your organization apply IT architecture techniques to define the platform's technical architecture and infrastructure to run the platform secure and in a scalable way? Did you include (enough) security mechanisms (i.e., controls, tools, procedures, etc.) to make sure the data stays safe?

- **On Customer Experience Design:** did your organization redesign the services with the citizen (i.e., the customer) in mind when putting them out on the platform model? Was the new experience tested with citizens before going live? What are your experiences here?

- **On Business Agility:** Is your organization capable of acting fast in terms of the evolution of the platform model? Can it adapt to the changing needs of the citizens? Did this require new skills?

The questions follow the defined practices as attributed to the level of responsibility in an organisation. The first questions were aimed at the strategic level. Then the tactical practices are addressed, and we ended with the question on the operational level. In the sections below, only the answers will be provided, but these follow the above set of questions.

4.4 Case Organization 1: VDAB - Matching Platform

The below text is the result of different interviews with key leaders at VDAB. The first interviews were at the time the vision to use the platform model for the core process of Matching (see below) was acknowledged by VDAB leadership. The last interview was one year later when a new CEO arrived, and some first reflections on the realizations were expressed.

4.4.1 Introduction to the Case Organization

The VDAB is the public employment service of Flanders. VDAB employment services offer a wide range of services to job seekers, employers, and employees through various channels. Their core mission is to help facilitate the search for a job and offer trainings to improve the individual's chances of finding a relevant job. In case no job can be found, the agency can facilitate providing unemployment benefits to the unemployed.

Figure 19. VDAB Matching logo - Source: VDAB.be

VDAB is an external and autonomous agency under the supervision of the Flanders Secretary for Work, managed by an independent executive board. VDAB is a very progressive organization, fully embracing technology to improve the execution of its mission.

The **VDAB Open Services Platform for Matching** is one of the core programs for achieving the strategic (business) objectives of the VDAB. This program links job seekers efficiently and quickly to open job openings, even if not all requirements are met. In this case, an extra set of short and targeted training can lead to the right match. The Matching algorithm also incorporates experiences, competencies, and other skills the candidate has acquired during his / her career. This could mean that a job seeker is offered a job in a completely different sector than the one the person has experience in. This is seen as a win-win for the job seeker and the employer who can hire faster than ever the desired capabilities to perform the job.

VDAB wanted to expand the reach of the competency-based matching of employees and employers using a **platform model.** The idea is that an ecosystem of partners has access to the vacancies and can do the matching themselves and directly. To help realize this, VDAB created a new department in 2018 that operates as the platform provider to the ecosystem. This program is known as "**Matching 2.0**". The idea was supported by a vision note that was approved by the VDAB board and set the basis for the new department to host and exploit the platform model, including managing the ecosystem.

During 2019, work was done to translate the vision into an organizational design and to refine the platform model. As we enter 2020, it is interesting to see what the lessons learned are versus the practices as defined in the research.

4.4.2 Meeting Minutes of the Interviews

At the time of the first interview, a joined business – IT platform team had been formed. At the time of the second interview, a new CEO had arrived while the team was getting prepared to launch the platform.

As such, it was valuable to capture not only the initial reflections, but also the progress and evolution of the platform model.

Below is a summarized version of the meeting following the defined practices:

On (Political) Leadership

The vision has been confirmed in the new Flemish government declaration (2019 - 2024), this is a good basis to ensure it will happen. The minister is fully aware of the objectives and supports the realization of the platform. The CEO of the VDAB has been tasked to execute the strategy.

On Vision and Strategy

There is a new CEO since Summer 2019. The CEO still supports the realization of the Matching 2.0 platform but is putting additional perspectives on the table:

- The vision is still OK but needs continuous communication and fine-tuning. Sometimes, other operational issues tend to get higher priority. In a context where the leadership is a politician, it takes continuous repeating that realization of the vision takes longer than a day.
- The original idea to provide the platform for free is being reconsidered, so more work is done to create a good overview of the costs versus the benefits that the platform will bring.
- The new CEO also wants to understand better what is in it for VDAB. What data and value, other than facilitating the transaction on the platform, can VDAB extract from the platform?

In general, the platform model is still acknowledged by the new VDAB leadership to support the core process of matching. There is a more economic perspective emerging, namely what the cost and benefits will be of this platform approach. This is in line with the defined practices, where we advocate capturing cost and benefits early in the business case.

On Value

VDAB has spent quite a long time defining the value for all stakeholders. Even today, this is still being refined. Value has been investigated from different angles: internal to the VDAB, to the citizens, to the ecosystem providers, but also partners like employers, other business units to the VDAB, and political level. Even the value for VDAB as a brand was determined. Value is well captured and documented.

There is more focus on ROI than in the early days of the project. The realization has come that the government cannot set up and manage the platform for free. All benefits are now for the suppliers, by using the data of VDAB. There is more attention to ROI models. This does not mean that VDAB needs to make money on this, but at least recuperate operating costs. The main question is: what is in it for VDAB?

IT has been very supportive in building out certain building blocks to be made available through the platform. For example, the provisioning of a data lake that allows exploration of all transaction data is well received. It was recognized again that IT plays a vital role in setting up the platform and balancing costs versus benefits. Costs are often internal, while revenue is for the external agencies using the data.

On Governance

The agile way of working is well established at the VDAB. Most projects already run in an agile way, with business and IT team both working closely together. Teams are empowered to drive innovation within the SAFe framework.

The innovation cell is well matured and integrated into the business. Several ideas have been put into production, and even the algorithms for the matching process (using AI technology) will be exposed on the platform.

On Ecosystems

The current collaboration model is based on gentlemen agreements with the partners, often on an individual basis. This needs to be straightened out towards a consistent model where costs and benefits are standard, or some scenarios are available. Up till now, the model was such that the partners could get the vacancies from the VDAB for free, but in return would send data back with statistics on processing time, how many matches, success rate, turnaround times, needed skills for a 100% match, etc. The model seems to be in favour of the partners, and the ROI for VDAB will be refined and improved.

VDAB also wants to avoid they become elite by choosing which partners get access and which not and the associated pricing model. Therefore, they are working on selection norms and criteria that the partner should undersign when getting the matching data. In the past, this was based on judgment, and now this must mature towards real, verifiable criteria.

There are ongoing talks with ecosystem partners on what the perfect collaboration model should be.

VDAB acknowledges this to be a very important practice and deserves a lot of attention and collaboration to get it straight. It also needs continuous effort to keep the partners interested and motivated to participate.

There is a need for dedicated marketing and communication, both inside and outside of the VDAB:

- How to motivate partners to join the platform and become part of the ecosystem
- Clear price setting
- Clear agreements

On (Innovation) Culture

Like in any organization, some old cultural habits still prevail, trying to cast doubt that this project will not succeed. Therefore, the interviewed acknowledged there is much attention ongoing and still needed to communicating vision, status, objectives, milestones, etc. to everyone in VDAB.

It requires continuous stakeholder management internally to keep everyone on the same line - specially to avoid going back to silo thinking inside a team or part of the organization.

On Risk Management

First, the legal department was referred to. Legal is still investigating where the boundaries of using privacy-related data are for the partners, how they need to handle this, and remain GDPR compliant. Discussions are ongoing. VDAB is the legal entity responsible for employee data, so it must be watertight that partners do not use this data for other purposes than intended via the matching.

Some items on traceability on data processing - who is accountable for reading the data sent by VDAB - there must be a legal counterpart to undersign a clause of confidentiality.

Next, the topic of security was addressed. Here the interviewed stated that the IT department had done a special study how to mitigate attacks on the platform. As there are already other citizen facing applications live that contain sensitive data (e.g., My Career Application), the IT department was not on uncharted terrain here.

On Platform Architecture and Infrastructure

It was acknowledged that a good API strategy is key to get the functionality interaction and integration sorted. This required specific expert IT skills that were sometimes hard to get on the local market. Nevertheless, together with the IT team, this was work in progress.

IT has the necessary expertise to help build out the Platform technically applying the defined standards and technologies at the VDAB.

Integration with legacy IT remains a point of attention as it can slow down the rollout of API or platform functionality.

On Customer Experience Design

The interviewed confirmed that this was an attention point from day one and that experience from other projects helped. However, it requires constant attention to make sure the user experience remains good, from the website of VDAB to the matching services.

A new app is also being built by VDAB, where users can add or remove competencies when searching for a job. This way, job seekers can apply for training where there is a gap in skills required.

On Business Agility

It was recognized that the organizational change of becoming a platform organization is a long and uphill one; the risk is that the ambitions are set too high - in the end, it must be realizable.

Summary of the interview

In summary, the interviews confirmed that setting up a platform model requires good preparation, communication, involvement, and leadership to pursue beyond the basic ideas to real implementation.

4.4.3 Evaluation of case versus defined practices

After the interview and based on the transcript, an evaluation was done on how VDAB complies with the defined practices. Table 40 was filled in by the author, discussed, and agreed with the case owner and acted as a validation of the interviews.

Based on the above table and given the number of "Yes" answers, it can be stated that the VDAB is covering all defined practices. Enough evidence was provided during the interview to come to this conclusion. However, this does not imply that it will be easy sailing as of now. The Interviewee clearly stated that keeping the Platform Dream alive is ongoing and continuous work.

Table 40. Does VDAB comply with the defined practices - Own work

Level	Practice	Considered	Comments
Strategic Level	Vision	Yes	The Matching platform vision is well articulated inside the organization but also anchored in the new Flemish Government declaration (2019 - 2024)
	(Political) Leadership	Yes	The new CEO confirmed the Matching Platform as the way to go. Leadership is in place to drive the formation of the Matching platform organization and the platform itself. Certain elements of the business case are being re-evaluated, and more focus on cost versus benefits was needed.
	Strategy	Yes	Based on the leadership of the CEO of VDAB, a team has been put in place to draw the strategy how to realize the Matching Platform. This includes the business and IT strategy of realizing the platform model.
	Governance	Yes	Processes for the Matching Platform organization are still refined, but the basis is defined and documented
	(Innovation) Culture	Yes	Agile way of working is well established and proven inside the VDAB and applied to the design of the platform model
	Value	Yes	Work has been spent on determining the value for all stakeholders. Refinement is ongoing. Costs for designing, hosting, and exploiting the platform are calculated but are under revision and refinement
Tactical Level	Business Agility	Yes	VDAB has the necessary business agility, but it needs attention to keep up the momentum of innovation
	Business Ecosystem Design	Yes	This is well defined, under control, and established inside VDAB. It is recognized to be an ongoing journey and requires vigilance to keep the collaboration going and growing stronger.
	Platform Architecture & Infrastructure	Yes	The IT team has the necessary standards and technology available to design and roll out the platform. An API strategy is being refined, and special attention goes to interoperability, especially with legacy IT systems.
	Data Driven	Yes	Through the innovation cell, many ideas on how to use data better in the business decision process have started, and some have made it into production (e.g., Matching algorithm using AI technology)
Operational Level	Talent and Skills	Yes	IT has the right skills, and business is evolving and adapting. HR is involved in attracting the right expertise to help build the platform model. VDAB has an extensive network of partners where they can source the skills needed to build the platform model.
	Customer Experience Design	Yes	So far, discussions with the end-user have been facilitated via the ecosystem partners. Little contact was directly with the unemployed to see how to provide a good experience. This is because providing a good experience is being delegated to the ecosystem partners. However, for the connecting VDAB websites (e.g., "My Career"), appropriate techniques were used to provide citizens a good experience. The interviewed added that this is to be an ongoing exercise.
	Risk Management	Yes	For example, the legal department is involved in helping to straighten out the ecosystem models.
	Technology Ecosystem Design	Yes	The IT team is very supportive and aligned with the platform organization. This project is only one of the many for IT, but the right alignment and objectives have been agreed upon.

Vice versa, the interviewed confirmed that the defined practices were representing a good 360° of all facets to cover. Based on this feedback, the defined list of practices is valid, but this is not a general conclusion yet.

4.4.4 Evaluation of Governance Mechanisms

In this section, a validity check was done on the existence of governance mechanisms to manage the practices. Based on the interview, it was understood that governance mechanisms were in the process of being continuously refined.

4.4.4.1 Process Checklist

Together with the case owner, the results of the interview and above list of evaluated practices, the checklist for processes was filled in (Table 41).

As stated, the above table is not the result of an audit, but the interviewed gave verbal assurance of the existence of these processes. The interviewed added that although the above table was mapped to corresponding COBIT2019 processes, the VDAB did not design their processes according to COBIT2019, so sometimes some interpretation was needed to understand the meaning of COBIT2019 versus the VDAB reality.

4.4.4.2 Organizational Structures Checklist

Together with the case owner, the results of the interview and above list of evaluated practices, the checklist for organizational structures was filled in (Table 42).

In summary, VDAB has established the Platform team within the standard organizational structure and is reusing existing structures to address the platform model.

It is to be noted that the above table is not the result of an audit, but the interviewed gave verbal assurance of the existence of these structures.

4.4.4.3 Relational Mechanisms Checklist

Together with the case owner, the results of the interview and above list of evaluated practices, the checklist for relational mechanisms was filled in (Table 43).

Note the above table is a condensed version of the checklist introduced earlier. For completeness, one is referred to the chapter where the governance mechanisms were established

Also, for the relational mechanisms, the Platform Team is reusing the existing VDAB mechanisms, but has defined its own mechanisms to specialize in acquiring and spreading knowledge and understanding on the platform model, inside and outside the VDAB.

It is to be noted that the above table is not the result of an audit, but the interviewed gave verbal assurance of the existence of these mechanisms.

4.4.5 Mapping Onto Organizational Capability & Impact Quadrant

Next, we asked the case owner if the organization had the right capabilities in place for turning the practices into practice. After all, the practices might have been identified before, but the question is off course how they were turned into practice at the VDAB. For that, the defined organizational Capability & Impact quadrant model was used.

The below table was filled in by the interviewed. The scores are attributed based on the experience of and reflection by the case owner.

Table 41. Process Checklist as applied to VDAB Case - Own Work

COBIT2019 Process	In place Y/N?	Comments
APO02 - Managed Strategy	Y	See practice of Vision & Strategy – a documented, and managed strategy exists
APO12 - Managed Risk	Y	See practice of Rik Management
APO13 - Managed Security	Y	Handled within the IT department
DSS05 - Managed Security Services	Y	Handled within the IT department
MEA03 - Managed compliance with external requirements	Y	The legal department is involved
EDM01 - Ensured Governance Framework Setting and Maintenance	Y	The mission of the Platform team is further refining its own governance mechanisms
EDM03 - Ensured Risk Optimization	Y	Risks are handled within the platform team
EDM05 - Ensured stakeholder engagement	Y	As stated, both internally and to the ecosystem
APO03 - Managed Enterprise Architecture	Y	Handled within the IT department, but extra is to be noted that the platform is rolled out with the standard technology of VDAB
APO04 - Managed Innovation	Y	The Matching process is being innovated and the platform team is responsible for rolling out the redefined services on the platform
APO08 - Managed relations	Y	The platform team does spend time with the ecosystem
APO14 - Managed Data	Y	The VDAB innovation team has worked on new technology using the matching data to predict who is best eligible for the job. This data is being shared with the ecosystem to optimize service delivery (i.e., targeting the right candidate)
BAI05 - Managed Organizational Change	Y	The platform team is a separate team composing of business and IT people, so a result of an organizational change
BAI06 - Managed IT Changes	Y	The Platform team has a list of projects and technology roadmap with the IT Team
DSS06 - Managed Business Process Controls	N	The platform team consists of business representatives, but this group is not directly responsible for service design. Service Design is handled by another central team at VDAB.
EDM02 - Managed Benefits delivery	Y	The platform team does provide insights in the benefits the platform through monthly reports to the VDAB management (e.g., how many people were given a job)
APO01 - Managed I&T Management Framework	Y	VDAB has standard framework and this is applied in the Platform Team
APO05 - Managed Portfolio	Y	The platform team has a budget allocated and handles new requests for additional investments. These are brought to the board, after aligning with business and IT managers.
APO07 - Managed Human Resources	Y	The platform team operates according to the HR processes of VDAB
APO09 - Managed Service Level Agreements	Y	The platform team has defined a set of SLAs with the IT team for ensuring availability of the platform
APO10 - Managed Vendors	Y	The platform team maintains a relationship with ecosystems (i.e., providers), but also with possible technology vendors to optimize the matching process
APO11 - Managed Quality	Y	Through the standard VDAB process
BAI08 - Managed Knowledge	Y	Through the standard VDAB approach, but the platform team has its internal knowledge sharing as well
MEA01 - Manage Performance and Conformance Reporting	Y	Reports go out to management regularly, i.e., monthly
MEA04 - Managed Assurance	Y	Through the standard VDAB processes
EDM04 - Ensure Resource Optimization	Y	Through the standard VDAB processes
APO06 - Managed Budgets and Costs	Y	The Platform team reports on budget and spent to the board

Table 42. Organizational structures at the VDAB case - Own work

Platform related organizational structures	In Place (Y/N)?	Comments
Platform Strategy committee at the level of board of directors	N	The platform team has this function, but there is no separate platform strategy committee. This is handled on the regular agenda of the VDAB board.
Digital Platform Transformation Leader, reporting to CEO	N	The platform team has its own leader, but this person is not reporting the CEO, but was fitted in the current organization of VDAB.
Platform Steering Committee	Y	The platform team organizes its own meetings and liaises with business and IT withing the existing steering committees.
Platform Governance Officer	Y	The leader of the platform team has the authority to define the governance, but within the established standards of the VDAB
Platform Security / compliance / risk officer	Y	These responsibilities lie within other organizations at the VDAB, but the Platform Team does liaise or consult with them (e.g., legal team)
Platform Project steering committee	Y	The platform team organizes its own meetings and liaises with business and IT withing the existing steering committees.
Platform Security steering committee	Y	This is handled in the meetings with IT
Platform Architecture steering committee	Y	This is handled in the meetings with IT

Table 43. Relational mechanisms at the VDAB case - Own work

Are the following relational mechanisms in place for Platform Model?	In Place (Y/N)?	Comments?
1. Co-location	Y	The platform team shares the same buildings, not always the same offices
2. Knowledge management	Y	The platform team has specific attention to understanding the platform model. Trainings were given on defining the value (Note: part of this training was given by Prof. Dr Steven De Haes in February 2019).
3. Ongoing awareness training on Platform model strategy	Y	The platform team does this, but recognizes this to be an ongoing duty, not only internally, but also and foremost with the ecosystem partners
4. Articulate the vision on Platform Model	Y	The vision is well articulated, and the platform team has good knowledge of it. The people specifically joined the platform team based on the vision and wanting to realize it.
5. Internal Communication	Y	The Platform team communicates through the VDAB standards (i.e., section in the VDAB newsletter)
6. Network outside the organization	Y	The Platform team (or members of the team) are often present at conferences to understand the latest trends, or even presenting the VDAB case (e.g., AWS Public Sector conference in Brussels, 2019 edition)
7. Platform Account Manager	Y	Within the platform team, there is a person dedicated to growing the ecosystem.

- On the X-Axis: **Capability** indicators answers (shortened version):

Table 44. VDAB scoring on Capability indicators - Own Work

Indicator	#	Question	Score (1...5)
STRATEGY	1	Clear, coherent, and actionable platform strategy	5
	2	Strategy focuses on transforming the whole business	4
	3	Strategy strives for maximum reconfiguration and availability of the services	5
BUSINESS ECOSYSTEM DESIGN	4	Active process with (potential) ecosystem providers	4
	5	My organization has a clear business roadmap	4
	6	My organization orchestrates the ecosystem providers in an open and co-creative approach.	4
PLATFORM ARCHITECTURE & INFRASTRUCTURE	7	My organization funds and resources the platform services and technology transformation adequately.	4
	8	Platform investment considers an organization-wide approach	4
	9	Integrate new technologies with older 'legacy technologies'	3
	10	Technology and solutions in place for analytics	5
	11	Systems, applications, and tools in place to interact (e.g., API approach)	3
	12	Technology in place to support service execution on the platform.	4
TECHNOLOGY ECOSYSTEM DESIGN	13	We connect with our partners digitally (APIs).	4
	14	Leveraging functional reusable components from our government	4
	15	Roadmap available on new functionality or components	3
DATA DRIVEN	16	Business intelligence system for timely decisions available	3
	17	Process and technology in place to understand & act upon citizen behaviour	4
	18	My organization is data focused to provide proactive service advice	5
TALENT AND SKILLS	19	Skills and competencies to design and build of services	4
	20	Easy to attract technical staff	3
	21	Investments in developing digital skills of employees	4
RISK MANAGEMENT	22	Technology in place to prevent cyber-attacks and other security risks	4
	23	Regular assessments on technical, business, and social risk factors	4
	24	Scalability of platform infrastructure meets the demand	4
	25	Proactive risk management approach	5
CUSTOMER EXPERIENCE DESIGN	26	User Experience research is conducted	3
	27	Design and deliver a tailored product to fulfil specific citizen's needs.	3
	28	Citizens get fully integrated experience	5
	29	Continuous improving digital and physical experiences on the platform	5
	30	Citizens can effectively communicate with my organization	3

- On the Y - Axis: **Impact** indicators answers (condensed version):

Table 45. VDAB scoring on Impact indicators - Own Work

Indicator	#	Question	Score (1...5)
VISION	1	Long-term goal for platform model	5
	2	We are customer centric	5
	3	The platform business model is part of our vision	5
	4	The platform strategy is aligned with the overall business strategy.	5
(POLITICAL) LEADERSHIP	5	Leaders have goals and objectives how to use the platform model	5
	6	Leaders can communicate their future foresight	4
	7	Leaders identify and realize opportunities for platform efficiency.	4
	8	Leaders empower cross-functional collaborative teams	5
GOVERNANCE	9	Agile processes are used to react to rapid business change.	5
	10	Employees can self-organize to address problems	5
	11	Staff is empowered to work autonomously	5
INNOVATION CULTURE	12	Everyone can think creatively, innovating the platform	4
	13	Service design is done in rigorous and systematic approach to innovation	4
	14	Process for introducing new technologies exists	5
	15	Small iterative experiments to realize innovation	5
	16	Innovation activities are a regular task	5
	17	Employees feel empowered and take calculated risks to be successful	5
	18	Employees work in interdisciplinary teams	5
	19	Teams work collaboratively on projects and share developments	5
VALUE	20	Citizens can effectively communicate with us	4
	21	Providers can effectively communicate with us to co-create value	4
	22	All staff work in sync towards implementing our (platform) vision	4
	23	My organization fosters an integrated digital ecosystem	4
	24	Everyone knows of, understands, and can act on our platform strategy	3
	25	Very few technical issues in the delivery of our platform services	3
	26	Technical issues are resolved within an acceptable period	4
	27	Platform is generating efficiencies versus the defined metrics	4
BUSINESS AGILITY	28	Proven ability to identify citizen's latent needs	4
	29	Ability to pivot its approach based on analysis of citizen insights	3
	30	Technical teams implement services fast	4

Next, the scores were added up. Based on the earlier defined scoring reference table, we come to the following totals for the VDAB:

Table 46. VDAB Platform Quadrant Scoring - Own Work

Quadrant	Capability Score	Impact Score
Initiate	0 - 75	0 - 75
Competent	76 - 150	0 - 75
Purposeful	0 - 75	76 - 150
Transformative	76 - 150	76 - 150
>> ACTUAL Score (Max Score is 150 per column)	**119**	**132**

With these scores, we can draw where VDAB fits in the model. This was visualized as follows:

Figure 20. VDAB Platform Magic Quadrant Visualization - Own Work

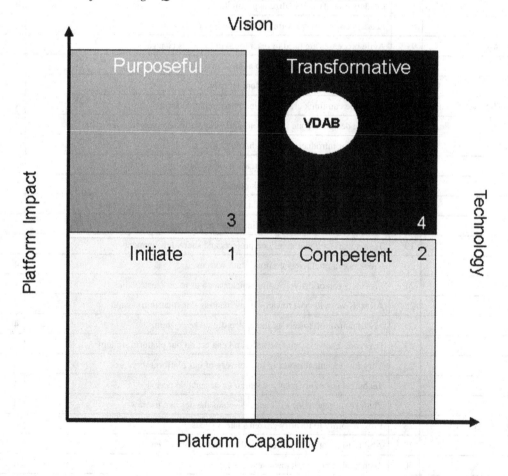

Based on the scores, VDAB is positioned in the top right quadrant and is thus applying the platform model in a transformative way. The artifact that was used proved its value in this case.

4.4.6 Mapping on Maturity Level

Next it was time to verify if the VDAB is considered mature in its desire to position the platform model to improve value and citizen's experience. For that, we used the maturity model, that was created earlier as a validation method. This way, an organization can see where they are today and wants to be going forward, which is expressed in the defined maturity levels on the model.

The below score was given by the author and discussed with the interviewed.

Based on the evidence presented before, VDAB was mapped to Level 5, i.e., Leading the platform model, certainly when compared to other organizations in the Flemish Government. This is because VDAB has demonstrated experience, has established an ecosystem, acknowledged the platform to be part of the business strategy, and - most importantly - the platform is up and running and provides value. In other words, all top practices are demonstrated in real life.

Table 47. VDAB Scoring on Maturity Model - Own Work

Specific	Level 1: Exploring	Level 2: Developing	Level 3: Embracing	Level 4: Using	Level 5: Leading
Ambition	**Exploring** the Business Platform Model	**Developing** a business Platform model	**Embracing** it for first experimental cases or well-defined scope	**Using** a platform model for certain services	**Leading** the model, actively putting services out to the ecosystem
VDAB					*****

4.4.7 Summary of the VDAB Case

As the VDAB was already experienced with rolling out a platform model, the validation methods served as a checklist where still to improve and mature. Through the list of defined Practices, the case owner was able to identify the areas to focus on when further rolling out the platform model. For example, the continuous effort in building out the ecosystem was recognized as key to the success. Also, the practice referring to political leadership was acknowledged by the case owner as one of the key items to take care of.

The governance mechanisms at VDAB were evaluated through the process checklists. The observation here was that most of the governance mechanisms were established within the defined and existing standards, committees, and lines of responsibility of the VDAB.

Through discussions with the interviewed, reflections were made on the new platform team that VDAB set up. It was recognized to keep communicating the vision and strategy to all stakeholders at VDAB. Also, to continuously refine the "one business – IT platform team" at VDAB and allow more autonomy as the platform is further rolled out in the market.

For this organisation, the artifacts served as reflection towards the current set up and experience but were also indicative for improving the VDAB platform organization towards providing better citizen services.

The artifacts were validated as useful by the case owner, through informal feedback. This implies that the artifacts can be used for government organizations that are already mature in rolling out the platform model as well. Obviously, more similar case organizations are needed to generalize this conclusion.

4.5 CASE ORGANIZATION 2: LB365

The below text is the result of different interviews with key leaders at LB365 over a period of twelve months from late 2019 till late-2020.

At the time of the first interviews (late 2019), the initiative was still being developed and rolled out. At the same time, the team was already looking to increase the number of services defined. The road-map of LB365 was particularly interesting to be investigated as a case as the team plans to evolve from "Opening Functionality" to "Opening Services".

A second series of interviews were done in December 2020. At this stage, the platform was live, and the ecosystem was growing. The lessons learned were captured and documented as updates to the original answers. That way, the reader can clearly see the evolution.

4.5.1 Introduction to the Case Organization

Figure 21. LB365 Logo - Source: Internet

LB365 is an initiative supported by V-ICT-OR, the organisation that unites local municipalities and IT. V-ICT-OR that develops and promotes expertise in the field of IT, at every level in the local govern-ment. Central to its mission is to achieve more efficiency and effectiveness in IT by using and reusing the same standards. By promoting collaboration across all levels of local municipalities, the consortium wants to lower the threshold for creating synergy significantly. Besides supporting local initiatives, the association also wishes to act as an intermediary to support cooperation between the various other levels of government in Flanders.

The concept of LB365 is based on the VLAVIRGEM concept which is further explained in the sec-tion on Vision and Strategy below. V-ICT-OR formed a consortium that joins approximately 20 local municipalities in the joined goal to optimize citizen services at the local level and reuse components for similar processes.

Often the service of a government starts and ends with an interaction at the local level (e.g., you visit the town hall to request a service, permit, license, passport, etc.). Equally often, the regional administra-tive level has little feeling with that interaction point, thus pushing down complex procedures that each municipality tries to solve and implement in their own way. It is in that perspective that LB365 was created: how can we create a set of generic and reusable components that each municipality can use instead of building them? The objective is to open functionality – through the components - to an ecosystem that can build applications or services on top of the components.

At first sight, this fits the business platform model of "Opening Functionality" level perfectly. During the discussions, it was noticed how LB365 could evolve to an "Opening Services" model as defined in this research project. As such, the case is more around "what if you use the artifacts of this research project (i.e., Practices and Design Criteria) and extend the roadmap of LB365 towards the "Opening Services" model - what would that mean?"

4.5.2 Meeting Minutes of the Interviews

Below is a summarized version of the meetings.

On (Political) Leadership:

One of the founding municipalities is Brasschaat, where the current prime minister of the Flemish Government, Mr. Jan Jambon, was a member of the council when this idea was presented. In a sense, we have a high political figure supporting this.

However, as LB365 represents a coalition of multiple local communities, one will have to ensure that they all remain on board, and thus continuing to provide leadership was acknowledged as key.

The team is working more and more closely with political leadership to help communicate the need for adjusting the operating model when going live. This goes hand in hand with managing change, culture, and innovation.

Late 2020, the Flemish government announced an ambitious digitalization plan, with more attention to technology to improve citizen services. A new IT agency has been created to centralize all digital efforts, with substantial extra budget to spend. Although a platform model is not explicitly mentioned, the first political signs of the importance of redesigning citizen services are on the horizon. The prime minister is expressing the desire to streamline the digital experience on all levels of government and mentioned LB365 as a cornerstone project for local municipalities. As such, we see that political leadership is recognizing technology to optimize citizen services.

On Vision and Strategy

The idea behind LB365 emerged already in 2003, with the vision of building a virtual new municipality from the ground up, and 100% digital. The initiative is called VLAVIRGEM (V-ICT-OR, 2020). As the scope is on local municipalities, the parallel with Smart Cities comes quickly to mind.

VLAVIRGEM idealizes a city/municipality in which information and technology are used to manage and control the city. This encompasses both the administration and its digital services as well as the facilities such as education, energy, environment, urban development, health care and prevention, transport and mobility systems, local economy, and all other possible utilities. A smart city / municipality aims to improve the quality of life by organizing the city more efficiently and reducing the distance between the inhabitants and the administration. All parts of the city/municipality will be connected via a network of sensors, internet, and high-quality technological devices with the "internet of things" as a motor, a connected world of technology to achieve the stated objectives.

Figure 22. VLAVIRGEM Vision - Source: V-ICT-OR website

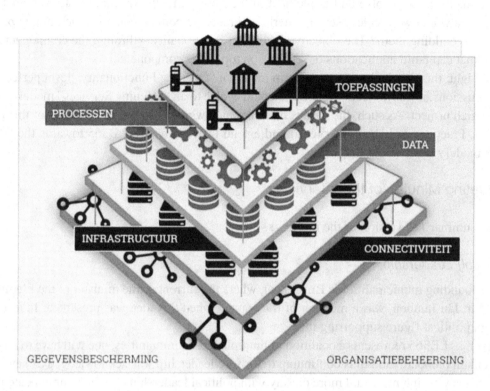

The idea is that there is a platform established that delivers generic building blocks of functionality that every municipality needs, also known as components. That platform is referred to as LB365.

First, a study was done on what these components should be. The observation was that sometimes many versions of the same information float around in different applications and that not every department knows or can react to the information that the other department has. For example: when a person dies, not all departments are informed at the same time or in the same way. This has strange consequences: the family still gets (local) tax letters, communication still goes to the address at the care centre, while the bed has been taken by another person.

Next, the architecture was established based on creating small reusable building blocks of functionality that can be reused. As back office, it was chosen to standardize all technology on the Microsoft Dynamics CRM toolset, that runs in the Cloud.

In-line with different levels of Government as a Platform defined earlier in the research, LB365 has made sure the right technology and standards were defined upfront to ensure "Opening Data" and "Opening Functionality" features to be available on the platform. For example, data from local municipalities can be linked more easily with authentic data sources on the regional and federal level, due to the introduction of standard data models and exchange protocols, including the usage of APIs from regional systems. Because the data is correctly exchanged and kept up to date, this approach facilitates an end-to-end quality citizen service, from request to delivery.

During the last interview in December 2020, it was confirmed the above still to be the guiding strategy. The platform is now live and rolling out to more municipalities and more functionality becomes available. The defined architectural principles have resulted in less data errors across the municipalities, regional government data sources and suppliers using the standard data models. The LB365 team confirmed the idea of moving to "Opening Services" remains the goal and even new use cases have emerged. See governance section how this is being decided and prioritized.

On Value

The value for the municipalities is:

- First, there is value in the technical reuse. Common functionality has been developed only once and can be integrated many times into the back-office applications. This way, no new application needs to be written, just connected to the generic service.
- Second, there is value in streamlining operations and service delivery as the process is the same, and data is stored similarly. This will increase the efficiency of the caseworkers at the local level, even across silos in the same local communality.
- By standardizing the data models and integrating directly with the master databases at regional or federal level, the quality of the data increases, making the service faster to start or change.

There is a cost for developing and maintaining the generic functionality. The experts made a business case that shows savings versus each municipality developing its own similar software. The cost for building the reusable functionality is spread over the participating municipalities, so the higher the number of participants, the lower the cost per participant as they each share the code produced and put on the platform.

During the last interview, the LB365 team confirmed that the business model had proven itself. For example, the total cost for one of the cities (i.e., Brasschaat) is lower than anticipated. This is because new cities have joined making the total cost of ownership lower as more parties absorb the investment. So, the value for the consortium member has materialized in terms of lower cost of operations by using the platform model.

On Governance

V-ICT-OR acts as the facilitating organization that unites the municipalities and brings them together in the LB365 project. The project has established a governance structure to consult with the stakeholders (i.e., the participating local communities that will act as consumers of the services offered), and a technical group of experts works together with Microsoft to ensure the services get deployed in the right way.

Over the last months, the steering group has institutionalized itself further, for example, by setting up more frequent meeting schedules and closer alignment of demand and supply. As the number of participants has grown, the budget for building new features grew alongside. The team frequently gets together to follow up on performance of the live platform and set the roadmap for new features to be developed or rolled out. Decisions are made in group. As the group grows, this might require a more formal approach to take decisions.

There was no evidence of a designed governance model in place for the specific situation of LB365. Given more progress and roll out, LB365 might reach a point where this needs to be considered, but so far, the current governance structures seem to be enough.

On Ecosystems

In this case, the ecosystem is on the consumer side. LB365 has established a participation model that local municipalities in Flanders can subscribe to. The investment costs (i.e., for the initial development) were spread across the initial participants, each getting the right to consume all created services. The original idea was centred around "Offering Functionality" on the platform, but the platform has evolved so that the applications and services that are built by the producers can also be shared amongst the participants. This way, we see an evolution towards the "Opening Services" level.

Regarding business ecosystem: during the last interview, it was stated that the number of municipalities (i.e., new consumers) as the number of components offered (i.e., available to producers) has been growing steadily over the last 12 months. The LB365 team is actively scouting a promoting both the business as technology ecosystem.

The domains in which LB365 is operating is broadening as well. Originally, generic common processes like complaints, registrations, etc., were put on the platform. Recently, processes for a specific domain like local healthcare, mobility, environment, etc., services are being added. The below drawing shows the expansion of the scope:

Figure 23. LB365 - Expanding scope of local services - Source: V-ICT-OR site

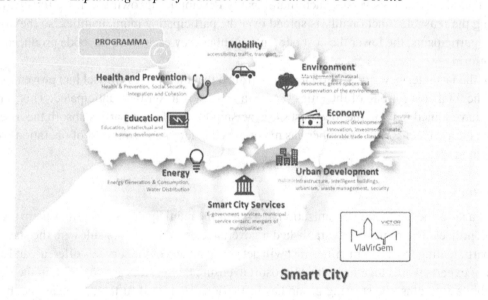

Regarding technology ecosystem: the LB365 technical team first defined a base technical platform with the help of Microsoft and defined a standard to use and deploy APIs to exchange data and functionality with all stakeholders. The standardisation effort has continued and now includes ways to interact with regional or federal authentic data sources as well. All IT providers now have access to both the core functionality APIs as well as extra APIs that allow data exchange.

On (Innovation) Culture

The interviewed reported in the first interviews an uphill battle for municipalities to accept reusing functionality instead of writing and maintaining it themselves (or through their local IT suppliers). Therefore, much attention went to communication, making sure the municipalities understood the value. There was much pragmatism at the local government level and that sometimes caused a "wait and see" attitude. Once the initial results of the platform became clear, then a wave of new interest was noted.

LB365 has gone thru its own platform chicken and the egg experience, i.e., creating enough services (i.e., supply) in the beginning to attract municipalities (i.e., consumers) so that the original goal of offering citizen services based on reusable functional building blocks became a reality.

During the last meeting, LB365 acknowledged again that creating network effects are crucial for the success of the platform. The team has done extra efforts to communicate the approach and the value to municipalities and new ones have been joining the consortium. This means that new participants are not just users of the services offered, but are part of the committee that owns, funds, and builds out the components. The number of services is growing, which gives a feeling of trust to the participants.

The interviewed stated that they must remain vigilant and make sure that all participants, big or small, get an equal say in the roadmap of new features. Bigger municipalities tended to put pressure to prioritize their own unique needs. The team has put governance in place to mitigate this and treat all features based on their value and reusability. The LB365 team coordinates all of this and acknowledged this is an ongoing effort. Likewise, the attention of the LB365 has shifted from getting the platform running in the beginning, to now mitigating further expansion, involving different processes to be used to monitor progress.

On Risk Management

When LB365 started, there were little legal obstructions identified to any of the anticipated services on the platform. After all, the various IT suppliers were already exchanging data securely in their own software products, be it per municipality. The real test came once common services were being deployed on the central platform, and data about citizens – residing in whichever municipality - was being stored, saved, and exchanged, certainly with the regional and federal master databases. Extra attention went to securing data security on the platform, both from a data modelling as technology point of view.

The increasing legal obligations on GDPR and data privacy has brought forward extra requirements to the LB365 data security approach. The team realized that security is not a one-off approach but required constant attention and adaptation.

Security is now an item on the agenda of the LB365 committee. As more municipalities have joined, extra attention goes to identifying risks and managing security.

On Platform Architecture and Infrastructure

From day one, a choice was made to install and deploy all the services on a Microsoft Dynamics CRM "as a service" environment. Although LB365 was acting as the platform owner, much of the platform architecture was built around the CRM tool. By using this CRM tool, the team could also apply a standard approach to APIs and security, to name a few, as this is offered alongside – but integrated with – the CRM tool.

As the platform is not yet deployed, attention will have to go to running and maintaining the platform. This applies to both the architectural but also the operational point of view. This point was not addressed in the interviews.

As the LB365 platform is deployed on AZURE, the platform benefits from new features and extra integration options from the underlying Cloud infrastructure. This made it easier to scale the platform as new municipalities joined. Also, the Microsoft CRM application itself has been evolving. Microsoft now supports a more API approach to CRM functionality. It fits the LB365 vision, where common components will be shared by all IT suppliers.

As such, both the LB365 as AZURE technical architecture has grown and lesson learned have resulted in changed how the LB365 platform is technically managed and IT roadmaps have been adapted. Technology is a fast moving forward business, the LB365 team recognized that this requires constant attention and vigilance. It was recognized that a platform architecture is never finished, certainly as the platform is now live and technical monitoring data becomes available.

On Customer Experience Design

At the start of the LB365 project, there was no specific attention yet to designing a good citizen experience. The assumption was that the IT supplier would be able to do this, when deploying applications on the platform. Also, the CRM tool chosen underneath was already offering quite some good experiences out-of-the-box.

The last interview revealed that the LB365 team is now devoting more attention to the citizen experience. Civil servants using the applications on the platform are now being asked feedback about their experience. As they are often the primary interface to citizens, this feedback captures well how efficient a service request is managed on the platform. This feedback is brought to the LB365 steering committee for review and action, including feedback to the IT suppliers. As such, it was recognized that offering a good experience is part of the success of a platform and the LB365 steering committee is increasingly focused on it.

On Business Agility

The interviewees were initially focused on getting the functionality programmed but realized soon that change management would be needed to manage the new operating model in the municipalities of the consortium. Rolling out the new components implied a new way of working for the civil servants in how to deal with the citizen request. It also was a change for the IT suppliers, that needed to adapt their software to use the components.

The LB365 steering committee managed this well via monthly change management meetings with the IT suppliers. Issues were agreed and prioritized. To handle the changes, an agile approach to development of the components was adopted along the journey. This is now the standard for all enhancements.

In the last interview, it was mentioned that working in an agile way had not only brought technical benefits. Also, the mechanisms for funding and investments were being handled and decided in an agile way. This means that the LB365 committee has a better control mechanism to see the return on investment of the platform.

New Citizen Services

During the first and later interviews, several ideas (see bulleted list here below) were brainstormed whereby the LB365 could be uplifted to the "Opening Services" level. About every local municipality is currently offering local services that are supported by individual applications and operating staff. It was stated that moving these services to the LB365 platform model and open them for all citizens, then even more efficiency in both technical support as operating processes would be achievable.

The mentioned ideas were:

- **Sports camps**: About every local municipality has its own website to provide information on how to book sports camps. For an organization looking to book a camp, this is a laborious process as many websites must be scanned to find a suitable location. Then contact must be initiated with that local communality. Offering available space on a platform model would facilitate more efficiency and transparency.

- **"Vrijetijdskampen"** (i.e., Leisure Facilities): like sports camps but not focused on sport, rather on leisure, thus requiring different facilities to be made available (e.g., Sleeping, washing facilities).

- **"Klusjesdienst"** (i.e., Casual Labour services). This is a mandatory service from the Flemish Regional Government but organized locally per local communality. The current process is still different and sometimes even archaic as one must fill in forms and go to the counter to subscribe to the service. Moving this process to a platform approach would be beneficial to all. One local communality already applies the concept of a platform. Bonheiden has set up a website where producers (i.e., people that are endorsed by the local municipality to provide services) are listed, and consumers (i.e., residents of Bonheiden) can look for a match and execute an order.

- **"Choose Your School"** (www.kiesjeschool.be) Still today, we see images on TV News where parents must camp days in front of a school to be able to register their kids in that school. Many attempts from the regional government have been undertaken to provide a platform where the local government offers on-line registration, but this is not a big success yet. That is why certain local governments are developing their platform where parents can select the school and register their kids. In this case, the cities of Bonheiden, Mechelen, Sint-Katelijne-Waver en Zemst co-developed a platform to achieve this.

- **After school care for children at school**. This case was mentioned specifically during the last interview session. A regional Flemish decree was published in 2019 for organizing after-school care for children. The organisation of this activity has been delegated to local municipalities and goes live in 2021. The Flemish government has created a business platform model style website (opvang.vlaanderen) whereby citizens can look for locally approved providers of after school care. However, the last element of the puzzle, the facilitation of a transaction is not available. For this, the citizen must contact the local municipality and use the local application or process to request the service. This proves that the platform model is live but realized case by case and without a proper platform strategy. This was recognized by the LB365 team as an interesting feature to be put on the LB365 platform.

However strong the cases might read, it is the nature of the local municipalities to offer localized services, i.e., services that are meant for the citizens living in that city or communality only. A critical note was expressed in the interview that too much openness would cause an influx of citizens from other cities to participate in the other city. So, work is also needed on vision and (political) leadership to break the silos across the borders of a city.

Summary of the Interviews

As a summary of the interviews, LB365 started with the idea of creating reusable functionality for similar business processes across municipalities. This fits the "Opening Functionality" model as described at the start. The platform model was established with a clear vision, strategy, business case, and expected results. The platform is now live, new municipalities have joined and IT suppliers have put applications live that use the standard components, but also secure and standard ways of exchanging data.

First lessons learned are injected in adapting the governance approach as well as in the business and technical ecosystem handling. The team constantly discovering what the options could be if this platform is extended to the definition of "Opening Services".

Although early, it seems the practices and governance models defined in this research are coming to surface as the LB365 steering committee grows and has more duties to cover in advancing the platform. LB365 is clearly underway towards an "Opening Services" model.

4.5.3 Evaluation of Case Versus Defined Practices

After all interviews and based on the transcript, an evaluation was done on how LB365 complies with the defined practices. Table 49 was filled in by the author, discussed, and agreed with the case owners and acted as a validation of the interviews.

Based on the interviews and given the number of overarching "Yes" answers in the summary table above, it can be stated that the LB365 team is covering most defined practices. Only one practice (i.e., data driven) was not reflected in the interviews as it is not yet in the design of LB365. This might be a point of attention, going forward. For some practices, and "Yes, but" consideration was given. This is not because the practices are not addressed, but to make sure the LB365 team understands and mitigates the recommendation given.

LB365 provided enough evidence during the interviews to come to this conclusion. The LB365 interviewed confirmed that the list of defined practices was representing a good 360° of all facets to cover. The team indicated that they learned something out of this exercise and that this will be taken up in the project. Based on this feedback, the validation of the artifact seems to hold up in the case of LB365, but this is not a general conclusion yet.

4.5.4 Evaluation of Governance Mechanisms

In this section, a validity check was done on the existence of governance mechanisms at LB365 to manage the practices. Based on the interviews, it was understood that no governance model was specifically designed for purpose, it was an ongoing evolution of best practices that is being applied.

4.5.4.1 Process Checklist

Based on the interviews and above list of evaluated practices, the checklist for processes was filled in as follows:

Table 49. Does LB365 comply with the defined practices - Own work

Level	Practice	Considered	Comments
Strategic Level	Vision	Yes	Against the overarching vision of VLAVIRGEM, LB365 has a clear vision of what to achieve and how the platform model can help achieve this. However, the platform transaction was not immediately focused on "Opening services", but first on realizing reusable components. Today, there are fine examples that are growing towards the "Opening Services" model.
	Strategy & (Political) Leadership	Yes, but	There is a good thought-out strategy behind the vision, and the team is in place and working hard to realize the vision. There are isolated cases of political support, but we can only advise to increase the awareness of (political) leadership to support this idea, both at the local level as at the regional Flanders government level. We mention this as a risk to be managed as, without that support, the rollout of the platform might be hampered.
	Governance	Yes	LB365 has set up a governance model with V-ICT-OR to coordinate development and enlarge the ecosystem.
	(Innovation) Culture	Yes	The idea of leveraging functionality as an answer to common business process is innovative. The usage of agile methods to steer new features both on the business as technical side is well aligned with the practices of this research.
	Value	Yes	After one year live, value has been demonstrated for all stakeholders. There is value established by providing a better experience to civil servants using the applications on the LB365 platform. The value for citizens is that the service request is handled more efficiently. And finally, there is value for municipalities as the initial investment is lowering IT costs due to the reuse of components amongst all.
Tactical Level	Business Agility	Yes	This was demonstrated in the last interviews whereby the LB365 steering committee now interacts with civil servants and IT suppliers to prioritize new functionality.
	Business Ecosystem Design	Yes	A coalition of participating local communalities has been established and is growing since going live. The LB365 steering committee is monitoring progress.
	Platform Architecture & Infrastructure	Yes	A choice for platform technology was made early in the project. Microsoft is on board as a technology partner, and the CRM software will be the foundation of the new functionality. As everything runs in the Cloud and "As a Service" with well-designed integration options like API, this is a smart move. During the first year of operations, lessons were learned how to improve the platform technical architecture as also Microsoft has evolved the CRM application towards an API based approach.
	Data Driven	No	There was no evidence yet that the platform will be data driven and adjust its services based on the behaviour of the consumer/user. This was acknowledged in the last interviews as future requirements.
Operational Level	Talent and Skills	Yes, but	The LB365 team has assembled the necessary skills to start designing the platform architecture and the services. As the platform is now live, extra capabilities were needed to both run and change the platform technicalities.
	Customer Experience Design	Yes	Although this was not an initial focus, the platform is now live and the LB365 Steering committee is liaising with end-users to see efficient the applications are performing.
	Risk Management	Yes	As the platform is now live, many of the risks are identified and controlled by the LB365 Team.
	Technology Ecosystem Design	Yes, but	Reference is made to the choice of using Microsoft Dynamics as a technology base. The LB365 team will have keep on guaranteeing that any integration of third-party software or APIs is highly standardized, but also open and transparent. Imposing proprietary integration mechanisms to vendors wanting to use the functionality in their applications should be avoided.

Table 50. Process Checklist as applied to LB365 - Own Work

COBIT2019 Process	In place Y/N?	Comments
APO02 - Managed Strategy	Yes, but	Although handled informally, the deliverables of this process are in places and metrics are defined to measure benefits and return on investments.
APO12 - Managed Risk	Yes, but	Handled by and via the LB365 Steering Committee.
APO13 - Managed Security	Yes, but	Handled by and via the LB365 Steering Committee.
DSS05 - Managed Security Services	Yes, but	Handled by and via the LB365 Steering Committee.
MEA03 - Managed compliance with external requirements	Yes, but	Handled by and via the LB365 Steering Committee.
EDM01 - Ensured Governance Framework Setting and Maintenance	Yes	The LB365 team is continuously refining its own governance mechanisms.
EDM03 - Ensured Risk Optimization	Yes, but	Handled by and via the LB365 Steering Committee.
EDM05 - Ensured stakeholder engagement	Yes	The LB365 team is evaluating and monitoring all stakeholders on the platform and takes appropriate actions in the Steering committee.
APO03 - Managed Enterprise Architecture	Yes	
APO04 - Managed Innovation	Yes	The LB365 Steering Committee is actively managing the ongoing innovation of the platform, for example by introducing agile methods for enhancing the functionality.
APO08 - Managed relations	Yes	The LB365 Steering Committee does spend time with the ecosystem partners, both the municipalities as the IT providers.
APO14 - Managed Data	Yes, but	A lot of attention goes to defining and applying standard data models to the data exchanged on the platform
BAI05 - Managed Organizational Change	Yes, but	Handled via the LB365 Steering Committee.
BAI06 - Managed IT Changes	Yes, but	Handled via the LB365 Steering Committee.
DSS06 - Managed Business Process Controls	Yes, but	Handled via the LB365 Steering Committee.
EDM02 - Managed Benefits delivery	Yes, but	Handled via the LB365 Steering Committee.
APO01 - Managed I&T Management Framework	Yes	Handled via the LB365 Steering Committee. In this project a specific consortium was set up to split costs amongst participants and share the benefits to all. V-ICT-OR is managing this aspect.
APO05 - Managed Portfolio	Yes, but	Handled via the LB365 Steering Committee.
APO07 - Managed Human Resources	Yes, but	Handled via the LB365 Steering Committee.
APO09 - Managed Service Level Agreements	Yes, but	Handled via the LB365 Steering Committee.
APO10 - Managed Vendors	Yes, but	Handled via the LB365 Steering Committee.
APO11 - Managed Quality	Yes, but	Handled via the LB365 Steering Committee.
BAI08 - Managed Knowledge	Yes, but	Handled via the LB365 Steering Committee.
MEA01 - Manage Performance and Conformance Reporting	Yes, but	Handled via the LB365 Steering Committee.
MEA04 - Managed Assurance	Yes, but	Handled via the LB365 Steering Committee.
EDM04 - Ensure Resource Optimization	Yes, but	Handled via the LB365 Steering Committee.
APO06 - Managed Budgets and Costs	Yes, but	Handled via the LB365 Steering Committee.

The above table is not the result of an audit, rather following the interviews with the LB365 team. As stated, the LB365 team has not designed a specific governance model for their platform and not used COBIT2019 as a reference framework. This means that most used processes above were given an "yes, but" rating, which does not imply a negative connotation, rather that eventually more guidance can be found in COBIT2019 to refine the controls.

4.5.4.2 Organizational Structures Checklist

Based on the interviews with the LB365 team and above list of evaluated practices, the checklist for organizational structures was filled in as follows:

Table 51. Organizational structures at LB365 - Own work

Platform related organizational structures	In Place (Y/N)?	Comments
Platform Strategy committee at the level of board of directors	Yes, but	Handled via the LB365 Steering Committee
Digital Platform Transformation Leader, reporting to CEO	Yes, but	Handled via the LB365 Steering Committee
Platform Steering Committee	Yes, but	Handled via the LB365 Steering Committee
Platform Governance Officer	Yes, but	Handled via the LB365 Steering Committee
Platform Security / compliance / risk officer	Yes, but	Handled via the LB365 Steering Committee
Platform Project steering committee	Yes, but	Handled via the LB365 Steering Committee
Platform Security steering committee	Yes, but	Handled via the LB365 Steering Committee
Platform Architecture steering committee	Yes, but	Handled via the LB365 Steering Committee

The same remark as for the processes is valid here: most organisational structures in the table above are given a "yes, but" score as there is an opportunity to further refine the current organisational structures as the platform matures and more municipalities join. At the time of the last interview, the LB365 Steering Committee was the central organ to manage all the organisational structures.

It is to be noted that the above table is not the result of an audit, but the interviewed gave verbal assurance of the existence of these structures.

4.5.4.3 Relational Mechanisms Checklist

Based on the interviews with the LB365 team and above list of evaluated practices, the checklist for relational mechanisms was filled in as shown in Table 52.

Note the above table is a condensed version of the checklist introduced earlier. For completeness, one is referred to the chapter where the governance mechanisms were established

Also, for the relational mechanisms, the LB365 Steering Committee is again the central organ to ensure smooth interactions inside and outside the consortium. Using the V-ICT-OR network ensures a broad reach to all municipalities in Flanders and to the political leadership authorities.

It is to be noted that the above table is not the result of an audit, but the interviewed gave verbal assurance of the existence of these mechanisms.

Table 52. Relational mechanisms at LB365 - Own work

Are the following relational mechanisms in place for the Platform Model?	In Place (Y/N)?	Comments?
1. Co-location	N	Due to the wide coverage of the partners and municipalities in the LB365 Team, not all people involved are constantly sitting together. V-ICT-OR has an office were the LB365 Steering Committee meets.
2. Knowledge management	Y	The LB365 Steering Committee has a specific attention to understanding the platform model.
3. Ongoing awareness training on Platform model strategy	Y	The LB365 Steering Committee keeps itself up to date on the evolution of the platform model. For example, by attending the V-ICT-OR conferences on platform models, where both busines as technical ecosystem partners meet.
4. Articulate the vision on Platform Model	Y	The vision is well articulated, and the platform team has good knowledge of it. Even political leadership is targeted to understand the value and operating model of LB365.
5. Internal Communication	Y	The LB365 Steering Committee communicates through various channels to local municipalities, for example, by reusing the network of the V-ICT-OR organisation.
6. Network outside the organization	Y	The LB365 Steering Committee is attending the various V-ICT-OR conferences on IT in general and platform models specifically, where both busines as technical ecosystem partners meet.
7. Platform Account Manager	Y	The LB365 Steering Committee is handling this.

4.5.5 Mapping Onto Organizational Capability & Impact Quadrant

Next, we asked the LB365 case owner if the organization had the right capabilities in place for turning the practices into reality. The below table was initially filled in when the platform was not yet operational but refined after the last interview when the platform had proven itself. As such, we have a unique evaluation here to see how the scores evolved after going live and incorporating the first lessons learned.

Here as well, the earlier defined organizational Capability & Impact quadrant model was used to capture the feedback through scoring the questions.

The below table was filled in by the interviewed. The scores are attributed based on the experience of and reflection by the case owner.

- On the X-Axis: **Capability** indicators answers (Condensed version) (Table 53)
- On the Y - Axis: **Impact** indicators answers (Condensed version) (Table 54)

Next, the scores were added up. Based on the earlier defined scoring reference table, we come to the following totals for LB365:

With these scores, we can draw where LB365 fits in the model, before and after going live. This was visualized as follows:

Based on the scores, LB365 was in the lower end of the top right quadrant before going live. The vision and strategy were defined correctly, so this adds up the horizontal score. After going live, the scores go up, due to progressive insights and adaptation of the organization to reality. New municipalities have joined, the technical platform has grown and the LB365 Steering Committee is handling all current and future roadmaps in a controlled approach – although it is advisable to further institutionalize the governance model as complexity will grow along with the platform.

Table 53. LB365 scoring on Capability indicators - Own Work

Indicator	#	Question	Score (1...5)	Score after 1 year (1...5)
STRATEGY	1	My organization has a clear, coherent, and actionable strategy	4	5
	2	Strategy focuses on transforming the whole business	4	
	3	Strategy strives for maximum reconfiguration and availability of the services	4	
BUSINESS ECOSYSTEM DESIGN	4	Active process with (potential) ecosystem providers	4	
	5	My organization has a clear business roadmap	4	
	6	My organization orchestrates the ecosystem providers in an open and co-creative approach.	4	
PLATFORM ARCHITECTURE & INFRASTRUCTURE	7	My organization funds and resources the platform services and technology transformation adequately.	4	
	8	Platform investment considers an organization-wide approach	4	
	9	Integrate new technologies with older 'legacy technologies'	2	4
	10	Technology and solutions are in place for analytics	4	
	11	Systems, applications, and tools in place (e.g., API approach)	4	
	12	Technology in place to support service execution on the platform.	3	4
TECHNOLOGY ECOSYSTEM DESIGN	13	We connect with our partners digitally (APIs).	4	
	14	Leveraging functional reusable components from our government	5	
	15	Roadmap available on new functionality or components	4	
DATA DRIVEN	16	Business intelligence system for timely decisions available	3	
	17	Process and technology in place to understand & act upon citizen behaviour	2	
	18	My organization is data focused to provide proactive service advice	2	
TALENT AND SKILLS	19	Skills and competencies to design and build of services	4	
	20	Easy to attract technical staff	1	3
	21	Investments in developing digital skills of employees	2	3
RISK MANAGEMENT	22	Technology in place to prevent cyber-attacks and other security risks	3	4
	23	Regular assessments on technical, business, and social risk factors	3	
	24	Scalability of platform infrastructure meets the demand	4	
	25	Proactive risk management approach	3	
CUSTOMER EXPERIENCE DESIGN	26	User Experience research is conducted	1	4
	27	Design and deliver a tailored product to fulfil specific citizen's needs.	3	
	28	Citizens get fully integrated experience	3	4
	29	Continuous improving digital and physical experiences on the platform	2	4
	30	Citizens can effectively communicate with my organization	2	4

Table 54. LB365 scoring on Impact indicators - Own Work

Indicator	#	Question	Score before going live (1...5)	Score after 1 year (1...5)
VISION	1	Long-term goal for platform model	4	
	2	We are customer centric	5	
	3	The platform business model is part of our vision	3	4
	4	The platform strategy is aligned with the overall business strategy	5	
(POLITICAL) LEADERSHIP	5	Leaders have goals and objectives how to use the platform model	4	
	6	Leaders can communicate their future foresight	4	
	7	Leaders identify and realize opportunities for platform efficiency.	4	4
	8	Leaders empower cross-functional collaborative teams	4	
GOVERNANCE	9	Agile processes are used to react to rapid business change.	2	4
	10	Employees can self-organize to address problems	2	3
	11	Staff is empowered to work autonomously	3	4
INNOVATION CULTURE	12	Everyone can think creatively, innovating the platform	4	
	13	Service design is done in rigorous and systematic approach to innovation	2	4
	14	Process for introducing new technologies exists	4	
	15	Small iterative experiments to realize innovation	4	
	16	Innovation activities are a regular task	3	
	17	Employees feel empowered and take calculated risks to be successful	2	3
	18	Employees work in interdisciplinary teams	5	
	19	Teams work collaboratively on projects and share developments	4	
VALUE	20	Citizens can effectively communicate with us	1	4
	21	Providers can effectively communicate with us to co-create value	1	4
	22	All staff work in sync towards implementing our (platform) vision	2	4
	23	My organization fosters an integrated digital ecosystem	1	3
	24	Everyone knows of, understands, and can act on our platform strategy	3	
	25	Very few technical issues in the delivery of our platform services	2	3
	26	Technical issues are resolved within an acceptable period	3	
	27	Platform is generating efficiencies versus the defined metrics	1	4
BUSINESS AGILITY	28	Proven ability to identify citizen's latent needs	3	4
	29	Ability to pivot its approach based on analysis of citizen insights	1	
	30	Technical teams implement services fast	1	3

4.5.6 Mapping on Maturity Level

Next it was time to verify if the LB365 is considered mature in its desire to position the platform model to improve value and citizen's experience. For that, we used the maturity model, that was created earlier

Table 55. LB365 Platform Quadrant Scoring - Own Work

Quadrant	Capability Score	Impact Score
Initiate	0 - 75	0 - 75
Competent	76 - 150	0 - 75
Purposeful	0 - 75	76 - 150
Transformative	76 - 150	76 - 150
>> Total Score before going live (Max Score is 150 per column)	**96**	**87**
>> Total score after going live (Max Score is 150 per column)	**112**	**112**

Figure 24. LB365 Platform Magic Quadrant Visualization - Own Work

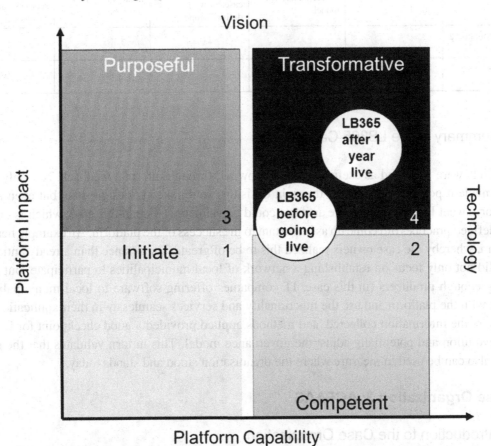

as a validation method. This way, an organization can see where they are today and wants to be going forward, which is expressed in the defined maturity levels on the model.

The below score was given by the author and discussed with the interviewed. Here again, the case is unique as LB365 could be evaluated in the condition before and after going live.

Based on the evidence presented, LB365 was originally mapped to Level 3, i.e., Embracing the platform model. This was mainly because LB365 was then still in development mode. However, during the last interview, the platform was live and the scope of services, municipality and IT suppliers using the platform has grown.

As such, there is no doubt LB365 was operating at Level 5 and leading the adoption of citizen services at local level. The only reason to attribute 3 stars is because there is still room for further institutionalizing and formalizing the processes of managing the platform.

Table 55. LB365 Scoring on Maturity Model - Own Work

Specific	Level 1: Exploring	Level 2: Developing	Level 3: Embracing	Level 4: Using	Level 5: Leading
Ambition	**Exploring** the Business Platform Model	**Developing** a business Platform model	**Embracing** it for first experimental cases or well-defined scope	**Using** a platform model for certain services	**Leading** the model, actively putting services out to the ecosystem
LB365 before going live			*****		
LB365 after going live					***

4.5.7 Summary of the LB365 Case

The artifacts were validated as useful by the case owner, through informal feedback. Not only did the artifacts made it possible to reflect if the right decisions were taken at design time, but even more as the platform went live. Therefore, the artifacts could be validated twice in this case, which is unique.

The defined practices did come up as important to the success of the platform. An example here is the ecosystem, whereby the case owners realized this to be of greater importance than already anticipated. LB365 did not only focus on establishing a network of local municipalities to participate, but also on attracting enough producers (in this case, IT companies offering software to local municipalities) to integrate with the platform and use the functionality and services seamlessly in their applications.

As such, the information collected, and methods applied provided a good checkpoint for LB365 to see the evolution and potentially adjust the governance model. This in turn validates that the defined artifacts also can be used to measure where the organisation stood and stands today.

4.6 Case Organization 3: ACPAAS

4.6.1 Introduction to the Case Organization

ACPAAS (i.e., "Antwerp City Platform as a Service") is the initiative from the IT Department of the City of Antwerp (Digipolis, 2020a) to provide standardized IT functionality to developers. The IT department has created several building blocks to be used during application development for different administrations of the city of Antwerp. ACPAAS allow the administrations to build and deploy tailor-made applications but based on reusable functional components that are standardized and delivered "as a service", meaning

Figure 25. ACPAAS Start Screen - Source: <u>acpaas.digipolis.be</u>

that the infrastructure complexity is hidden away from the developers. The benefits are that applications are more flexible to maintain, cheaper to build, and are delivered a much shorter time to market.

Some of the functionality is made available to third parties, meaning App-builders outside of the administrations (i.e., companies, universities, citizens, …). The city of Antwerp organizes competitions whereby start-ups can create the application. This way, the modernization of the application estate is sped up, and reuse of the functionality increases. The software components underneath the functionality is maximally based on open sources and free of paid licenses.

ACPAAS fits in the category of government business platforms on the borderline between "Opening Functionality" and "Opening Services". For now, most of the applications are still internal to the administrations, but public examples emerge without deliberate platform planning or strategy behind it. For example, the "Find a meeting room" application (Digipolis, 2020b) fits the criteria of "Opening Services" quite well. For an explanation of this case, see earlier in this research project.

4.6.2 Meeting Minutes of the Interviews

Unfortunately, only IT responded to the request for an interview. The business declined to participate and stated they are not yet discussing or planning for platform business models on the "Opening Services" level.

Below is a summarized version of the meetings.

On (Political) Leadership

The council of Antwerp was involved in every step of the development. The mayor of Antwerp endorsed the use of new technology to propel the city of Antwerp towards a Smart City, making valuable digital applications available to its citizens, continuously reiterating this vision since he was appointed mayor in 2012.

To the CIO of the City of Antwerp and CEO of Digipolis, this meant a shift of emphasis from social to economic issues(EU Joinup, 2018) as he had to rethink the entire IT procurement policy. The CIO wants to shift away from big tenders that were submitted by big companies with bid budgets, thus excluding many small and innovative players from expressing their ideas.

This case is thus a fine example where a political leader expresses the need for using more technology to improve the life of citizens living in Antwerp.

On Vision and Strategy

The driver for ACPAAS was to bring the cost of IT down by standardizing the components on a lower technical level and enabling reuse on a functional level. Similar business functionality could be deployed much faster than writing it from scratch again. This created leverage and speed at the administration to quickly turn over older legacy applications to newly designed ones.

The vision and strategy were entirely based on cost-saving and standardization of IT components, and the business case was endorsed by the council of Antwerp in 2015. The idea was to bring costs down by building reusable IT components.

"The target set in 2015 was to become ten times less expensive, ten times faster, and ten times more innovative", the CIO stated.

On Value

On the IT side, the value sits in the speed and low cost of developing new applications through an ecosystem of start-ups (see below). By leveraging the IT components that can be activated dynamically and automatically, development speed is reduced to coding the specifics of the business process.

For the ecosystem – essentially the smaller firms that are still in a start-up phase – this provides a means to win business and start delivering.

For the business, this provides quick access and turnaround time to have business processes modernized or renewed, using new technologies. This turns the classical relationship between business and suppliers from a waterfall approach into an agile way of working. This requires a different culture at the business as well.

The value is also increasing per project as lessons learned are turned into best practices. This results in a success rate of over 90%. As large projects do not exist anymore, there are also no large project failures anymore, resulting in value for the agility of the organization.

One of the most important drivers starting this project was to bring the IT budget down and avoid similar functionality to be invented repeatedly.

A calculation was done in 2018 what the costs were since starting in 2015. The IT department made a total investment in that period of roughly 4M €, which delivered 55 IT components as part of the ACPAAS platform. This makes a cost of +/- 70K € per component. The 55 components were deployed in more than 300 business applications. This would never be achievable if the IT department would have chosen standard applications like SAP or Oracle.

On Governance

The adoption of standard functionality and IT components required the City of Antwerp to overturn the way its IT was organized completely. The focus of IT has become to deliver standard IT components and foster the relationship with start-ups that become an extended community of application developers. To facilitate this, the procurement process had to be changed as well. Requests for new applications are published on the portals, and every interested party can come and pitch their idea. Selected parties can come and do a second pitch, including an actual offer within two weeks. This reduced the whole procurement cycle to a maximum of 6 weeks. Furthermore, since the budget per contract is about EUR 100,000, risks in both time and money are low.

On Ecosystems

Here the City of Antwerp took a specific decision: to open the functionality to start-ups that can subscribe to a contest. The idea to buy from start-ups was born in January 2015, Tobias stated, but it was not an immediate success. It proved not so easy to find start-ups that were interested in helping the city of Antwerp. External help was called in, and several start-up gurus provided tips and promised to help up set up a network and promote the contests. It was referred to a testimonial that was published by the EU on these early days and the ecosystem building (EU Joinup, 2018).

On (Innovation) Culture

Processing all the pitches can be time-consuming for the staff of the City of Antwerp. Becoming agile or working at the speed of a start-up was challenging for the business in the beginning. But the enthusiasm of the start-ups seems to have inspired the businesspeople that this way of working is the new baseline or the new normal.

This also meant that new skills and capabilities are needed at the business ideas. As an example, the number of traditional business analysts is reduced and replaced by architects that know how to mix and match the IT components to suit the need of the business and make sure the start-up can deliver.

On Risk Management

As stated earlier, it was the procurement department that underwent many changes. Instead of negotiating a big pile of documents detailing out every requirement, the dynamic is now shifted in procuring a sprint in which a new version of the application is delivered. Start-ups get paid by sprint, meaning every two weeks a new demo is done.

Figure 26. Overview of ACPAAS Building blocks - Source: Digipolis Website

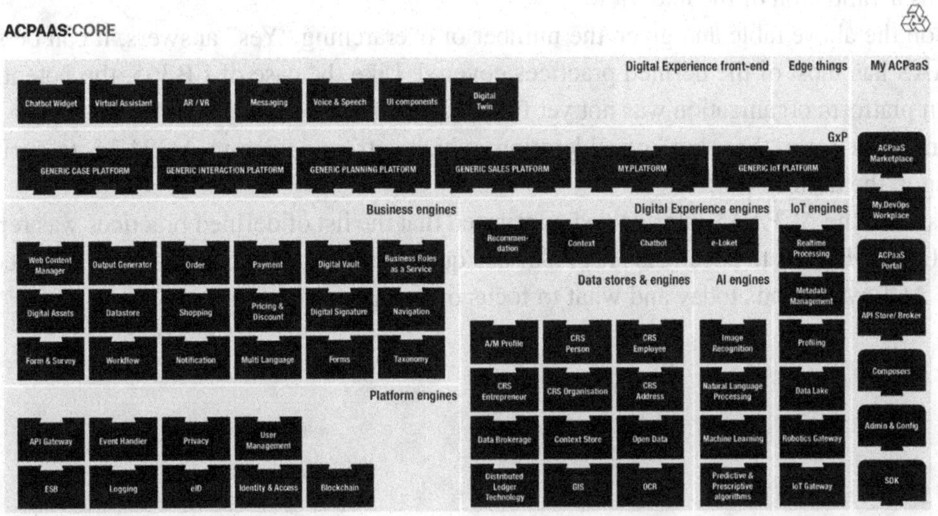

On Platform Architecture & Infrastructure

All the stated business advantages come from the standardization of IT components that allow quick assemblage in applications. There is a standard defined for most common IT needs, like building a chatbot, integrating with a database, or using AI technology. This did not fall out of the sky and took much preparation at the IT side.

The build-up of the components continues today, and new technology is being included per new release. The latest list is displayed In Figure 26.

On Customer Experience Design

ACPAAS has also invested in the creation of IT components that improve the user experience. Every business entity can tweak the site or application with its colour palette, but the functionality is the same for each.

On Business Agility

It was emphasized that rolling out a platform model was not only about IT change, but both an organizational, procurement, and operating model change that was set in motion with the availability of IT components.

Summary of the interviews

This case is particularly interesting as it has touched most of the practices that were defined in this research project, certainly the top ones like (political) leadership, vision, and strategy. The technicalities have been well worked out and business is seeing value of the platform. Although the business was not represented, the case owner had enough detail and experience to cover their point of view as well.

4.6.3 Evaluation of Case versus Defined Practices

After the interview and based on the transcript, an evaluation was done on how ACPAAS complies with the defined practices. Table 56 was filled in by the author, discussed, and agreed with the case owner and acted as a validation of the interviews.

Based on the above table and given the number of overarching "Yes" answers, it can be stated that the ACPAAS has most of the defined practices covered. Like the case of LB365, the potential to be a data driven platform organization was not yet fully explored.

Two practices got a "Yes, but" consideration, which acts as advice to ACPAAS to verify how to improve upon the defined practice.

Here as well, the ACPAAS interviewed confirmed that the list of defined practices was representing a good 360° of all facets to cover. As ACPAAS has quite some experience, it served as a checkpoint to see where ACPAAS stands today and what to focus on or improve.

Table 56. Does ACPAAS comply with the defined practices - Own work

Level	Practice	Considered	Comments
Strategic Level	Vision	Yes	ACPAAS based its platform on a vision of IT component standardization and innovation, using start-ups as an engine to deliver fast results to the business.
	Strategy & (Political) Leadership	Yes	Driven by a mayor with a keen interest in improving the life of Antwerp citizens with (new) technology, ACPAAS contributes by rapidly turning over ideas into applications that can be released, and components reused.
	Governance	Yes	Business, IT, but also procurement have been overhauled to embrace agility and deliver in small steps
	(Innovation) Culture	Yes	Business is now getting used to new demos or delivery every two weeks by the start-ups, quite a change than the original and classical waterfall approach
	Value	Yes	Value is demonstrated on the IT and business side. There is a clear ROI on the IT side and agility on the business side.
Tactical Level	Business Agility	Yes	Driven by the two-weekly release schedule from start-ups, the business team had to change their approach of receiving new functionality and put in production.
	Business Ecosystem Design	Yes, but	This circles mostly around the buy from start-ups approach, whereby business listens to the start-ups defend themselves in pitches and then select the party to engage with. This is different from the "Opening Services" model as here, it is not the citizens that make the transaction, but the businesspeople of the city of Antwerp in a coordinated way.
	Platform Architecture & Infrastructure	Yes	ACPAAS was designed from day one with the right architecture elements, i.e., standardization if functional components wrapped in a standard infrastructure solution. There was specific attention to the standardization of APIs so that third parties could rely on using these in their applications. There is a management console that oversees what developers can do, but also to design new components or spin up infrastructure.
	Data Driven	No	We have not seen this element being covered, other than operational reporting on usage and number of business processes modernized with start-ups (versus cost). The platform does not cater for proactive services to suppliers, so we did not find evidence on advanced analytics for internal service optimization. Note there is a component in the library for advanced analytics, but this is meant for developers to be used in their application building projects.
Operational Level	Talent and Skills	Yes	New skills on the business and IT side were introduced to handle the IT components and changing the relationship with suppliers towards a two-weekly schedule. There are also new skills in the procurement division introduced to cope with start-ups and smaller cycles of delivery
	Customer Experience Design	Yes	The customer experience mostly comes from reusing IT components that have been developed on ACPAAS and made available to start-ups. This drives standardization but also allows creativity for the start-ups to propose new elements to increase the customer experience. Every application uses the same IT components for customer experience, and this is a two-way street, meaning the IT team continuously watches what new elements to introduce in the relevant IT components.
	Risk Management	Yes, but	Risk is being transferred and mitigated in the business projects, where delivery is centred around a two-weekly rhythm with external suppliers like start-ups. This avoids big risks before going live. On the technical front, the IT estate is being monitored but can be scaled out easily.
	Technology Ecosystem Design	Yes	Start-ups cannot only use APIs but can also contribute APIs to the ACPAAS API Store to be reused by others. This feeds an entire ecosystem whereby the latest supplier can build upon the previous supplier.

4.6.4 Evaluation of Governance Mechanisms

In this section, a validity check was done on the existence of governance mechanisms to manage the defined practices within the ACPAAS Team. As ACPAAS has been up and running for at least 2 years, it was interesting to reflect upon how the governance mechanisms were established and evaluated today.

4.6.4.1 Process Checklist

Together with the case owner, the results of the interview and above list of evaluated practices, the checklist for processes was filled in (Table 57):

The above table is not the result of an audit but obtained through discussions with the case owners. It is interesting to see how the ACPAAS platform model is in this case organization fully entangled inside the IT team and applying existing processes. Linking the existing processes to COBIT2019 was done arbitrarily as ACPAAS has not used a framework like COBIT2019 to guide them.

4.6.4.2 Organizational Structures Checklist

Together with the case owner, the results of the interview and above list of evaluated practices, the checklist for organizational structures was filled in (Table 58).

Here as well, we can conclude that most of the platform structures are established inside the IT team and reusing the practices of Digipolis, the IT company working exclusively for the City of Antwerp. The CIO is on the City Council and thus also represents how the vision of the Mayor of Antwerp is executing the strategy to realize the vision.

It is to be noted that the above table is not the result of an audit, but the interviewed gave verbal assurance of the existence of these structures.

4.6.4.3 Relational Mechanisms Checklist

Together with the case owner, the results of the interview and above list of evaluated practices, the checklist for relational mechanisms was filled in (Table 59).

Note the above table is a condensed version of the checklist introduced earlier. For completeness, one is referred to the chapter where the governance mechanisms were established

Similar as for the organizational structures, most of the relational mechanisms for ACPAAS are reusing and are interconnected with the IT team.

It is to be noted that the above table is not the result of an audit, but the interviewed gave verbal assurance of the existence of these mechanisms.

4.6.5 Mapping Onto Organizational Capability and Impact Quadrant

Next, we asked the ACPAAS case owner if the organization had the right capabilities in place for turning the practices into reality. Here as well, the earlier defined organizational Capability & Impact quadrant model was used to capture the feedback through scoring the questions.

The below table was filled in by the interviewed. The scores are attributed based on the experience of and reflection by the case owner.

Table 57. Process Checklist as applied to ACPAAS Case - Own Work

COBIT2019 Process	In place Y/N?	Comments
APO02 - Managed Strategy	Y	The IT team does manage the strategy of ACPAAS, reusing the existing process at the IT department.
APO12 - Managed Risk	Y	The IT team does manage the risk of ACPAAS, reusing the existing process at the IT department.
APO13 - Managed Security	Y	The standards of security applicable at Digipolis are applied to the ACPASS team as well.
DSS05 - Managed Security Services	Y	The standards of security applicable at Digipolis are applied to the ACPASS team as well
MEA03 - Managed compliance with external requirements	Y	Digipolis Legal department is involved in managing contracts and projects that are executed with external resources.
EDM01 - Ensured Governance Framework Setting and Maintenance	Y	The existing governance mechanisms at Digipolis were applied to the ACPAAS team.
EDM03 - Ensured Risk Optimization	Y	The IT team ensures risk optimization of ACPAAS, reusing the existing process at the IT department.
EDM05 - Ensured stakeholder engagement	Y	The IT team is liaising with external parties directly, with the help of legal to assign contracts.
APO03 - Managed Enterprise Architecture	Y	The IT team has developed the reference architecture for ACPAAS. It is also maintained in the IT team.
APO04 - Managed Innovation	Y	Innovation on ACPAAS is handled inside and together with the IT team. There is only one roadmap. Note that innovation at the business is realized through letting the third parties work with the standardized functionality in rewriting legacy or even new applications.
APO08 - Managed relations	Y	The IT team handles relations inside and outside of Digipolis
APO14 - Managed Data	Y	The IT Team operates data for the platform as any other data, i.e. applying the same processes and technology.
BAI05 - Managed Organizational Change	Y	The IT team is the main liaison to the business, even to help the business adjust to agile approach of delivery of new functionality.
BAI06 - Managed IT Changes	Y	ACPAAS is managed by the IT Team, there is one IT budget and roadmap, including the growth of new building blocks.
DSS06 - Managed Business Process Controls	Y	The IT team has direct contacts with the business to propose and conduct projects with third parties.
EDM02 - Managed Benefits delivery	Y	The IT team has its own mechanisms to calculate the ROI. The business provides details on how the perceived value is turned into tangible and intangible benefits. Example of tangible benefits: the user can work through the request in less steps and now also on a mobile device Example of intangible benefit: the citizen provides good feedback on the experience of using the new application
APO01 - Managed I&T Management Framework	Y	Digipolis is a long-standing company providing services to the City of Antwerp. Its management framework is well established and even approved at the City Council. ACPAAS is operating inside this framework
APO05 - Managed Portfolio	Y	The IT team has a backlog of new building blocks and has associated budget requests to the board of Digipolis
APO07 - Managed Human Resources	Y	All resources working on ACPAAS are recruited from within Digipolis or hired externally via the established HR processes
APO09 - Managed Service Level Agreements	Y	ACPAAS has a set of SLAs it guarantees to users. These are documented on the website.
APO10 - Managed Vendors	Y	The IT Team manages vendors themselves, including the suppliers of technology or expertise to ACPAAS
APO11 - Managed Quality	Y	ACPAAS is using the same quality framework as applicable to other IT projects
BAI08 - Managed Knowledge	Y	Although handled inside the IT team, the ACPAAS members have expressed and shared their knowledge extensively externally. Example is that ACPAAS was a winner of the Gartner Innovation Award in 2018.
MEA01 - Manage Performance and Conformance Reporting	Y	All reporting on ACPAAS is consolidated in the standard and existing reports of the IT team
MEA04 - Managed Assurance	Y	This is also handled within the current standards of the IT team and Digipolis
EDM04 - Ensure Resource Optimization	Y	This is also handled within the current standards of the IT team and Digipolis
APO06 - Managed Budgets and Costs	Y	ACPAAS has its dedicated budget lines, but these are also consolidated into the overall IT Budget.

Table 58. Organizational structures at the ACPAAS case - Own work

Platform related organizational structures	In Place (Y/N)?	Comments
Platform Strategy committee at the level of board of directors	Y	There is no separate platform strategy committee. This is handled on the regular agenda of the Digipolis board, this part of the IT Steering Committee.
Digital Platform Transformation Leader, reporting to CEO	Y	The CIO of Digipolis is also representing the ACPAAS to the City Counsel. As such, the CIO is the transformation leader
Platform Steering Committee	Y	There is no separate Platform Steering Committee. ACPAAS is a topic on the IT Steering Committee.
Platform Governance Officer	Y	There is no separate Platform Governance Office. The Chief Enterprise Architect together with the CIO of Digipolis hold this responsibility.
Platform Security / compliance / risk officer	Y	This is handled in the regular IT meetings and within the standards of Digipolis
Platform Project steering committee	Y	This is handled in the regular IT meetings and within the standards of Digipolis
Platform Security steering committee	Y	This is handled in the meetings with IT
Platform Architecture steering committee	Y	This is handled in the meetings with IT

Table 59. Relational mechanisms at the ACPAAS case - Own work

Are the following relational mechanisms in place for Platform Model?	In Place (Y/N)?	Comments?
1. Co-location	Y	People working on ACPAAS are in the same building as all the IT people of Digipolis.
2. Knowledge management	Y	The IT team has specific attention to understanding the platform model. This started from an IT point of view.
3. Ongoing awareness training on Platform model strategy	Y	The IT team is responsible for all these items, including ACPAAS.
4. Articulate the vision on Platform Model	Y	The vision is defined by the Mayor of Antwerp and translated into a strategy that the IT team is executing upon.
5. Internal Communication	Y	All ACPAAS communications are handled via the IT team.
6. Network outside the organization	Y	The IT team has done a great job in presenting ACPAAS to the outside world, even winning Gartner Awards. As a result, there is an extensive network established, far beyond the borders of Flanders.
7. Platform Account Manager	Y	The Chief Enterprise Architect of Digipolis has this responsibility.

- On the X-Axis: **Capability** indicators answers (Condensed version):

Table 60. ACPAAS scoring on Capability indicators - Own Work

Indicator	#	Question	Score (1...5)
STRATEGY	1	My organization has a clear, coherent, and actionable strategy	4
	2	Strategy focuses on transforming the whole business	4
	3	Strategy strives for maximum reconfiguration and availability of the services	3
BUSINESS ECOSYSTEM DESIGN	4	Active process with (potential) ecosystem providers	5
	5	My organization has a clear business roadmap	2
	6	My organization orchestrates the ecosystem providers in an open and co-creative approach.	3
PLATFORM ARCHITECTURE & INFRASTRUCTURE	7	My organization funds and resources the platform services and technology transformation adequately.	4
	8	Platform investment considers an organization-wide approach	5
	9	Integrate new technologies with older 'legacy technologies'	5
	10	Technology and solutions are in place for analytics	3
	11	Systems, applications, and tools in place (e.g., API approach)	5
	12	Technology in place to support service execution on the platform.	5
TECHNOLOGY ECOSYSTEM DESIGN	13	We connect with our partners digitally (APIs).	5
	14	Leveraging functional reusable components from our government	5
	15	Roadmap available on new functionality or components	3
DATA DRIVEN	16	Business intelligence system for timely decisions available	3
	17	Process and technology in place to understand & act upon citizen behaviour	3
	18	My organization is data focused to provide proactive service advice	3
TALENT AND SKILLS	19	Skills and competencies to design and build of services	4
	20	Easy to attract technical staff	3
	21	Investments in developing digital skills of employees	4
RISK MANAGEMENT	22	Technology in place to prevent cyber-attacks and other security risks	5
	23	Regular assessments on technical, business, and social risk factors	4
	24	Scalability of platform infrastructure meets the demand	4
	25	Proactive risk management approach	4
CUSTOMER EXPERIENCE DESIGN	26	User Experience research is conducted	2
	27	Design and deliver a tailored product to fulfil specific citizen's needs.	4
	28	Citizens get fully integrated experience	3
	29	Continuous improving digital and physical experiences on the platform	3
	30	Citizens can effectively communicate with my organization	4

- On the Y - Axis: **Impact** indicators answers (Condensed version):

Table 61. ACPAAS scoring on Impact indicators - Own Work

Indicator	#	Question	Score (1...5)
VISION	1	Long-term goal for platform model	5
	2	We are customer centric	5
	3	The platform business model is part of our vision	5
	4	The platform strategy is aligned with the overall business strategy	1
(POLITICAL) LEADERSHIP	5	Leaders have goals and objectives how to use the platform model	5
	6	Leaders can communicate their future foresight	4
	7	Leaders identify and realize opportunities for platform efficiency.	4
	8	Leaders empower cross-functional collaborative teams	4
GOVERNANCE	9	Agile processes are used to react to rapid business change.	3
	10	Employees can self-organize to address problems	4
	11	Staff is empowered to work autonomously	4
INNOVATION CULTURE	12	Everyone can think creatively, innovating the platform	3
	13	Service design is done in rigorous and systematic approach to innovation	2
	14	Process for introducing new technologies exists	4
	15	Small iterative experiments to realize innovation	4
	16	Innovation activities are a regular task	4
	17	Employees feel empowered and take calculated risks to be successful	4
	18	Employees work in interdisciplinary teams	4
	19	Teams work collaboratively on projects and share developments	4
VALUE	20	Citizens can effectively communicate with us	5
	21	Providers can effectively communicate with us to co-create value	4
	22	All staff work in sync towards implementing our (platform) vision	3
	23	My organization fosters an integrated digital ecosystem	2
	24	Everyone knows of, understands, and can act on our platform strategy	4
	25	Very few technical issues in the delivery of our platform services	4
	26	Technical issues are resolved within an acceptable period	4
	27	Platform is generating efficiencies versus the defined metrics	5
BUSINESS AGILITY	28	Proven ability to identify citizen's latent needs	4
	29	Ability to pivot its approach based on analysis of citizen insights	2
	30	Technical teams implement services fast	3

Next, the scores were added up. Based on the earlier defined scoring reference table, we come to the following totals for ACPAAS:

Table 62. ACPAAS Platform Quadrant Scoring - Own Work

Quadrant	Capability Score	Impact Score
Initiate	0 - 75	0 - 75
Competent	76 - 150	0 - 75
Purposeful	0 - 75	76 - 150
Transformative	76 - 150	76 - 150
>> ACTUAL Score (Max Score is 150 per column)	114	113

With these scores, we can draw where ACPAAS fits in the model. This was visualized as follows:

Figure 27. ACPAAS Platform Magic Quadrant Visualization - Own Work

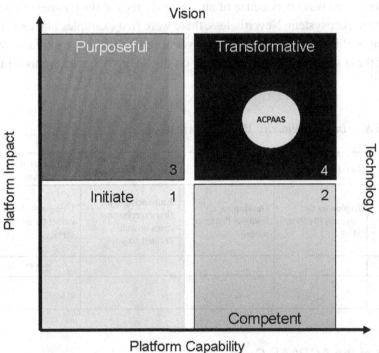

ACPAAS is positioned in the top right quadrant. This is because the vision and strategy are right, the practices are considered, capabilities are in place, and impact in execution is proven. However, we marked it in yellow as the business was not reachable to comment on these scores.

It is to be noted that ACPAAS started as an "Opening Functionality" platform and is still evolving. The "Opening Services" is within reach, but not yet endorsed by the business. In other words, the focus is still on improving the current application landscape and quickly modernize the estate. Although the business has adopted an agile approach, it remains within the borders of traditional government business, whereby the administrations are the actor. The ecosystem is a network of suppliers like start-ups that can quickly develop applications. Services and service execution are not yet transferred to the ecosystem.

However, a first emerging example like Room Finder(Digipolis, 2020b) is a testimonial that ACPAAS has the extra characteristics at least identified to be able to grow to an "Opening Services" approach. This requires the city hall to show another round of (political) leadership and embrace the next step in the journey to become an orchestrator of services.

4.6.6 Mapping on Maturity Level

Based on the defined maturity model, we map ACPAAS in its totality to be at Level 4, i.e., Using the platform model. This is because of the proven execution and realizations. ACPAAS has received credits from Gartner at its 2017 Gartner Symposium ITxpo for its leading and innovative way of redesigning applications using start-ups.

The main reason for Level 4 was because the realizations were aligned with the "Opening Functionality" model. However, there was no evidence of an active strategy at the business to put service delivery and execution out to the ecosystem. Nevertheless, there were first examples like Room Finder emerging, so an extra mention of this service was made at Level 5, i.e., Leading the platform model. We hope in the future more of these services become available on the ACPAAS platform as all the practices were in place to realize this.

Table 63. ACPAAS Scoring on Maturity Model - Own Work

Specific	Level 1: Exploring	Level 2: Developing	Level 3: Embracing	Level 4: Using	Level 5: Leading
Ambition	**Exploring** the Business Platform Model	**Developing** a business Platform model	**Embracing** it for first experimental cases or well-defined scope	**Using** a platform model for certain services	**Leading** the model, actively putting services out to the ecosystem
ACPAAS				*****	
Theatre finder ("Zalenzoeker")					*****

4.6.7 Summary of the ACPAAS Case

ACPAAS is a mature platform model. In the interviews, evidence was shown that the platform is operating well and that the results are impressive. However, the initial focus of ACPAAS was on "Opening Functionality" and thus on improving the quality of the applications for both civil servants as residents of the city of Antwerp.

The defined practices were all recognized by the case owners, and deemed necessary, even for operating on the "Opening Functionality" level. As an example, the political leadership was well established from day one, and the case owner acknowledged this to be a key differentiator in convincing the business and IT teams to execute the strategy and roll out the model.

When we reflect on the usefulness of the governance mechanisms defined and validated through the organizational capabilities' method, we need to remark the absence of business leaders in wanting to comment on this. Here, more awareness is needed to get the business in the driving seat. The governance mechanisms should be further explored and the ambition of ACPAAS should be uplifted to a "Opening Services" model. As the first examples of "Opening Services" are emerging (i.e., Room Finder), the timing might be right to consider uplifting the vision.

The artifacts proved valuable in validating that the current ACPAAS platform is considering the defined practices but in an "Opening Functionality" model rather than an "Opening Services" model. The case owner indicated that especially the one business – IT team approach was a good suggestion.

As conclusion, the artifacts were deemed useful in today's situation of ACPAAS. The artifacts could also act as guidance in evolving the platform model to an "Opening Services" approach. For that to happen, a new wave of political leadership is needed to reset or refine the vision and start a new strategy exercise.

This case proved that the artifacts can also be used for case organization that are not yet operating on the "Opening Services" level. Here as well, the practices seem applicable, at least the political leadership as a condition to uplift the vision. It is suggested to investigate more case organizations to be able to make a more general view on this.

4.7 OVERALL CASES COMPARED

We have investigated three cases in Flanders that are embracing the platform model in some form or another. When combining the results of the individual case organizations, an overall picture was drawn and displayed for informational purposes only:

Figure 28. Platform Magic Quadrant Visualization for all cases - Own Work

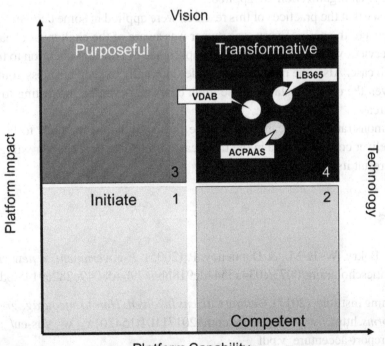

All three organizations map to the Transformative quadrant and stay relatively close together. This was expected as the organizations were selected because of their desire and proven journey to becoming an "Opening Services" model.

4.8 SUMMARY OF THE CASE EVALUATION

Information on three selected government cases was collected through interviews with case owners. The purpose was to verify if the validation methods were deemed useful, effective, repeatable, and deliver consistent data across the selected case organizations in a transparent way.

This approach can be considered a lightweight evaluation (Johanesson & Perjons, 2014). It does not assess how well an artefact works, but only shows that it could be used meaningfully in the three cases.

It is opiniated that the three cases demonstrated that the artefacts and the associated validation methods provided value to the interviewed, which means:

- For **VDAB**, it was mainly a validation of where they are today and where and how they can improve the maturity of the organization.
- For **LB365**, the artifacts demonstrated the evolution as the team was interviewed twice, namely before rollout and after going live.
- For **ACPAAS**, it was the realization that more is possible given the current well-established technical foundations. Given an uplift of the vision, the business can focus on moving to an "Opening Services" orientation.

The exercise of demonstrating the artifacts is hereby considered repeatable, meaning all checklists and techniques can be reused, no matter wat the state of the case organization is. Certainly, when the criteria for selecting an organization are applied.

It was good to see that the practices of this research were applied in some degree. Political leadership is at the base of each of the cases. Every case owner was aware of the challenges ahead and continue to work to roll out services with an ecosystem. The importance of constant attention to the ecosystem was recognized in each case. Also, the need act as a leader internally to keep business going in the direction of the vision. Given the different status, these three cases can even be interesting to learn from when their maturity increases.

Given this demonstration, one can now assume these validation methods to be useful, effective, repeatable, and deliver consistent data in the three cases used. The next chapter will go in dept into the evaluation of the result itself.

REFERENCES

Abdelghaffar, H., Bakry, W.-E. M., & Duquenoy, P. (2005). *E-government: a new vision for success.* https://pdfs.semanticscholar.org/0073/c034a5544dc918bb6c79b4d8049c282e9148.pdf

Accenture Governing Institute. (2017). *Getting Citizens Involved: How to encourage greater use of citizen engagement platforms.* https://www.accenture.com/t20171101t154201z__w__/us-en/_acnmedia/pdf-64/accenture-gov17-report-accenture_v.pdf

Alreemy, Z., Chang, V., Walters, R., & Wills, G. (2016). Critical success factors (CSFs) for information technology governance (ITG). *International Journal of Information Management, 36*(6), 907–916. doi:10.1016/j.ijinfomgt.2016.05.017

Anthony, R. N. (1965). *Planning and Control: a Framework for Analysis*. Harvard University Press. https://www.diva-portal.org/smash/get/diva2:326878/FULLTEXT01.pdf

Australian Digital Transformation Agency. (2019). *Digital Service Platforms Strategy*. https://www.dta. gov.au/our-projects/digital-service-platforms-strategy/six-keys-success/4-strengthen-digital-leadership-governance-and-accountabilities

Axelos. (2012). Management of Risk (M_o_R®): Guidance for Practitioners (3rd ed.). Axelos.

Axelos. (2015). *PRINCE2 Best Practice Solutions*. Axelos.

Axelos. (2016). *What is ITIL® Best Practice?* https://www.axelos.com/best-practice-solutions/itil/what-is-itil

Axelos. (2020). *Managing Successful Programmes*. https://www.axelos.com/best-practice-solutions/msp-4th-edition

Barcevičius, E., Cibaitė, G., Gineikytė, V., Klimavičiūtė, L., Matulevič, L., Misuraca, G., & Vanini, I. (2019). *Exploring Digital Government transformation in the EU*. https://doi.org/ doi:10.2760/17207

Benlian, A., Hilkert, D., & Hess, T. (2015). How open is this platform? The meaning and measurement of platform openness from the complementors' perspective. *Journal of Information Technology, 30*(3), 209–228. doi:10.1057/jit.2015.6

Berkman. (2005). *Roadmap for Open ICT Ecosystems*. https://cyber.harvard.edu/epolicy/

Boorsma, B. (2017). *A New Digital Deal: Beyond Smart Cities. How to Best Leverage Digitalization for the Benefit of our Communities*. Boekscout BV. https://anewdigitaldeal.com/

Boynton, A. C., & Zmud, R. W. (1984). *An Assessment of Critical Success Factors. Sloan Management Review, 25(4)*.

Brown, A., Fishenden, J., & Thompson, M. (2014). Digitizing Government - Understanding and Implementing New Digital Business Models. Digitizing Government. https://doi.org/ doi:10.1057/9781137443649

Brown, A., Fishenden, J., Thompson, M., Brown, A., Fishenden, J., & Thompson, M. (2014). *Organizational Structures and Digital Transformation*. Digitizing Government. doi:10.1057/9781137443649_10

Calder, A. (2008). ISO/IEC 38500: The IT Governance Standard. IT Governance Publishing.

Calder, A. (2009). *IT Governance Implementing Frameworks and Standards for the Corporate Governance of IT*. http://books.google.com/books?id=E1x5AgAAQBAJ&pgis=1

Calder, A. (2013). ISO27001/ISO27002 A Pocket guide. ISO27001 / ISO27002.

Choudary, S. P. (2019). *The Three Design Elements for Designing Platforms*. https://drive.google.com/file/d/0B36EfI3U2bSMOHdyU0lhWmNuTjg/view

CMMI Institute. (2018). *What Is CMMI?* Author.

Cordella, A., & Paletti, A. (2019). Government as a platform, orchestration, and public value creation: The Italian case. *Government Information Quarterly, 36*(4), 101409. doi:10.1016/j.giq.2019.101409

Cram, A. (2007). The IT balanced scorecard revisited. *Information Systems Control Journal, 3*, 1–4. http://www.isaca.org/Journal/Past-Issues/2007/Volume-5/Documents/jpdf0705-the-it-balanced.pdf

Danneels, L. (2017). *Transforming government: The way towards digital era governance.* https://lirias. kuleuven.be/bitstream/123456789/597211/1/PhD+final.pdf

De Haes, S. (2017). *Het herdefiniëren van governance-benaderingen voor digitale transformatie in gevestigde bedrijven.* https://blog.antwerpmanagementschool.be/nl/herdefinieren-van-digitale-trans-formatie-in-gevestigde-bedrijven

De Haes, S., Van Grembergen, W., Anant, J., & Huygh, T. (2020). Enterprise Governance of Information Technology. Enterprise Governance of Information Technology. https://doi.org/ doi:10.1007/978-0-387-84882-2

Dickinson, R. A., Ferguson, C. R., & Sircar, S. (1984). Critical Success Factors and Small Business. *American Journal of Small Business, 8*(3), 49–57. doi:10.1177/104225878400800309

Digipolis. (2020a). *Antwerp City Platform as a Service (ACPaaS).* https://acpaas.digipolis.be/nl/home

Digipolis. (2020b). *Zaalzoeker.* https://zaalzoeker.antwerpen.be/zaalzoeker

Diirr, T., & Santos, G. (2014). *Improvement of IT service processes: a study of critical success factors.* https://doi.org/ doi:10.1186/2195-1721-2-4

Engin, Z., & Treleaven, P. (2019). Algorithmic Government: Automating Public Services and Supporting Civil Servants in using Data Science Technologies. *The Computer Journal, 62*(3), 448–460. doi:10.1093/comjnl/bxy082

EU - DG Joint Research Centre. (2018). *Digital Platform for public services.* https://joinup.ec.europa. eu/collection/elise-european-location-interoperability-solutions-e-government/document/report-digital-platform-public-services

European Commission Joint Research Centre. (2018). *Digital Platform for public services.* Author.

European Commission. (2018). *Open data | Digital Single Market.* https://ec.europa.eu/digital-single-market/en/open-data

Forum, T. M. (2020). *GB921 Business Process Framework (eTOM).* https://www.tmforum.org/resources/suite/gb921-business-process-framework-etom-r17-0-1/

Gaikema, M., Donkersloot, M., Johnson, J., & Mulder, H. (2019). *Increase the success of Governmental IT-projects.* http://www.healthpayrollinquiry.qld.gov.au/__data/assets/pdf_file/0014/207203/Queensland-

Gartner. (2016). *Digital Leadership is a Team Sport.* https://www.gartner.com/smarterwithgartner/digital-leadership-is-a-team-sport/

Gawer, A. (2014). Bridging differing perspectives on technological platforms: Toward an integrative framework. *Research Policy, 43*(7), 1239–1249. doi:10.1016/j.respol.2014.03.006

Harrison, T. M., Pardo, T. A., & Cook, M. (2012). Creating Open Government Ecosystems: A Research and Development Agenda. *Future Internet, 4*(4), 900–928. doi:10.3390/fi4040900

Hein, A., Weking, J., Schreieck, M., Wiesche, M., Böhm, M., & Krcmar, H. (2019). Value co-creation practices in business-to-business platform ecosystems. *Electronic Markets, 29*(3), 503–518. doi:10.100712525-019-00337-y

Hoogervorst, J. (2017). *Foundation of Enterprise Governance and Enterprise Engineering.* Springer International Publishing. doi:10.1007/978-3-319-72107-1

ISACA. (2019a). *COBIT 2019 Design Guide.* https://www.isaca.org/bookstore/bookstore-cobit_19-digital/wcb19dgd

ISACA. (2019b). *COBIT 2019 Framework Introduction and methodology.* www.icasa.org/COBITuse

ISO/IEC. (2005). *ISO/IEC 20000-2: Information technology—Service management.* Service Management.

Janssen, M., & Estevez, E. (2013). Lean government and platform-based governance-Doing more with less. *Government Information Quarterly, 30*(Suppl. 1), S1–S8. doi:10.1016/j.giq.2012.11.003

Jayaram, M. N. (2003). *What is Capability Maturity Model (CMM)?* https://searchsoftwarequality.techtarget.com/definition/Capability-Maturity-Model

Johanesson, P., & Perjons, E. (2014). Design Science. In *An introduction to Design Science.* Springer. doi:10.1007/978-1-4899-6331-4_35

Johnson, J. (2018). *CHAOS Report: Decision Latency revisited.* http://jtonedm.com/2012/11/21/decision-latency-revisited/

Joinup, E. U. (2018). *"Buy from startups" strategy pays off for City of Antwerp.* https://joinup.ec.europa.eu/collection/open-source-observatory-osor/document/buy-startups-strategy-pays-city-antwerp

Kaplan, R. S., & Norton, D. P. (1992). The Balanced Scorecard - Measures That Drive Performance. *Harvard Business Review.*

Kaplan, R. S., & Norton, D. P. (1996). *Translating Strategy into action: The Balanced Scorecard.* https://www.amazon.co.uk/Balanced-Scorecard-Translating-Strategy-Action/dp/0875846513

Kohlegger, M., Maier, R., & Thalmann, S. (2009). Understanding maturity models results of a structured content analysis. *Proceedings of I-KNOW 2009 - 9th International Conference on Knowledge Management and Knowledge Technologies and Proceedings of I-SEMANTICS 2009 - 5th International Conference on Semantic Systems,* 51–61. http://www.sei.cmu.edu/cmmi/

Lang, N., von Szczepanski, K., & Wurzer, C. (2019). *The Emerging Art of Ecosystem Management.* Boston Consulting Group. https://www.bcg.com/publications/2019/emerging-art-ecosystem-management.aspx

Leading Edge Forum. (2019). *Are platform organizations a wicked problem.* https://leadingedgeforum.com/research/are-platform-organizations-a-wicked-problem/

McFadden, E. (2016). Innovation in the public sector: Is risk aversion a cause or a symptom? *Nesta*, 1–9. https://www.nesta.org.uk/blog/innovation-in-the-public-sector-is-risk-aversion-a-cause-or-a-symptom/

Mettler, T. (2009). *A Design Science Research Perspective on Maturity Models in Information Systems*. https://www.researchgate.net/publication/44939433_A_Design_Science_Research_Perspective_on_Maturity_Models_in_Information_Systems

Müller, S. D., & Skau, S. A. (2015). Success factors influencing implementation of e-government at different stages of maturity: A literature review. *International Journal of Electronic Governance, 7*(2), 136–170. doi:10.1504/IJEG.2015.069495

Mureddu, F., Osimo, D., Misuraca, G., & Armenia, S. (2012). A new roadmap for next-generation policy-making. *ACM International Conference Proceeding Series*, 62–66. https://doi.org/10.1145/2463728.2463743

Napitupulu, D. (2014). The Critical Success Factors Study for e-Government Implementation. *International Journal of Computers and Applications, 89*(16), 23–32. doi:10.5120/15716-4588

Napitupulu, D., Sensuse, D. I., & Sucahyo, Y. G. (2017). Critical success factors of e-government implementation based on meta-ethnography. *2017 5th International Conference on Cyber and IT Service Management, CITSM 2017*, 1–6. https://doi.org/10.1109/CITSM.2017.8089300

New Zealand. Department of Internal Affairs. (2017). *Government as a Platform Value Proposition Discussion Paper*. https://www.digital.govt.nz/blog/gov-as-a-platform-a-value-proposition-discussion-paper

Nicholson-Crotty, S., Nicholson-Crotty, J., & Webeck, S. (2019). Are public managers more risk averse? Framing effects and status quo bias across the sectors. *Journal of Behavioral Public Administration, 2*(1), 1–14. doi:10.30636/jbpa.21.35

Nolan, R. L., & Gibson, C. F. (1974). Managing the Four Stages of EDP Growth. *Harvard Business Review*. https://hbr.org/1974/01/managing-the-four-stages-of-edp-growth

OECD. (2014). *Recommendation of the Council on Digital Government Strategies Public Governance and Territorial Development Directorate*. https://www.oecd.org/gov/digital-government/Recommendation-digital-government-strategies.pdf

Othman, M. H., & Razali, R. (2018). Whole of government critical success factors towards integrated e-government services: A preliminary review. *Jurnal Pengurusan, 53*. Advance online publication. doi:10.17576/pengurusan-2018-53-07

Ottlewski, L., & Gollnhofer, J. (2019). Private and Public Sector Platforms Characteristics and Differences. *Marketing Review St. Gallen, 2*. https://www.researchgate.net/publication/331980681%0APrivate

Parker, G. G., Van Alstyne, M. W., & Choudary, S. P. (2016). *Platform Revolution*. W.W. Norton & Company. doi:10.1017/CBO9781107415324.004

Paul, R., & Lawrence, J. W. L. (1967). *Organization and Environment*. Harvard University Press. https://www.hup.harvard.edu/catalog.php?isbn=9780875840642

Petrov, M., Burov, V., Shklyaruk, M., & Sharov, A. (2018). the Government As a Platform - a (Cyber) State for the Digital Economy. *Cyberleninka.Ru,* 1–48. https://cyberleninka.ru/article/n/the-government-as-a-platform-a-cyber-state-for-the-digital-economy

PMBOK. (2017). PMBOK Guide (6th ed.). Project Management Institute.

Poon, P., & Wagner, C. (2000). Critical success factors revisited: success and failure cases of information systems for senior executives. *Aphids as Crop Pests.* www.elsevier.comrlocaterdsw

Pope, R. (2019). *Playbook: Government as a Platform.* https://ash.harvard.edu/files/ash/files/293091_hvd_ash_gvmnt_as_platform_v2.pdf

PWC. (2007). *The Road Ahead for Public Service Delivery - Delivering on the Customer Promise.* www.psrc-pwc.com

Rantanen, M. M., Koskinen, J., & Hyrynsalmi, S. (2019). E-government ecosystem: A new view to explain complex phenomenon. *2019 42nd International Convention on Information and Communication Technology, Electronics and Microelectronics, MIPRO 2019 - Proceedings,* 1408–1413. https://doi.org/10.23919/MIPRO.2019.8756909

Ref, R., Heald, M., & Jankelovics, O. (2017). Your Role in the Ecosystem. *Accenture Strategy,* 1–11. https://www.accenture.com/_acnmedia/PDF-56/Accenture-Strategy-Your-Role-in-the-Ecosystem.pdf#zoom=50

Reponen, S. (2017). *Government-as-a-platform: Enabling participation in a government service innovation ecosystem* (Master's Thesis). https://aaltodoc.aalto.fi/bitstream/handle/123456789/24802/master_Reponen_Sara_2017.pdf?sequence=1

Rockart, J. F. (1978). *A new approach to defining the Chief Executive's information needs.* https://pdfs.semanticscholar.org/1b3a/7bf37b2728f137960bd21762e0dc98c6bfb1.pdf

Rosin, T. (2019). *Digital Dexterity 101: How to Develop a Digital-Ready Organization.* https://blog.walkme.com/digital-dexterity/

Shahiduzzaman, M., Kowalkiewicz, M., Barrett, R., & McNaughton, M. (2017). *Digital Business: Towards a value-centric maturity model.* https://chairdigitaleconomy.com.au/wp-content/uploads/2018/04/Digital-Business-Part-B.pdf

Smith, B. (2002). Six-sigma design (quality control). *IEEE Spectrum.* Advance online publication. doi:10.1109/6.275174

Symons, C., Cecere, M., Young, G. O., & Lambert, N. (2005). IT Governance Framework - Best Practices. *Forrester,* 1–17. http://i.bnet.com/whitepapers/051103656300.pdf

Tapscott, D. (2009). Grown up digital: how the net generation is changing your world. Choice Reviews Online. https://doi.org/ doi:10.5860/choice.47-3242

Taylor, F. W. (1911). *The principles of scientific management.* https://archive.org/details/principlesof-scie00taylrich

Techzine België. (2019). *API-beleid in Vlaanderen: 'We willen de bol.com van de overheid zijn.'* https://www.techzine.be/blogs/data/25944/api-beleid-in-vlaanderen-we-willen-de-bol-com-van-de-overheid-zijn/

Tiwana, A. A. (2014). *Platform Ecosystems, aligning Architecture, Governance and Strategy.* https://www.amazon.com/Platform-Ecosystems-Aligning-Architecture-Governance/dp/0124080669

Uludag, Ö., Hefele, S., & Matthes, F. (2016). Platform and Ecosystem Governance. *Digital Mobility Platforms and Ecosystems State of the Art Report,* 1–24. https://doi.org/ doi:10.14459/2016MD1324021

V-ICT-OR. (2020). *VlaVirGem.* https://v-ict-or.be/vlavirgem/wat-is-vlavirgem

VDAB. (2020). *Open services bij VDAB.* https://partners.vdab.be/werkgevers/openservices

Vitalari, N., & Shaughnessy, H. (2012). *The Elastic Enterprise: The New Manifesto for Business Revolution.* Telemachus Press, LLC. https://eiexchange.com/content/38-the-elastic-edge-a-better-way-to-work-innovate-c

Westerman, G., Tannou, M., Bonnet, D., Ferraris, P., & McAfee, A. (2012). *The Digital Advantage: How digital leaders outperform their peers in every industry.* https://www.capgemini.com/wp-content/uploads/2017/07/The_Digital_Advantage__How_Digital_Leaders_Outperform_their_Peers_in_Every_Industry.pdf

Wikipedia. (2020a). *Henri Fayol.* https://en.wikipedia.org/wiki/Henri_Fayol

Wikipedia. (2020b). *Mary Parker Follett.* https://en.wikipedia.org/wiki/Mary_Parker_Follett

Chapter 3
Findings and Core Practices in the Domain of Business Platform Models:
Overall Evaluation of the Practices

Yves Vanderbeken
DXC, Belgium

ABSTRACT

This chapter provides an overall evaluation of the created practices. The author reflects on the conclusions that are based on research in general and applying the various methods in actual case organizations. The original hypothesis and research questions are answered and commented upon. All of this information is assembled in a set of best practices, meaning in what order to apply the practices and what to achieve. Finally, the author provides tips for further research and topics that might be useful to add in a next iteration.

1 INTRODUCTION

At this point, the practices and governance model were both defined and applied in three case organizations by using the associated validation methods. Therefore, it is now possible to evaluate if the artifacts contributed to determining where the organisation stands today in relation to the business platform model each is applying.

2 USEFULNESS OF THE DEFINED ARTIFACTS

This section is limited to evaluating the usefulness of the validation methods themselves. The evaluation of what was learned by applying the methods is described in the next chapter. The evaluation was based

DOI: 10.4018/978-1-7998-7367-9.ch003

on working with the identified experts of the case organisations. The following reflections were given versus the stated objectives earlier:

- Objective: To determine what **value** the platform model will bring to all stakeholders in optimizing citizen services – tangible or intangible
  >> This was acknowledged as one of the key practices and confirmed in all three cases:
- For VDAB, value is established by allowing an ecosystem to match vacancies with unemployed directly, meaning a faster return to employment. This is a win-win for all.
- For ACPAAS, value generated by allowing start-ups to use standard functionality to develop applications for internal and external usage.
- For LB365, value is obtained by standardizing common processes and allowing third parties to build applications on top of this.
- Objective: To evaluate the existence of **digital leadership, governance, and accountabilities** to set the vision and turn this into an executable strategy
  >> This was defined as one of the top practices and confirmed in all three cases as crucial before starting to define an execution strategy. For example, VDAB and ACPAAS demonstrated how political leadership gave the kickstart for the platform model. LB365 also testified how a local politician – now prime minister of Flanders – kept on supporting the initiative and links it to a renewed digitalization investment program Flanders will be launching in 2021.
- Objective: To measure how **trust and confidence** was enabled by the government to allow all stakeholders to function safely in the "Opening Services" model.
  >> Building out a successful ecosystem requires trust and confidence at the side of the government, the ecosystem providers, and the citizens to use the services. This was defined as one of the top practices and acknowledged in the cases. The VDAB acknowledged this is not a onetime effort but requires constant attention. For ACPAAS, this requires building out a good network with start-ups that want to engage with the City of Antwerp. LB365 took a different approach by first forming a coalition of participating municipalities and getting trust upfront. This coalition had an initial budget to invest in the first components, but as more are joining the return on investment becomes more interesting as the costs are shared across all.
- Objective: To measure how the government organization **transformed their culture, skills, and capabilities** to apply a platform model.
  >> Applying the platform model requires different mindset inside the organization. All three cases acknowledged this. VDAB has set up a united business – IT team for managing the platform but agreed that this required different mindset and capabilities. For ACPAAS, this was already part of the DNA of the IT team. New skills were attracted to build the components of the ACPAAS platform. The transformation on the business side in Antwerp was equally challenging, to work in 2-week delivery schemes for accepting new applications or functionality. LB365 spent quite some time setting up the consortium discussing what functionality would be needed both on the business and IT side to get the benefits of the platform introduced at the local municipalities.
- Objective: To measure how **new technology and data** were used to build the platform model, connect stakeholders, and allow the unification of government services towards the platform.
  >> Setting up new technology to host the platform and act responsible with behaviour data was one of the key practices defined. This was only partially acknowledged. For VDAB, this did

not mean getting much new technology, as the IT team already had the technology capabilities in house. In the VDAB Innovation lab, advanced analytics techniques were already in use for improving the matching algorithms. For LB365, choosing a standard technology from Microsoft Azure implied that many of the features were available out of the box and that IT vendors could use the many standard integration mechanisms out of the box. It is to be noted that this was not enough, the LB365 steering committee had to spend quite time in defining the standard data models and ways to interact with regional and federal authentic data sources.

- Objective: To address **risk** and **legislative barriers** in dealing with the ecosystem and allowing the government organization to transfer the responsibility of the service execution to the ecosystem
  >> It was identified by all 3 case owners that embracing a platform model did require changing in risk management. This was mostly acknowledged at the VDAB, where the legal department was closely involved in setting the rules for the ecosystem partner to be able to access the job matching functionality. For ACPAAS, this meant changing the procurement approach from big contracts to small and agile ways of working, meaning there was not one budget to approve, but often budgets were allocated per sprint or deliverable. For LB365, this was defined upfront via the consortium approach. Each municipality knew upfront the costs and the return on investment as more would join.
- Objective: To foster **collaboration and innovation** across government and beyond by redesigning the government services to become self-service, proactive, and personalized on the platform
  >> This requirement was translated into a series of practices like culture, but also in the establishment of the right governance mechanisms. Although acknowledged by the three case organizations, the way it was handled is different. For VDAB and ACPAAS, a culture of collaboration and innovation already existed, so it was easy to reuse that mentality to apply it on a platform model. LB365 had a tougher hill to climb as they had to convince each local municipality to participate in the consortium. As LB365 assembled a coalition of about 15 municipalities kickstarting the project, one can state there was evidence that collaboration had been applied and made the difference.
- Objective: To measure how the **governance mechanisms** were put in place to ensure the platform model is managed and controlled in the best possible way.
  >> One of the artifacts of this research project was the definition of governance mechanisms. On governance, it was noted that none of the three cases designed governance mechanisms in the beginning or following the roll out of the platform model. For VDAB and ACPAAS, existing governance structures were mainly reused. For LB365, most of the governance is concentrated in the newly formed Steering committee, that brings all stakeholders together. This is not necessarily wrong and can be attributed to a pragmatic approach that is prevailing, but at a certain level of maturity, a more formal and designed governance mechanism will be needed. All case owners found guidance in the artifacts how to do this but testified that this was not very high on the agenda.

EGIT (De Haes et al., 2020) states that value can only be realized if the governance mechanisms are designed to execute the vision and strategy of where you want to be. One can thus argue if the defined governance mechanism as developed in this research project will make the difference. Or that there is still a lot of awareness needed at management level that "one size of governance does not fit all purposes".

Given the evidence and acknowledgement of the experts of the case organizations versus the objectives, it can be stated with reasonable (Weiss, 2003) that the practices and governance model did prove to be a good baseline for evaluating the status of a platform model. On top of that, the experts agreed that the validation methods were:

- **Useful** – the defined practices were confirmed representing a 360° view of the approach the case organization has taken to design, build and/or roll out their respective platform model. They helped in validating that the practices were applied or helped identify where actions or corrections were needed.
- **Effective** - the defined practices and governance mechanisms suggested will allow to build an effective platform organization and provide value to the stakeholders, although all agreed this is work in progress.
- **Repeatable** – as the validation methods were in used in the same way in three case organizations that are in different stages of deploying the platform model, the repeatability is conformed.
- **Consistent** - by applying the scoring and checklist approach, consistency in comparing results was obtained. For example, the organizational capability method and the resulting graph made it easier to understand where an organization stands today and even visualize results for one or all case organizations.

This brings us to the following conclusions on the practices and governance model:

Table 1. Evaluation of functional characteristics of artifacts - Own Work

Artifact	Validation Method	Requirements met?	Functional Evaluation
Defined practices	Method 1: Organizational Capability	Yes	Useful, effective, repeatable, and consistent
Governance Mechanisms	Method 2: Governance Compliance Checklist	Yes	Useful, effective, repeatable, and consistent
Practices and Governance Mechanisms	Method 3: Platform maturity model	Yes	Only used in a limited way but useful, effective, repeatable, and consistent
Practices and Governance Mechanisms	Method 4: Platform Balanced Score Card	Not tested	Not used with the case organizations

Summarizing, the validation methods met their requirements, and the underlying practices and governance model were recognized by the case experts as guiding for their respective organization.

3 CONCLUSION

3.1 Contribution of this Research

In the beginning of this project, literature research (Accenture, 2018) revealed that even the top countries adopting Government as a Business Platform model today still have significant room to improve and lag industry progress.

This research demonstrated clearly that this statement is highly relevant for the region of Flanders in Belgium. Government organizations embarking on a platform model have room for improvement. The cases indicated that organizations were indeed taking different routes along their journey, leveraging the strength of an ecosystem on a local level and bottom-up approach only. Although all government organizations investigated in Flanders were aware of the digital disruption and were embarking on a platform journey, scaling best practices remained a challenge. All struggled in evolving to the role of ecosystem moderator from their traditional role of government as an actor.

Although there was an increased budget allocated in the latest Flemish Policy declaration to digitizing the citizen services with technologies, there was still no evidence for a top-down led vision and strategy to realize these goals.

So, defining the practices and governance model was indeed needed to determine a baseline on how to set up a platform and be aware of how to run and maintain such a platform model. It was shown that adapting to the platform model in Flanders required a pronounced political vision, strategy, mindset, culture, and skills for the government employees to make it a success. The political vision was only expressed at local level, like the ACPAAS case in Antwerp or the VDAB.

This research project contributed to the body of knowledge as it proposed a set of useful practices and designed governance model for government organizations embarking on a platform-oriented model in Flanders. Top-down or bottom-up, every organisation can benefit from this knowledge.

The defined artifacts were turned into a set of validation methods that helped in the case research to determine where organizations stand today and what the next suggested step was. With these artifacts, gaps could be analysed, and actions identified on what organizations needed to do to improve their platform strategy.

With the artifacts from this research, governments got extra guidance that contributed to the following activities in defining, designing, or implementing the business platform model for citizen services:

- **Set the right vision and strategy** to set a clear expectation to the citizens what and how a platform model can help.
  >> The defined practices should contribute to a better understanding of setting the right vision and strategy, especially in getting the political support.
- **Determine the right value creation,** applying the platform model to tap into new sources of (digital) supply - de-linking assets from value where possible.
  >> This research examined in depth what value is and explained why this is key practice.
- **Enable value/service consumption** through new forms of citizen behaviour like self-service
  >> This research defined how a platform model operates and how a service consumption approach looks like. This is part of the defined practices.
- **Manage trust and quality control** through community-driven curation and guaranteeing the right ecosystem players to deliver the right services to the citizens

>> As part of the research, attention was given to how an ecosystem must be defined and managed. This is part of the top defined practices.

- **Define and execute the right governance model** to become the moderator of an ecosystem, constituting of a mix of government agencies, industry, and other stakeholders (e.g., education institutes).

  >> See the designed governance model, where guidance on the right processes to be implemented was provided.

- **Validating** if the organization was on the right track and determining what the actions might be to reach a higher maturity level

  >> See Validation methods that can be used by government organizations to get a sense of where they are today and what the next steps are. As a result of all the produced guidance, all government leaders should be inspired to define and implement the business platform model for citizen services to provide the following **value to citizens**:

- **Faster access** to proactive and tailored citizen services according to unique citizen's needs.
- Get a consistently delightful **experience** instead of being the errand boy of the government agency.
- **Involve** citizens in the evaluation of existing and/or design of new services.
- More **transparency** as more government organizations open their data and application to citizens and business.
- Faster creation and launch of new services.
- **Guaranteed authenticity** and privacy as governments control and govern the ecosystem and the manipulation of data associated with the transaction.
- **More innovation** as industry partners can create new value on top of the government provided data and functionality.

Governments need to better recognize and address the changing needs of citizens over their entire lifetimes, provide platforms to help them get the resources and make the connections they need, and see a whole set of public goods created by the sum of their many original parts (Slaughter Anne-Marie, 2017).

As far as we are aware, this is the first study that defined these practices and defined an associated governance model for governments. The biggest contribution of this study was that it delivered immediate value, and the defined set was directly applicable in today's reality.

1.3.2 Conclusions from literature research

Literature revealed that governments are embracing the platform model in a scattered approach and are running behind versus industry, where this model is already considered the new normal. Yes, experiments are going on everywhere, but most stick to the provisioning of (open) data and opening supporting functionality (e.g., Verify My ID in UK). Nevertheless, some interesting cases of citizen services based on matching an ecosystem with citizens already exist today, both internationally as in Flanders.

Here and there, we saw isolated cases of government organizations in Flanders like VDAB that dare to take the platform model to the next level to allow citizen services to be handled by an ecosystem – in the case of VDAB this was related to job matching services. This was and is courageous. Other initiatives were emerging like the Theatre Finder in Antwerp or "Surviving in Brussels" that have the best in mind. When digging a bit deeper, some seemed to be lacking a proper business platform vision or strategy behind it. Some were unintentional by-products of an "Opening Functionality" focus like the Theatre Finder.

In Flanders, experts interviewed indicated that there is not yet an ongoing political debate on how to optimize citizen services using technology and the platform model as a basis. In fact, politicians did not think they can score or win votes with proclaiming a platform model. Politicians in Flanders might not yet fully understand the value of the platform model and seemed hesitant to speak out on it – despite some EU reports setting the stage for platform models. Most likely we will continue to see different initiatives like VDAB emerge - all with good intentions and support from the organizational leadership level - but not top-down coordinated, not starting from a strategy on Flanders level, but rather bottom-up organized, per government organization.

One interviewed stated that it was strange that local municipalities were the furthest in adopting this model, that this is the world upside down. The point was that local municipalities should be leveraging a centrally deployed platform model, not invent one themselves on a case-by-case basis. This raised a very valid point: if every government organization, independent of their level in the government hierarchy, defines its own strategy and rolls out its own platform, it will be confusing for the ecosystem to select where to engage. Hence the chicken and the egg effect (Choudary, 2015) will play as different government organizations approach the same ecosystem partners to join different platforms.

Analysis of the literature (Shahiduzzaman et al., 2017), discussions with the experts and case owners revealed that government organizations that had established a good working business platform model ("Opening Services") tended to have the following digital capabilities (competence areas):

- A clear, coherent, and well-developed strategy for how digital will be incorporated into business.
- Leadership, management, and governance that supports its strategy.
- Investment in digital knowledge, skills, and competencies to realize its strategy.
- A proactive culture that embraces innovation.

The above bullets were reinforced in the interviews with case owners. The list of defined practices extended upon these four themes. Thus, the artifacts represent a fair interpretation of the literature research phase.

3.3 Conclusions from Working with the Case Organizations

Three government organizations using the platform model in Flanders were investigated in detail, fitting the defined requirements, and by applying the validation methods. For each, detailed discussions with the case owners were held to see where the organizations stand today and what their next move could be to increase the efficiency of using the platform model.

In general, the case organization acknowledged the defined practices. The top practices as value, ecosystem and leadership were all recognized and acknowledged relevant, i.e.,

- Each of the three case organizations stated that **political leadership** was crucial to express a vision that the organization then can translate into a strategy. But all also observed that Flanders was not showing much political leadership towards platform models from the top. This was in line with the findings from the literature research phase.

There was a desire expressed by all case organizations to see the regional government of Flanders defining a vision and even provide guidance how to build and operate a business platform model. At least, the case organizations were looking for a signal that government services could be delegated to an ecosystem of partners.

Additionally, there was also the wish to have a common platform developed once in Flanders whereby each organization can tap his services into. Now, each organization was building its own platform, with its own specifics. There was an opportunity here to increase efficiencies. But it all starts with a political vision, which was reported lacking in Flanders.

- Regarding **value**, it was observed that the organizations each have a different perspective on value. This was understandable as the platforms that each case organization operated, were individual initiatives, starting from the organizational objectives, not driven by an overarching Flemish Government vision. During the interviews with VDAB, an interesting new perspective on value emerged. With the arrival of new leadership at VDAB, there was an additional request to the VDAB platform team to describe what the value was for VDAB itself in applying the platform model. During the research, we stated that there must be value for all stakeholders. VDAB was now asking the question of "what does this mean it for us, VDAB", or "Will the public money that is invested in the Matching platform also bring value to VDAB?" Benefits realization management suddenly became a topic and was now being addressed by the platform team.

For LB365, the discussion with the interviewed quickly evolved to what possible services could be built upon the current foundations. The interviewed realized immediately that the platform had more potential than foreseen in the original vision. The practices and validation methods triggered a forward-thinking process.

- All three organizations also had a different view on the role and involvement of an **ecosystem**. ACPAAS focused on start-ups to help create new applications with provided reusable functionality. LB365 was looking to provide common functionality used in local municipalities to an ecosystem of software builders. Only VDAB came close to the real idea behind the business platform model: delegating service execution to the ecosystem. But in a platform model, it is the citizen (or the unemployed in the case of VDAB) that should have control over the transactions initiated with the ecosystem. Today, it is the ecosystem that is launching the transaction to the unemployed, meaning contacting the unemployed with a proposal for a matching job.
- In a sense, **innovation and culture** were evaluated as easy to kickstart. Although constant attention was needed to keep everyone aligned, VDAB and ACPAAS noted that getting an innovation mindset in place was not that difficult. Keeping it alive and keeping people excited and focused though did require constant attention.

The results on the defined **governance mechanisms** were surprising. There seemed to be little appetite to designing a specific governance model for the purpose of supporting a platform model, i.e.,

- VDAB and ACPAAS chose for reusing and integrating the platform model in the current governance mechanisms.

- VDAB did set up a new team but was still reusing the current processes and committees to discuss and handle the platform approach.
- ACPAAS was managed within the scope of the IT Department, with a leadership role for the Chief Enterprise Architect
- LB365 is currently using one structure, i.e., the LB365 Steering Committee, for managing all governance aspects related to the scope, technical matters, growing the consortium and delivering the platform itself.

As a summary, the problem statement was verified and confirmed with the case organizations through the demonstrated artifacts. However, there is much work to do in Flanders to steam ahead in the re-development and deployment of government services using a platform model.

The three cases demonstrated that the platform model did provide value and brought benefits to the government organization, to the ecosystem, and the end-users. The three cases were thus an example in Flanders, and all stated that more was possible, given clear (political) leadership. In other words, the platform is here to stay.

3.4 Conclusions on the Research Questions

We defined two research questions that drove the literature and case research, i.e.,

- **Research Question 1**: What are some of the best practices for defining a Government Business Platform model for citizen services?
- **Research Question 2**: What are the **criteria** that constitute **a well-governed design** for a Government Business Platform model for citizen services?

In the next section, each research question will be answered.

3.4.1 Conclusion for research question 1 - practices.

Based on literature research and interviews with experts, a list of success factors was analysed and documented as a set of practices. The list was mapped on the typical level of responsibilities in an organization, which did not imply that strategic mapped practices were more important than others. The mapping implied where the suggested responsibility for action laid in an organization.

The resulting table is hereby copied for convenience:

This list was first validated in discussions with experts and later converted into a series of validation methods. The validation methods were used with case organizations in determining if the defined practices were recognized and used. Based on the output of the interviews, the list was deemed useful, effective, repeatable, and consistent.

The top practices like value, political leadership, and the establishment of the ecosystem were all confirmed by the case owners as crucial to articulate what the core interaction or matching process of the platform should be.

Given the reflections from the research phase and in working with the case organizations there was a clear indication (Weiss, 2003) to determine that this artifact were correctly defined, validated, and tested in Flanders. The artifact and associated validation methods could be reused when evaluating other government organizations, in Flanders and anywhere else in the world.

To get to a higher level of certainty, it is advised to repeat the case research with similar type of governments organization in- or outside Flanders.

Table 2. Final List of practices mapped to organizational responsibility – Own Work

Strategic level	Tactical Level	Operational Level
Vision	Business Agility	Talent and Skills
Strategy	Business Ecosystem Design (Top)	Customer Experience Design
(Political) Leadership (Top)	Platform Architecture & Infrastructure	Risk Management
Governance	Data Driven	Technology Ecosystem Design (Top)
(Innovation) Culture		
Value (Top)		

3.4.2 Conclusion on research question 2 – governance.

For answering the second research question, the COBIT2019 Design Toolkit was used to design governance mechanisms matching the platform model. The Toolkit offered a set of 10 design factors to consider when determining the right COBIT2019 processes for the organization.

The values entered in the design factors represented Flemish government organizations embracing a platform model. The reason was that a governance model needed to be designed to its specific context to maximize the value of applying the governance processes, structures, and relational mechanisms(De Haes et al., 2020). The result was thus a customized set of processes for purpose.

Some recommended processes from the toolkit changed priority based on the literature research. Each action influencing the prioritization was documented as to why it was adjusted or not. The result was a prioritized list of processes to govern the defined practices.

The below table provides an overview of the prioritized processes:

This table was then merged with the defined practices and the levels of responsibility in an organization. Thus, a single overview was established that determines which processes contribute to handling what practices and where the typical responsibility sits in a government organization to handle this process. For example, for handling the practice of Vision and Strategy, COBIT2019 processes APO02 - Managed Strategy (Top Priority) and APO05 - Managed Portfolio are recommended and the responsibility sits with the strategic leadership team.

Next, a set of organizational structures and relational mechanisms were defined to match the designed processes and context of a government organization embracing the platform model. For example, it was suggested to have a dedicated Platform Steering Committee to report to the leadership to handle platform roll out progress. It was also suggested to establish a joined Platform Leadership Team where business and IT share the same goals and budget to realize the platform model.

Based on feedback with experts and case owners, we can also state there is clear indication(Weiss, 2003) that this artifact was defined properly. The artifact has proven valuable and useful in reviewing the selected government organizations in Flanders. More research is suggested to confirm if this artifact is equally valid for other organizations in Flanders, other regional or federal governments inside or outside of Belgium.

Table 3. Prioritized COBIT2019 processes - Own work

COBIT2019 Process	Assigned Priority
APO02 - Managed Strategy	Top
APO12 - Managed Risk	Top
APO13 - Managed Security	Top
DSS05 - Managed Security Services	Top
MEA03 - Managed compliance with external requirements	Top
EDM01 - Ensured Governance Framework Setting and Maintenance	High
EDM03 - Ensured Risk Optimization	High
EDM05 - Ensured stakeholder engagement	High
APO03 - Managed Enterprise Architecture	High
APO04 - Managed Innovation	High
APO08 - Managed relations	High
APO14 - Managed Data	High
BAI05 - Managed Organizational Change	High
BAI06 - Managed IT Changes	High
DSS06—Managed Business Process Controls	High
EDM02 - Managed Benefits delivery	Medium
APO01 - Managed I&T Management Framework	Medium
APO05 - Managed Portfolio	Medium
APO07 - Managed Human Resources	Medium
APO09 - Managed Service Level Agreements	Medium
APO10 - Managed Vendors	Medium
APO11 - Managed Quality	Medium
BAI08 - Managed Knowledge	Medium
MEA01 - Manage Performance and Conformance Reporting	Medium
MEA04—Managed Assurance	Medium
EDM04 - Ensure Resource Optimization	Low
APO06 - Managed Budgets and Costs	Low

3.5 CONCLUSION ON THE HYPOTHESIS

The hypothesis was drawn up as a statement that was subject to verification via research. In this case, this meant applying the validation methods in three government organizations. Using interviews, discussions and analysis, enough evidence was collected to reflect on the hypothesis.

The hypothesis was:

Citizens will be favouring to use a trusted and managed ecosystem for requesting and executing govern-ment services if the value is clear, and governments vow for a transparent design of the platform where safety and privacy of the citizen' data are guaranteed.

First, there was doubt that a lot of citizens would engage with a government through a platform model. Chapter "Are citizens waiting for the Platform model? (Consumers)" indicated that citizens all over the world still use predominantly traditional means to contact a government or request a government service. The level of trust that is established by talking to a civil servant (either physically or via phone) is still more appreciated than websites or platforms. But citizens are more and more getting digital, so it can be expected that the balance might shift. Governments can thus not sit and wait, but before all, must do a better job in communicating the availability and advantages of the digital channels and start positioning platform models.

Second, politicians in Flanders seemed hesitant to embrace the platform model. Not only because they thought there are little votes to be won, but also because there was a lack of vision towards a platform model. This was testified in the case organizations, that each have their own political support, but did not saw an overall vision being put forward by a politician. Interviews indicated also that politicians in Flanders were reluctant to express the desire to redirect the execution of government services towards an ecosystem.

The case organizations did reveal that it is possible to establish a trusted and managed ecosystem, but all three testified this does not come easy and does require ongoing effort. The value for each of the ecosystem partners must be clear and commercial parties are foremost looking to make a profit out of each transaction. Quite some examples were identified in and around Flanders of government-initiated platform models using an ecosystem of partners that operate in the non-profit sector. The government is mainly matching the needs of the people with the providers offering a service like meals, sleeping facilities or even homestays. The case of "Surviving in Brussels" is a prime example of this, but similar examples are found internationally.

What about value? Criteria to determine what drives value on a platform model for governments, ecosystem providers and consumers of services were presented earlier. For citizens, the value is time and convenience, as the services can be ordered anytime, anyplace, any device and are fully self-service. Citizens can choose the provider and even give feedback which provider was best. This feeds the network effects, meaning that good providers get more requests. For the government, platform models represent a faster way to provide proactive and tailored services. Using the latest technology, governments provide a means to access the services faster than before. Through delegation of authority, certain parts of the service can be executed to a non-governmental or commercial partner. This in turn increases operational efficiency of a government.

Again, the case of "Surviving in Brussels" brings all these value points of view together. A homeless person in need of a hot meal can use the smartphone to determine where a decent and subsidized eating facility is located. Based on the experience, the person can provide feedback. The provider gets clients and is paid by the government. A win-win for all.

Value is thus an important driver, but also well defined in literature and good examples were found. Value was confirmed by the case organizations as a top priority, although each organization had its own view of what the value exactly is and how to engage.

Offering government services on a platform must be simple, clear, and transparent. In chapter Data Related Practices, two practices emerged from using data and (advanced) technologies on a platform:

proper data management and ensuring transparency to all stakeholders what data to use, how it affects the service proposition, and where the red line is. Governments cannot afford a platform to be selling behaviour data to commercial players. Literature research revealed that safety and privacy of data must be guaranteed. Experts interviewed opiniated that this was one of the reasons governments in Flanders refrained from fully embracing a platform model, as no politician wanted to face the press describing the consequences of a data breach.

Research from literature and interviews indicated that citizens are still using traditional means to engage with governments and politicians in Flanders were reluctant to embrace the platform. On the positive side, there are excellent examples already out there in Flanders of government organization that do embrace the platform model, where there is (local) political leadership, where value for all stakeholders is demonstrated, and government operations are made more efficient in a secure and transparent way.

As such, the validity of the hypothesis is nuanced. If viewed top down from Flanders level, the hypothesis is not confirmed due to the lack of political leadership at the top. But, viewed bottom up from government organizations like VDAB, LB365 and ACPAAS, there was enough evidence that all elements of the hypothesis can be fulfilled, this making the hypothesis valid and confirmed.

The conclusion is therefore that the hypothesis seems to bring together the right attention points and is valid, but not all of points have been established in Flanders in a top-down way. Therefore, local government organizations are likely to continue to be the trendsetter in a bottom-up approach, hopefully and eventually triggering a political top-down reaction or standardization effort.

4 USING THE PRACTICES TO START A PLATFORM MODEL IN GOVERNMENT ORGANIZATIONS

Based on all the research, defined practices, and validation methods, a set of sequenced advice was derived that practitioners should consider when preparing for a government business platform ("Opening Services"). These are:

- **Express a clear vision** – what citizen services you want to improve? What is the target group? Why a platform model? What is your ambition: do you want to lead the pack or be a follower of other exemplary government organizations? Can the vision be linked to the government declaration? (>> Vision Practice)

- **Determine the anticipated value from day one** – Investigate what the potential value could be for all stakeholders. Translate value to each stakeholder, as value for citizens is different than for ecosystem partners that are looking for commercial opportunities. (>> Value Practice)

- **Get Political support** – translate the value into politically correct statements that politicians understand and can communicate to the public. This might involve different arguments than for a traditional business case. There should be more focus on intangible benefits like improving citizens' lives, allowing self-service, convenience, personalized services, more choice, etc. (>> Political Leadership practice)

- **Get local leadership in helping to define goals and set a strategy to realize the vision**. Be realistic in goals and determine criteria for benefits and success upfront. What network effects do you anticipate? How will you attract citizens? How will feedback be possible? What capabilities does the government organization need to have to be successful in setting up the platform model? What is the timeline? Are there compelling events that can trigger accelerating the deployment of the platform model? Start early on the business case and build it up over time. How much investment will be needed? What is the anticipated return on investment?

(>> Strategy practice)

- **Prepare for defensive reactions in your organization** when you express the idea. Continue to communicate to all levels and explain the vision. Communicate benefits to the organisation and the changing role of the civil servants. Get people involved and a sense of participating in shaping the new citizen services.

  (>> Culture practice)

- **Go for a separate organization that blends IT and businesspeople together**. Go for a one platform team, one vision approach. Agree that the platform is the product of both business and IT. Set up own governance in the team that embraces agility and change. Strive for a complete redesign of the citizen services, embracing an ecosystem approach. Embrace technology to make the difference. Do this in an incremental way.

  (>> Innovation practice)

- **Involve other departments** like legal and contracting from day one to set the stage for redesigning citizen services and determining the fees and rewards for ecosystem partners. Make it clear how data will be collected and made available to all stakeholders. Make agreements with the ecosystem how they can join, but also how they will be relieved from the ecosystem if they do not meet service level agreements or get bad feedback from citizens.

  (>> Governance practice)

- **Define the right governance mechanisms** to start light but think how this may and will evolve over time as the platform matures and more services are put on the platform. Use a framework like COBIT2019 to provide the right guidance. Use the predefined governance model to start but be ready to re-evaluate the design criteria to suit your own specific organisation.

  (>> Governance practice)

- **Start working on defining and establishing an ecosystem of partners from day one**. Define what an ecosystem partner must bring and contribute to the platform, what parts of the service can be delegated to the partner, how end-to-end responsibilities will shift, how you will measure this.

  (>> Business Ecosystem Design practice)

- **Work with IT** to define what technology to use or skills to acquire to host and operate the platform model. Work on application and infrastructure standards for sharing data and functionality with the ecosystem. Put security in the design from day one. Consider Cloud to quickly procure software or functionality to build the platform. IT capabilities might be needed from external suppliers to bring the necessary expertise (e.g., security, data handling) to design the technical platform layers. Use agile methods to deliver to the team. Remember that IT is as much in this as the business, you are both one team now.

  (>> Platform Architecture & Infrastructure practice)

- **Ensure (only) the right data is captured and analysed** to serve the goals of being proactive and personalized. Ensure data is handled secure and only visible to the right stakeholders. Be transparent to citizens to what data you captured and how you processed or stored it. If algorithms like Artificial Intelligence are used, be transparent to the citizens how these algorithms influence the selections, choices or partners provided to them.

(>> Data Driven practice)

- **The citizen experience is crucial.** When deploying the platform model, make sure you have thought enough how easy the service is presented and how a transaction can be established in a minimal number of steps. Involve citizens in the design, test ideas early. **Be an agile organization** that listens and acts upon the feedback.

(>> Customer Experience Design practice)

The **Platform Maturity model** can be used as a compass as it provides much of the answers to the above questions in an ascending sequence of maturity. This way an organization can also determine where they are today and where they want to be.

Once this is done, the **Balanced Score Card** can be used as a technique to link objectives and metrics together.

5 Suggestions for extra research

5.1 Platform models as part of a government digital transformation agenda

There was ample literature found that already provided recommendations to governments how to adopt the platform model. One particular study from the EU is hereby referenced for building out a business platform model (EU - DG Joint Research Centre, 2018). Basically, the EU recommended taken up the platform model in the overall digital transformation strategy of the government / country. This is well in line with the documented practices from this research, but there was additional guidance in the EU report worthwhile repeating, i.e.,

- Governments should adopt current economic, social, and public policies to regulate platform businesses fairly.
- Governments should focus on orchestrating and reusing existing government services as a starting point for developing their Digital Platforms.
- Governments should start building IoT capabilities.
- Governments should optimize the use of their open data services by defining a service delivery approach that matches the needs of their target consumers.
- Governments should invest in creating and designing ecosystems.
- Governments should consider Digital Platforms to be a cornerstone of their Digital Government transformation initiative.
- Governments should enhance digital public services with location information and location intelligence.
- Governments could benefit from supporting location data offerings to match providers and consumers.

Not all aspects of these guidelines were studied in-depth in this research but do fit in the defined practice of defining an overall strategy. The EU called out to governments to create an overall top-down oriented business platform model, which this case organizations also confirmed to be necessary. This is well inline with the top practices defined.

5.2 Platform (Corporate) Governance

It was realized during this study that strategizing and defining a government business platform model ("Opening Services") is a very broad area. Not all aspects have been addressed in full detail in this study, and therefore need further clarification.

For example, we addressed the need for a new form of corporate governance – referred to as Platform governance – to be drawn up and rolled out consistently across the stakeholders. Some of the outstanding corporate governance questions are hereby summarized:

- What should the units of political accountability be if we move beyond a traditional silo-based & department-shaped organization? (i.e., what is the impact of realizing the joined Platform Leadership Team on traditional organisational set up, the silos?)
- What will the government do with all the data it collects from the citizens in requesting services? Will it be deleted after a time? How to ensure transparency and demonstrate data is used in the best interest of the citizen, even if this includes more government money to be allocated to that group of citizens than budgeted?
- E-inclusion: how to make sure people without connectivity or access to the Internet get their equal share of the services offered on the platform?
- How to avoid a tax nightmare for the ecosystem if extra platform taxes are imposed on top of existing constructions? Will the government tax the ecosystem partners extra based on the number of transactions on the platform versus general tax?
- Delegating execution to a non-government agency or industry partner requires trust and transparency in selecting these partners. This is ongoing work. How to set up an ecosystem and maintain the focus on the government platform model?
- How does one ensure that the trust and consent layer operates as it should to protect citizens' interests?
- What strategies could be adopted to minimize the impact of data breaches at the data register layer? In a broader perspective, how does one secure the operations on a platform model?
- Is it a local play or international? How to stimulate or avoid international companies to offer services on a local/regional platform?
- Will the government ask a fair fee to ecosystem partners to be enlisted on the platform? Will there be a transaction fee when a citizen places an order?

It is recommended to study these questions carefully in a follow-up study. The results can be incorporated in refining the vision and strategy for a government business model ("Opening Services") and eventually used to refine the practices as compiled in this study.

5.3 Impact of new or Emerging Technologies Like IoT

In their book Platform Revolution, the authors state (Parker et al., 2016) that *the platform model will continue to shape transformations in the markets for labour and professional services as well as the operations of government*. The authors predict that IoT (Internet of Things) will bring new possibilities to the platform model by linking citizens and smart devices together.

These emerging technologies will allow many extra or new services to be deployed on the platform. For example, in healthcare, we can monitor a patient's health by collecting and sending signals to a central application running in the Cloud. There, computers can process and crosslink huge amounts of data and apply artificial intelligence techniques to provide recommendations to the patient or the healthcare authorities. Also, the results are sent back to the patient. This architecture is implemented partly in so-called self-care systems, which help people manage their wellbeing using many kinds of measurement instruments.

Further research is needed to determine the causes of/effects of/relationship between the defined practices and new technologies like IoT.

5.4 Security

Welby (Welby, 2019) stated that governments need to understand and guard against digital security threats in order to secure citizens and the services they require in order to safeguard their well-being.

In this research project, we did not dive deep into the topic of security for platform models but grouped all security aspects into governance and risk management as practices. The COBIT2019 processes APO13 - Managed Security and DSS05 - Managed Security Services were identified as top priority.

This was a narrow view of the many security threats that government platforms face. First, there is the technology aspect, whereby tools can be put in place to avoid hacking of personal data of the consumers. Then there is the process aspect, avoiding that services are hacked and abused to force transactions. Both need to be in balance.

Governments are extra vulnerable to the negative impact of a security breach as citizens (and the ecosystem) will lose trust in the platform if it is not well handled or mitigated.

To better understand the implications of security on a platform model, future studies should focus and address this topic and the meaning of these processes in detail.

6 FINAL NOTES

Throughout the whole journey of writing this research project, it became obvious that the platform model is here to stay, even in governments.

During the research, excellent examples were found all over the world, but also in Belgium and Flanders. Several countries, like the US, UK, Singapore, Dubai, Australia, or New-Zealand, are taking a fast-forward approach to adopting technology to build the platform model for improving citizen services. Many others, including Flanders, seemed a bit more hesitant to engage in a centralized and coordinated approach, certainly from a political leadership top-down point of view. This stands in stark contrast with the ingenuity and creativity that several government organizations were showing in a bottom-up approach, thus on their own. Platforms models cannot be stopped anymore, also not in governments.

At the time of writing the conclusions, the COVID-19 pandemic was in full swing. Lots of platforms emerged from nowhere and overnight in Flanders, mostly centred around social care, like supporting the healthcare system or – for example - helping the elderly with their shopping. Apart from the wonderful solidarity that was behind the initiatives, there was no indication of political leadership behind this, as these initiatives seemed not to be driven by a vision or a strategy.

Meanwhile, technology advancements are being launched faster than ever, and private companies are already busy designing the next generation of platforms, based on IoT or AI as novelties. This will put governments even more in a catch-up mode.

Platforms are here to stay and can be a success in Flanders, the case organizations have shown that. Citizens are getting used to the convenience these platform models bring. One can only look forward to a more consistent approach to finding and ordering government services on a well-regulated platform, where collected data is correctly used.

The ultimate question is thus: how long can governments and politicians (e.g., in Flanders) still wait to adopt at least some form of standardized approach to platform models? Because…

"A digital population cannot be well served by an analogue government"
Paul M. Tellier and David Emerson, Co-Chairs of the Prime Minister's Advisory Committee, on the evolution of Public Service, 2013

REFERENCES

Accenture. (2018). *GaaP 2018 Readiness Index*. https://www.accenture.com/_acnmedia/pdf-83/accenture-gaap-2018-readiness-index.pdf

Choudary S. P. (2015). *Platform Scale*. https://www.amazon.com/Platform-Scale-emerging-business-investment-ebook/dp/B015FAOKJ6

De Haes, S., Van Grembergen, W., Anant, J., & Huygh, T. (2020). Enterprise Governance of Information Technology. Enterprise Governance of Information Technology. doi:10.1007/978-3-030-25918-1

EU - DG Joint Research Centre. (2018). *Digital Platform for public services*. https://joinup.ec.europa.eu/collection/elise-european-location-interoperability-solutions-e-government/document/report-digital-platform-public-services

Parker, G. G., Van Alstyne, M. W., & Choudary, S. P. (2016). *Platform Revolution*. W.W. Norton & Company. doi:10.1017/CBO9781107415324.004

Shahiduzzaman, M., Kowalkiewicz, M., Barrett, R., & McNaughton, M. (2017). *Digital Business: Towards a value-centric maturity model*. https://chairdigitaleconomy.com.au/wp-content/uploads/2018/04/Digital-Business-Part-B.pdf

Slaughter Anne-Marie. (2017). 3 responsibilities every government has towards its citizens. *World Economic Forum*. https://www.weforum.org/agenda/2017/02/government-responsibility-to-citizens-anne-marie-slaughter/

Weiss, C. (2003). Expressing scientific uncertainty. *Law Probability and Risk*, *2*(1), 25–46. doi:10.1093/lpr/2.1.25

Welby, B. (2019). The impact of digital government on citizen well-being. *OECD Working Papers on Public Governance*, *32*. doi:10.1787/19934351

Chapter 4
Problems in the Area of Agile Methodologies

Tapan Kumar
Cognizant, The Netherlands

ABSTRACT

Working in distributed teams on digital platforms and software technology nowadays occurs in a more agile manner. This chapter will focus on the motivation for examining why agile methodology doesn't work efficiently in large enterprises having distributed agile software development model compared to small or medium enterprises. The chapter aims to unearth with the help of several known statistics and see if agile or distributed agile is always a success or not and explore the cascading impact on organizations when the agile implementation is not as per the expectation. Further on, the author tries to explore all factors and characteristics associated with distributed agile software development with the help of the literature, and thereafter, it intended to find an answer about the impact of distributed agile software development on team performance which has varying impact on organizations.

INTRODUCTION

"Agile or Distributed Agile" is Always a Success: An Overhype or Reality!

Agile methodology has played a vital role in recent days for software development by overcoming the drawbacks of other popular SDLC (Software Development Life Cycle) models like Waterfall, Iterative and Spiral, etc. The Agile methodology was initially developed for small co-located teams involving small projects, thereby introducing flexibility and quicker response to the customer requirements. The statistics proving the success stories involving Agile in small projects had also been illustrated in several reports like Standish Group Chaos Report 2020.

Following the path of applying agile for small co-located teams, agile practices had also caught interest in larger organizations, as well as to large-scale software development projects. However, it was observed that there had been some challenges in incorporating and practicing Agile to explore benefits as it was defined in the Agile Manifesto in 2001 and was also very well documented in the Standish Group Chaos

DOI: 10.4018/978-1-7998-7367-9.ch004

Report 2020, which referred that, big organizations were not as successful in practicing Agile for larger projects as compared to smaller projects (19% success in large size projects vs 59% success in small size projects). Besides, another survey done by Scott Ambler + Associates mentioned that the success ratio of the co-located agile team was higher than a distributed agile team. Despite such clear statistics in success ratio involving various dimensions, organizations still intended to do a huge investment in implementing a distributed agile software development model, which was surprising and exploratory.

In this chapter, the author has tried to evaluate and explore all factors and characteristics associated with distributed agile software development with the help of the literature and thereafter, intended to find an answer about the impact of distributed agile software development on team performance. This was done by conducting a case study within a leading bank in The Netherlands, where, at one end maturity assessment model was used for deriving the current maturity within the Bank, and on other hand, the maturity score using data statistics, helped to identify the gaps while using a single model for evaluation.

The outcome of the research did find a direct co-relation between distributed agile software development and team performance but in addition to this, it also unboxed various factors attributing to the usage of maturity models and challenges within the distributed agile model which might have been given less importance by organization thus not helping organization to do a good root cause analysis thus, negatively impacting team performance.

PROBLEM AREA

Agile has become the fastest growing IT development methodology with the majority of organizations doing agile implementations as stated in Harvard Business Review (Agile at scale (2018)). However when examined a leading bank within The Netherlands, which had transitioned from traditional software development methodology to Agile methodology, the return on investments was not up to the desired mark. This was visible in the velocity graph of several agile teams, which had not shown any sort of growth in the last 3 years, and besides, the completion ratio depicted via the burndown chart was around 50-70% instead of 100%. Prima facie it looked like big organizations were not as successful in practicing Agile for larger projects as compared to smaller projects (19% success in large size projects vs 59% success in small size projects). This statement has also been described by the Standish Group Chaos Report 2015 and 2020 under section Chaos resolution by Agile versus Waterfall.

Another observation for such large organizations having a geographically distributed delivery model, which, came with a list of challenges Paasivaara, M., Durasiewicz, S., & Lassenius, C. (2008) for example, communication problems Berczuk, S. (2007), lack of close physical proximity Agerfalk, P., & Fitzgerald, B. (2006), lack of team cohesion Ramesh, B., Cao, L., Mohan, K., & Xu, P. (2006), lack of shared context and knowledge Yap, M. (2005), and unavailability of team members Yap, M. (2005). Scott Ambler + Associates, in their 2014 survey found that 38% of teams were located geographically apart and had just shown a success rate of around 52% and the failure rate of around 54%. On the contrary, the success ratio was as high as 60% for co-located teams working in an agile development model with a failure rate of 25%. Despite such known challenges along with statistics, organizations still intended to implement the distributed agile software development model by doing huge investment in coaching, training, and setting a different infrastructural setup for better productivity yet, the issues/challenges seem to persist and can be observed in the retrospective session of agile teams were commonly addressed concerns for lower velocity were misunderstanding, miscommunications, unable to get support quickly, etc.

Another observation was the weak guideline in the agile manifesto and ambiguities within agile maturity models. At one end, where agile manifesto failed to give any direction based on the type of organizations or type of projects (small, medium, or large) and on the other end it contradicted other agile methodologies. For example, principle 6 of the agile manifesto illustrated the importance of face-face communication for working efficiently as an agile organization, thus challenging and contradicting the core principle of distributed agile methodology which promoted working at distinct locations to achieve the goal.

Another phenomenon that was observed in one of the leading banks within The Netherlands was the introduction of the various methodology in quick succession, unlike the traditional approach which had been there for decades. Such frequent implementation impacted team retention as people felt less motivated due to such frequent change in working culture and was easily visible in various associate survey reports which depicted unpleasant associate satisfaction index for the organization. It thus remains unclear on the need and objective for such experiments. Such trends of experimenting with agile methodology implementation within the organization also involved huge IT investment, thus impacting the financial aspect of an organization and team performance as the team needed to get accustomed to the new methodology which required time and effort.

Thus, it became very interesting to see the correlation between the distributed agile team and team performance and try exploring the problem, *"Distributed agile team can lead to team performance problems and thereby causing a negative impact on software development."*

RESEARCH QUESTION

The topic of the research focused on the problem statement, *"distributed agile team led to team performance problems and thereby causing negative impact to software development."*

The focus was to check the validity of this problem statement within one of the leading banks within The Netherlands. To do the initial assessment, preliminary interviews were done to get an opinion and direction on the research and gathered insights that could benefit the organization. Thus, after getting a direction from management and doing a preliminary examination of existing research and literature, the research was conducted and aimed at answering the research question (MRQ):

What is the impact of distributed agile software development on team performance?

As team performance is a subjective entity, the conclusion of the research was done with the help of a hypothesis by comparing the team performance of co-located agile teams and distributed agile teams within the Dutch Bank. This has further been extensively described in subsequent sections.

The conceptual model for the research has been depicted in the Figure 1 depicting the independent and the dependent variable

Scott Ambler + Associates, in their 2014 survey also found that 61% of agile teams were geographically distributed and had a return on investment scale at 3.2/10, quality at 4.4/10, delivery time at 2.6/10, stakeholder satisfaction at 4.8/10, and team morale at 4.8/10. These low figures did indicate a potential impact on team performance due to the distributed agile software model. Thus, researching validating the impact of distributed agile software development on team performance was very relevant for the organization working in a distributed agile software model.

Figure 1. Conceptual Model

Independent variable	Dependent variable
Distributed Agile Software Development	Team Performance

CONCEPTUAL MODEL

As the conceptual model used two distinct variables i.e. Distributed Agile Software Development and Team Performance, it was relevant to have an understanding of these variables to understand the research question, which has been described in subsequent sections.

Independent Variable: Distributed Agile Software Development

As Distributed Agile Software Development is an aggregation of Agile Software Development and Distributed Software Development, it became relevant to get a quick overview of these two terminologies to understand the terminology i.e. Distributed Agile Software Development which was the key parameter in the research question. This section tries to illustrate what has already been explored and documented in the literature.

AGILE SOFTWARE DEVELOPMENT

Agile manifesto[1] defined Agile as a better way of developing software by doing it and helping others do it and emphasizes on below goals:

Individuals and interactions over processes and tools
Working software over comprehensive documentation
Customer collaboration over contract negotiation
Responding to change over following a plan.

Agile Software Development Methodologies and Practices (Williams, L. (2010)), stated that the concept of agile was created and evolved in the mid-1990s. It emerged as an attempt to more formally and explicitly embrace higher rates of change in software requirements and customer expectations.

Agile encompassed various methodologies, including Adaptive Software Development (ASD) (Highsmith, J. (1997)), Agile Unified Process (AUP) (Ambler, S. (2006)), Crystal Methods (Cockburn, A. (2004)), Dynamic Systems Development Methodology (DSDM) (Tuffs, D., Stapleton, J., West, D. and Eason, Z. (1999)), eXtreme Programming (XP) (Anderson, A., Beattie, R., and Beck, K. (1998)), Feature Driven Development (FDD) (Coad, P., de Luca, J. and Lefebvre, E. (1999)), Kanban (Ladas, C. (2009)), Lean Software Development (Poppendieck, M. and Poppendieck, T. (2003)), Scrum (Schwaber, K. (1996)).

The agile methodology was based on the "iterative enhancement" (Basili, V.R. and Turner, A.J. (1975)) technique (Cohen, D., Lindvall, M. and Costa, P. (2004)). As an iteration-based methodology, each iteration in the agile methodology represented a small scale and self-contained Software Development Life Cycle (SDLC) by itself (Williams, L. (2010)). Unlike the Spiral model (Boehm, B. (1986)), agile methods assumed simplicity in all practices (Cohen, D., Lindvall, M. and Costa, P. (2004)).

Figure 2 [REMOVED REF FIELD]depicts a representation of the agile software development process (Lyoko, G., Phiri, J., & Phiri, A. (2016)) and comprises all phases of the same.

Figure 2. Agile Software Development (Lyoko, G., Phiri, J., & Phiri, A. (2016))

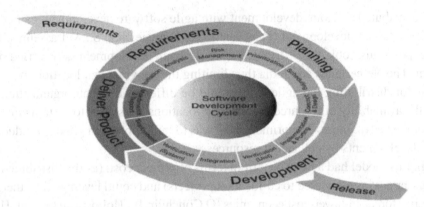

Distributed Software Development

A distributed software development or global software development (GSD) (Woodward, E., Surdek, S., & Ganis, M. (2010), Stadler, M., Vallon, R., Pazderka, M., & Grechenig, T. (2019)) comprised of software development teams, whose team members were sited in different, geographically dispersed, locations. Next to geographical separation, there were cultural, temporal, and configurational dimensions to consider.

Geographical Distance could be measured directionally and was described as "the effort required for one actor to visit another at the latter's home site" (Ågerfalk, P. J., Fitzgerald, B., Holmström Olsson, H., & Ó Conchúir, E. (2005)). The Allen-Curve stated that "the probability that people in a given organization would communicate with each other declined precipitously the farther away from each other they are situated and reached an asymptotic level at about 50 meters". This is not only valid for face-to-face communication, but the data from the study also showed a decline in the usage of all communication media with increasing distance (Allen, T. J., Henn, G. (2006)).

Socio-cultural Distance was one of the most complexes and therefore least understood dimensions of distance, but a critical element in a distributed setting (Casey, V. (2010)). It involved national as well as organizational culture, work ethic, and motivation, as well as spoken languages. Two individuals with different national backgrounds may share a common organizational culture and therefore have a low socio-cultural distance. A low socio-cultural distance lowered risks and increased communication (Ågerfalk, P. J., Fitzgerald, B., Holmström Olsson, H., & Ó Conchúir, E. (2005)). In the scope of the same company, sites still often differed in various aspects like informal habits, practices, or processes (Herbsleb, J. D. (2007)).

Configurational Dimension was defined by O'Leary and Cummings as "the arrangement of members across sites independent of the spatial and temporal distances among them (O'Leary, & Cummings.

(2007)). Especially situations with a concentrated core team and isolated members on different sites decrease awareness towards remote colleagues. A larger number of sites also boosts complexity for coordination and raises conflict potential (O'Leary, & Cummings. (2007)).

Temporal Distance, in terms of different time zones, impeded synchronous communication between individuals and therefore hurts communication (Carmel, E., & Agarwal, R. (2001)).

Distributed Agile Software Development

The pairing of distributed software development with agile software development can be referred to as distributed agile software development (Paasivaara, M., Durasiewicz, S., & Lassenius, C. (2008)). In terms of the scope of this book, the distributed agile software development was carried out in multiple locations within The Netherlands and India thus limiting the geographical location to two continents.

The reasons for distributing the agile team would be different for each organization because they might include the availability of resources in different locations, closeness to certain clusters, proximity to customers, or cost advantages. The primary goal was to build up worth goods at reduced prices than the co-located developments by enhancing resources.

Even though this model had a huge benefit in terms of project cost, (as the distributed model generally involved lesser cost as compared to co-location projects) and could leverage the emergence of large multi-skilled labor forces in lower-cost economies (Ó Conchúir, E., Holmström Olsson, H., Ågerfalk, P. J., & Fitzgerald, B. (2009)), this model primarily contradicted with the 6th principle of the agile manifesto[2] which stated:

The most efficient and effective method of conveying information to and within a development team is face-to-face conversation.

Thus, apart from the mentioned contradiction, the model also inherited challenges from distributed software development models (Carmel, E., & Agarwal, R. (2001)) and introduced a new set of challenges (Software Engineering Research and Practice, (2015)).

Dependent Variable: Team Performance

A team could be deðned as a social system of three or more people, which is embedded in an organization, whose members perceived themselves as such and are perceived as members by others(identity), and who collaborated on a common task (team-work) (Alderfer (1987), Hackman (1987), Wiendieck (1992), Guzzo and Shea (1992)). They are usually organized hierarchically and sometimes dispersed geographically, they must integrate, synthesize, and share information; and they needed to coordinate and cooperate as task demands shift throughout a performance episode to accomplish their mission (Salas et al., (2008)).

Teamwork on the other hand was deðned as the interdependent components of performance required to effectively coordinate the performance of multiple individuals (Salas et al., (2008)).

Team performance is conceptualized as a multilevel process (and not a product) arising as team members engage in managing their individual and team-level task work and teamwork processes (Kozlowski & Klein, (2000)). Team performance and team effectiveness (Singh, A. K., & Muncherji, N. (2007)) were often used synonymously in the literature; sometimes team performance was part of team effectiveness

(Cohen and Bailey (1997)), and sometimes team effectiveness was part of team performance (Hoegl, M., & Gemuenden, H. G. (2001)). Most of the models of team performance (or team effectiveness) originated from management science and psychology (Salas et al., (2008)).

By conducting this research with the outlined research question in mind, it would be possible to identify the impacts of a distributed agile team along with their associated characteristics and parameters on team performance.

Now that the conceptual model had been described briefly, the research question also mentioned the probable "IMPACT" of the two defined variables present in the conceptual model. To evaluate the impact, the necessity and understanding of the maturity model was needed which has further been described.

Maturity Models

Maturity could be regarded "as a measure to evaluate the capabilities of an organization regarding a certain discipline" (Rosemann & De Bruin (2005)). Wikipedia[3] stated the Maturity Model as "Maturity is a measurement of the ability of an organization for continuous improvement in a particular discipline. The higher the maturity, the higher will be the chances that incidents or errors will lead to improvements either in the quality or in the use of the resources of the discipline as implemented by the organization."

Maturity models are conceptual models that outline anticipated, typical, logical, and desired evolution path towards maturity (Becker & Knackstedt & Poeppelbuss (2009)). Maturity models could be understood as reference models (Herbsleb & Zubrow & Goldenson& Hayes & Paulk (1997)) and could be used to assess the as-is situation of an organization, to derive and prioritize improvement measures in terms of evolutionary stages (Iversen et al. (1999), Gottschalk (2009), Kazanjian & Drazin (1989)). These distinctive stages provided a roadmap for improvement where each stage was superior to the previous stage (Mehta, Oswald, Mehta (2007), Subba Rao, Metts, Mora Monge (2003)). This advancement on the evolution path meant a step by step progression on team performance (Katzenbach and Smith, (1993)).

Target Audience

There would be several who could have an interest in this research, the target audiences for this research would be organizations working in distributed agile software development model and aiming to improve team performance.

REFERENCES

Ågerfalk, P., & Fitzgerald, B. (2006). Introduction. *Communications of the ACM, 49*(10), 26–34. doi:10.1145/1164394.1164416

Ågerfalk, P., & Fitzgerald, B. (2006). Introduction. *Communications of the ACM, 49*(10), 26–34. doi:10.1145/1164394.1164416

Ågerfalk, P. J., Fitzgerald, B., Holmström Olsson, H., & Ó Conchúir, E. (n.d.). Benefits of global software development: The known and unknown. *Making Globally Distributed Software Development a Success Story*, 1-9.

Allen, T. J., & Henn, G. (2006). *The Organization and Architecture of Innovation: Managing the Flow of Technology* (1st ed.). Butterworth-Heinemann.

Berczuk, S. (2007). Back to basics: The role of agile principles in success with a distributed scrum team. *Proceedings of AGILE*, 382-388. 10.1109/AGILE.2007.17

Carmel, E., & Agarwal, R. (2001). Tactical approaches for alleviating distance in global software development. *IEEE Software, 18*(2), 22–29. doi:10.1109/52.914734

Casey, V. (2010). Imparting the importance of culture to global software development. *ACM Inroads, 1*(3), 51–57. doi:10.1145/1835428.1835443

Cohen, S.G., & Bailey, D.E. (1997). What Makes Teams Work: Group Effectiveness Research from the Shop Floor to the Executive Suite. *Journal of Management, 23*, 239-290. doi:10.1177/014920639702300303

Herbsleb, J. D. (2007). Global software engineering: The future of socio-technical coordination. *Future of Software Engineering (FOSE'07).*

Hoegl, M., & Gemuenden, H. G. (2001). Teamwork quality and the success of innovative projects: A theoretical concept and empirical evidence. *Organization Science, 12*(4), 435–449. doi:10.1287/orsc.12.4.435.10635

Hoegl, M., & Gemuenden, H. G. (2001). Teamwork quality and the success of innovative projects: A theoretical concept and empirical evidence. *Organization Science, 12*(4), 435–449. doi:10.1287/orsc.12.4.435.10635

Katzenbach & Smith. (1993). *The Wisdom of Teams.* Creating the High-performance Organization.

Kozlowski & Klein, (2000). *A multilevel approach to theory and research in organizations: Contextual, temporal, and emergent processes.* Academic Press.

Lindsjørn, Y., Sjøberg, D. I., Dingsøyr, T., Bergersen, G. R., & Dybå, T. (2016). Teamwork quality and project success in software development: A survey of Agile development teams. *Journal of Systems and Software, 122*, 274–286. doi:10.1016/j.jss.2016.09.028

Ó Conchúir, E., Holmström Olsson, H., Ågerfalk, P. J., & Fitzgerald, B. (2009). Benefits of global software development: Exploring the unexplored. *Software Process Improvement and Practice, 14*(4), 201–212. doi:10.1002pip.417

O'Leary, & Cummings. (2007). The spatial, temporal, and configurational characteristics of geographic dispersion in teams. *Management Information Systems Quarterly, 31*(3), 433. doi:10.2307/25148802

Paasivaara, M., Durasiewicz, S., & Lassenius, C. (2008). Distributed Agile development: Using Scrum in a large project. *2008 IEEE International Conference on Global Software Engineering.* 10.1109/ICGSE.2008.38

Paasivaara, M., Durasiewicz, S., & Lassenius, C. (2008). Distributed Agile development: Using Scrum in a large project. *2008 IEEE International Conference on Global Software Engineering.* 10.1109/ICGSE.2008.38

Ramesh, B., Cao, L., Mohan, K., & Xu, P. (2006). Can distributed software development be agile? *Communications of the ACM, 49*(10), 41–46. doi:10.1145/1164394.1164418

Stadler, M., Vallon, R., Pazderka, M., & Grechenig, T. (2019). *Agile distributed software development in nine central European teams: Challenges, benefits, and recommendations.* Academic Press.

Yap, M. (2005). Follow the sun: distributed extreme programming development. *Proceedings of the Agile conference*, 218-224. 10.1109/ADC.2005.26

ENDNOTES

[1]   See https://agilemanifesto.org/
[2]   See https://agilemanifesto.org/principles.html
[3]   See https://en.wikipedia.org/wiki/Maturity_model

Chapter 5
Research Findings in the Domain of Agile Methodologies

Tapan Kumar
Cognizant, The Netherlands

ABSTRACT

This chapter focuses on examining the impact of distributed agile software development on team performance. In order to get the answer, literature is examined to get necessary information associated with distributed agile software development and team performance along with the models/tools/frameworks, which were used to evaluate the team performance of agile teams. The chapter will then explain how the case study research design was used to get statistics using the data collection method and how the data has been analysed to derive any sort of conclusion and help in answering the research question in more quantified way. The chapter also elaborates with the help of literature analysis the characteristics, existing challenges, success factors, and practices associated with distributed agile software development, team performance, and existing maturity models. This chapter aims to provide insight into the core practices teams can apply when working on digital platforms.

INTRODUCTION

This Chapter focuses on the research approach done to get the answer to the research question "What is the impact of distributed agile software development on team performance?". In order to get the answer, the chapter as per author highlights the literature strategy being used to get necessary information associated with Distributed Agile software development and Team performance along with the models/tools/frameworks, which were used to evaluate the Team performance of Agile teams for this research.

The chapter will then explain how the case study research design was used to get statistics using the data collection method and how the data has been analysed to derive any sort of conclusion and help in answering the research question.

The chapter also elaborates with the help of literature the characteristics, existing challenges, success factors etc. associated with Distributed agile software development, Team performance and existing Maturity models.

DOI: 10.4018/978-1-7998-7367-9.ch005

Research Approach

The research fell under category applied research as the intention was to explore the impact of the distributed agile software development on team performance and gain a deeper understanding of the topic. The first choice of conducting this applied research was by using the design science research method. The goal of design science was to deliver an artifact as a practical contribution to science (Peffers, Tuunanen, Rothenberger, & Chatterjee, (2008)). However, it was not very clear what the impact of distributed agile software development could have on team performance, hence the idea of an artifact creation (Hevner, 2004) was very unlikely to be realized and might not have been useful as organizations would prefer to know improvements in existing approach rather than practicing new artefact.Thus, the choice of using the design science research method was not appropriate for this research.

The choice was made to use a qualitative research method and to perform an exploratory case study, where the co-relation of distributed agile software development and team performance was explored (Recker, (2013)). With the help of a literature study, the features and characteristics associated with distributed agile software development methodology were examined along with the factors, that collectively defined team performance (Dingsoyr, T., Faegri, T. E., Dyba, T., Haugset, B., & Lindsjorn, Y. (2016)). In the sequel to this, the research also explored the possible impacts of distributed agile software development on team performance and thus evaluated the outcome by testing the results across one of the leading bank within The Netherlands.

A set of predefined questionnaires from existing assessment models were shortlisted based on a detailed analysis of 87 assessment models and frameworks. This assessment were carried out with the inputs from agile coaches of one of the leading bank within The Netherlands, understanding the organizational culture and its way of working. These set of questionnaire helped in assessing the team performance and thereby understanding the impact of distributed agile framework features on the team performance parameters. The results had also been cross-validated with the agile team as well as with the management of the Leading bank within The Netherlands to draft the conclusion of this research.

Figure 1. Exploratory case study (Recker, (2013))

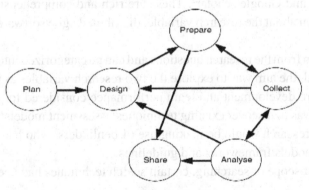

Using this approach, two different analysis on the same case were performed, framing this exploratory research further to an embedded single-case study as seen in Figure 1.

First, a plan was drafted to approach this research. Thereafter the design phase focused on doing the literature research and extracting information needed to conduct the case study. As part of the preparation phase, questionnaires for conducting interviews with expert panels were selected. The experts were given all necessary background about the objective of research and sufficient time to provide their answers. The questionnaires for agile teams were also prepared and circulated to get their set of answers necessary for answering the research question. The results were then collected and analyzed to evaluate the validity of the research question and provided necessary recommendations as part of practices inferred as an outcome of the research.

Literature Strategy

The literature strategy was based on two variables of the research: distributed agile software development and team performance. The literature study had been set up based on Systematic Literature Review (SLR) which, according to Shahin, Babar, and Zhu (2017), is the most used research method in "Evidence-Based Software Engineering". According to the researchers, SLR aims to provide a well-defined process for identifying, evaluating, and interpreting all available data that is relevant to a specific research question or topic.

To answer the main research question, relevant peer-reviewed publications were searched for. The author also searched for relevant literature to scientifically substantiate the introduction and problem definition in the preliminary research where possible. During the literature search, systematic search had been mainly used by searching search engines based on combined search terms and the snowball method by searching articles that were used as a reference in the articles found.

Because most publications and the terms themselves are in English, the search was mainly done with English search terms. EndNote was used to sort and categorize the articles found. To reduce the search results found to a manageable selection, the search process in Figure 2 below had been followed (Budgen, D., Burn, A. J., Brereton, O. P., Kitchenham, B. A., & Pretorius, R. (2010)).

Phase 1: In this phase, the aim was to find a comprehensive and unbiased collection of primary studies from the literature related to the research questions. For the electronic search, major databases were leveraged that had been widely used for software engineerings (SE) research such as the University of Antwerp (UA) Library[1] and Google scholar[2]. These are rich and comprehensive databases containing bibliographic information about the research variable, distributed agile software development, and team performance.

The search terms stem from the research questions and can be categorized into 2 dimensions, as shown in Figure 3. On one hand, the aim was to explore the two research variables as part of 2 dimensions i.e. distributed agile software development and team performance considered for the primary studies. On the other hand, the aim was to discover existing techniques/assessment models/frameworks to test these research variables. The research would be making use of benlinders[3] who had consolidated more than 80 of such assessment models/frameworks and guidelines.

To ensure a sufficient scope of searching, certain search techniques had been used such as;

- Boolean logic for combined search or exclusion with AND, OR, NOT
- Exact phrase, to enable searching by exact word combination
- Field-specific, search by title and/or summary

Figure 2. Stages of Systematic Literature review

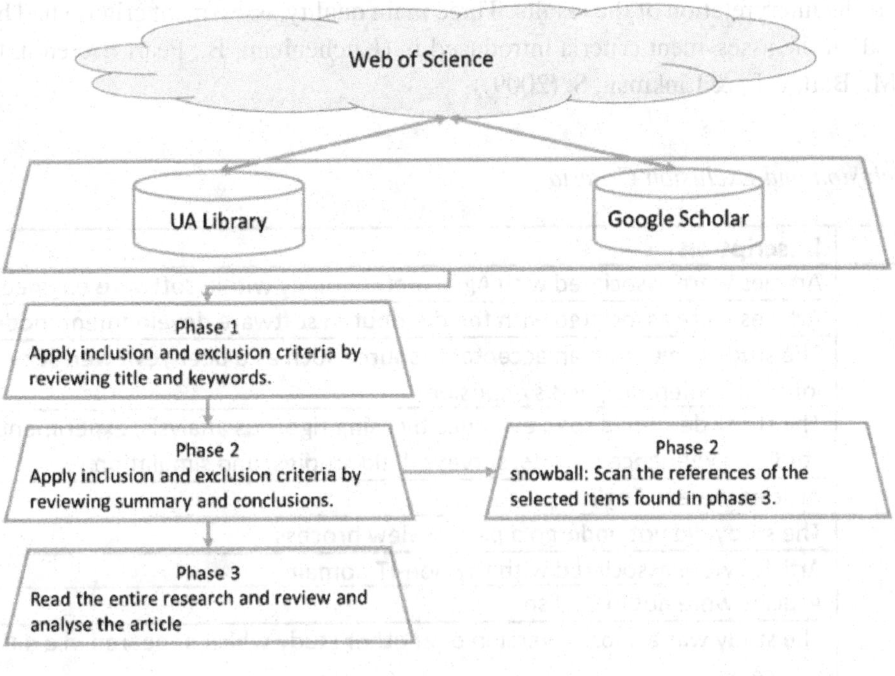

Phase 2: In this phase, the aim was to implement inclusion and exclusion criteria to identify the suitability of primary studies and making decisions for inclusion or exclusion of an article in the SLR based on the addressed research questions. The inclusion and exclusion criteria are shown in Figure 4.

In this phase, the aim was also to check the references of the filtered literature selected as an outcome of phase 2 and reassess them and let it undergo inclusion and exclusion criteria implementation and identify the suitable literature for this research.

Phase 3: In this phase, all the selected non-duplicate articles as an outcome of phase 2 were analyzed and synthesized such that, they were suitable and sufficient for answering the main research question.

For executing phase 2 and phase 3, the quality assessment criteria were being used to determine the rigorousness and credibility of the used research methods and the relevance of the studies. This assess-

Figure 3. Literature Search Dimensions and Keywords

Dimension	Search Keywords
Distributed agile software development	"distributed agile software development", timeline 2001 onwards, Peer-Reviewed, agile software development, distributed software development, agile development, software development, scrum, agile
Team performance	"team performance", timeline 2001 onwards, team performance, computer software development, agile software development, "team performance in software development"

ment was important to limit bias in conducting this SLR, to obtain insight into potential differences, and to support the interpretation of the results. Three main quality assessment criteria had been applied that was based on the assessment criteria introduced in (Kitchenham, B., Pearl Brereton, O., Budgen, D., Turner, M., Bailey, J., & Linkman, S. (2009)).

Figure 4. Inclusion and Exclusion Criteria

Type	Description
Inclusion	Articles were associated with Agile methodology within software engineering
	Articles were associated with the distributed software development model
	The study came from an acceptable source such as a peer-reviewed scientific journal, conference, and symposium
	The study described solid evidence by using rigorous analysis, experiments, case studies, experience reports, surveys, field studies, and simulation
	Articles were in English
Exclusion	The study did not undergo a peer-review process
	Articles were associated with the non-IT domain
	Articles were not in English
	The study was a shorter version of another study which appeared in a different source

The 3 quality assessment criteria were:

- **Relevance**: As part of this assessment criteria, the focus was to accumulate only those studies which had enough information to answer the research question. This quality assessment had been done according to the inclusion and exclusion criteria.
- **Coverage**: As part of this assessment criteria, all the selected articles were reviewed in iterations so that no relevant studies were missed. To achieve that, a search was performed on the entire list of relevant studies following the screening of the titles, abstracts, keywords, and conclusion. Moreover, a snowballing process was conducted to broaden the scope of selected studies
- **Validation**: As part of this assessment criteria, an analysis of the searched studies was done and checked if the primary studies contained the necessary information to answer the main research question.

Having done the literature strategy to gain relevant insight and information already present in existing literature, it was important to draft the theoretical model for this research and to enlist all factors playing vital role in getting answer to the research question.

Theoretical Model

The theoretical model depicted in Figure 5 illustrates the research question i.e. exploring the impacts of distributed agile software development on team performance.

Here, levels (also known as observed variables) defined the factors which played a vital role in determining how the team performance (depicted under dependent variable) would be impacted in case of distributed agile software development (mentioned under independent variable). Here measures which is also referred as observed variables illustrated few observations which were the prime reason to try getting answer to the main research question.

Figure 5. Theoretical model

Levels	Independent variable	Dependent variable	Measures
Multiple geographical location Cultural diversity Experience diversity	Distributed Agile Software Development	Team Performance	✓ Low Velocity ✓ Inability to finish requirements in committed timeframe ✓ Lower IT retention ✓ Miscommunication and misunderstanding within team

RESEARCH DESIGN

As described by (Robson, 1993), a case study was defined as a strategy of doing research which involves an empirical investigation of a particular contemporary phenomenon within its real-life context using multiple sources of evidence.

Case Study Research

To answer the research question, exploratory case study (Recker, (2013)) was performed and applied in one of the domains within a leading bank within The Netherlands. This research design helped in gaining further insight into the impacts of distributed agile software development on team performance.

The next step in the research design was to identify the type of case study design and its corresponding unit of analysis (Recker, (2013)) as depicted in Figure 6.

For this research a single-case, a holistic approach was most appropriate to design the case study. The choice of a single case had been taken since the execution of this case study was limited to the just one bank. In addition to this, it was interesting to see all variants within the bank's domain concerning the distributed agile team. Thus, the choice of using a holistic approach had been selected to see if there was any impact of a distributed agile team on team performance by comparing various aspects while working in a distributed software development model.

Figure 6. Case Study Design Recker (2013)

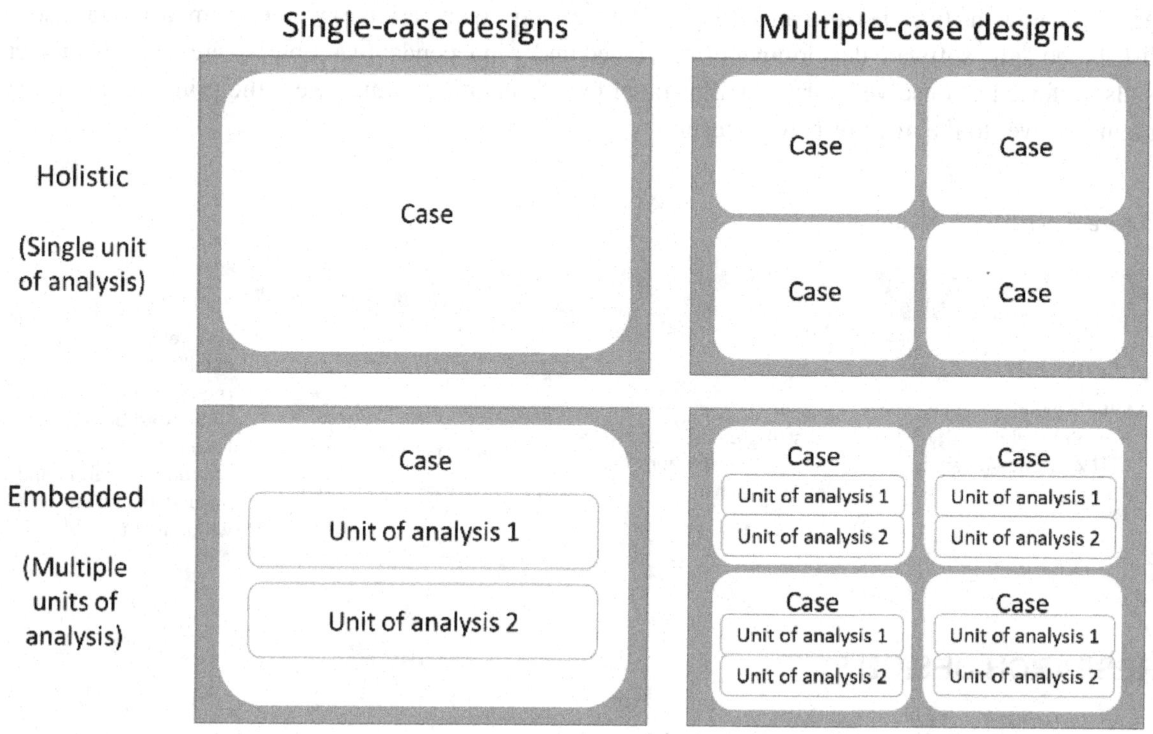

EXPLORATORY RESEARCH

Exploratory Research is research conducted for a problem that has not been studied more clearly, establishes priorities, develops operational definitions, and improves the final research design (Shields & Rangarajan, 2013). In this case, it was important to explore the correlation between distributed agile software development and team performance. For this research, in addition to literature, data was collected with the help of interviews with team management, questionnaires answered by agile team members, documentation, and observation of several agile teams (as an observer and also as a participant).

Data Collection

As the teams of the Dutch Bank were located in 2 geographic location, with the management team being located in the Netherlands and the agile teams distributed across 2 geographical location i.e. India and Netherlands, it was very convenient to have interviews with management located in the Netherlands and have the questionnaires answered with the help of online surveys. In addition to the data collected via interviews and questionnaires, observation and documentation were the other data collection techniques used for my research.

Figure 7. Data Collection (Documentation)

Document type	Description
Burndown chart	It showed the ideal progress of a release or a sprint from start to finish compared with the actual daily progress
Velocity Graph	A velocity chart showed the sum of estimates of the work delivered across all iterations
Confluence Page	This Atlassian's content collaboration tool helped to know about various agile teams and their composition of team members. It also comprised of the overall hierarchy of team structure within the domain
JIRA Board	The board would depict apart from the user stories[4]. All the issues highlighted during retrospective were also enlisted which acted as a hurdle for the team.

Interviews

The qualitative research interview for my research relied on purposive sampling, where the participants who were part of the interview process as well as for responding to online surveys were already working in the distributed agile software model and had in-depth and detailed information about the same. As the objective of this data collection technique was to define questions, propose new theory constructs, and/or build new theories (Recker, (2013)), the choice was to use both unstructured and semi-structured exploratory interviews.

During the start of the research, unstructured interviews were planned with agile coaches and delivery managers IT to understand various aspects associated with distributed agile software development within the organization along with in-depth information around the topic like associated training, processes and practices proposed and followed within the organization. After the unstructured interviews, semi-structured interviews were planned with the management involving pre-defined questionnaires. These questionnaires helped in calculating the maturity assessment of the agile teams which in turn translated to the team performance as per the perception of the management.

In addition to conducting the interviews with management, data was collected with the help of online surveys which comprised questionnaires from existing agile assessment models. These questionnaires, which needed to be answered by the agile team members helped in understanding the facts associated with the maturity of the team from team's perception, processes, and practices followed by team and some vital information on how the distributed agile software model has been implemented within the organization.

Documentation and Observation

The other data collection technique used in the research were documentation and observation. At one end where documentation helped in getting the facts associated with team functioning as mentioned in Figure 7.

On the other hand, there were few observations which also clearly showed existing challenges within the distributed agile software development teams and was evident as an action item during the retrospective session within Jira Board and has been enlisted in Figure 8 which mentions the observed parameters and the participants for the observation.

Figure 8. Data Collection (Observation)

Observation	Participants
Miscommunication, misunderstanding	Agile team and individual
Team coordination	Agile Team
Openness during the retrospective session	Individual
Supporting each other	Individual
Accountability	Agile team
Constructive feedback	Individual

Scoring

In the research, the data collection method such as interviews, online surveys, documentation, and observation were scored using Likert scale (Joshi, A., Kale, S., Chandel, S., & Pal, D. (2015)) to depict the maturity of the agile team. The overall score of individual participants were then be used in the data analysis phase to draw conclusions.

Data Analysis

As the research was qualitative, data collection and data analysis were interwoven or even dependant on each other (Recker, (2013)). The research question was tested with the help of existing agile assessment models[4].

For the selection of the assessment model used for testing, coding technique was used to categorize evaluated assessment models into frameworks, guideline and assessment tools, thereafter mapping the results with the bank's culture and way of working and thereafter selected the frameworks suitable for the case study to deduce the research question.

As the research question focused on the impact of distributed agile software development on team performance, which could be analyzed efficiently using statistical methods (Eisele, P. (2015)). Thus, for the research, the data analysis included both descriptive statistics (Spriestersbach, A., Röhrig, B., Prel, J. D., Gerhold-Ay, A., & Blettner, M. (2009)) and inferential statistics (Vieira, E. T. (2017)). At one end, where descriptive statistics was used to calculate the maturity of the teams working in distributed agile software development model as well as co-located agile software development model within the bank. At the other hand inferential statistics was used to answer the main research question with the help of a hypothesis by comparing the team performance of co-located agile teams and distributed agile teams.

Descriptive Statistics

Descriptive statistics (Spriestersbach, A., Röhrig, B., Prel, J. D., Gerhold-Ay, A., & Blettner, M. (2009)) are brief descriptive coefficients that summarizes a given data set, which could be either a representation of the entire or a sample of a population. Descriptive statistics are broken down into measures of central tendency and measures of variability (spread). Measures of central tendency includes the mean, median, and mode, while measures of variability include the standard deviation[5], variance[6], the minimum, and maximum variables, and the kurtosis[7] and skewness[8].

Inferential Statistics

Inferential statistics (Vieira, E. T. (2017)) are often used to compare the differences between the treatment groups. Inferential statistics use measurements from the sample of subjects in the experiment to compare the treatment groups and make generalizations about the larger population of subjects. In inferential statistics, data are analyzed from a sample to make inferences in the larger collection of the population. The purpose is to answer or test the hypotheses. A hypothesis (plural hypotheses) is a proposed explanation for a phenomenon. Hypothesis tests are thus procedures for making rational decisions about the reality of observed effects.

In inferential statistics, the term 'null hypothesis' (H0 'H-naught,' 'H-null') denotes that there is no relationship (difference) between the population variables in question (Travers, J. C., Cook, B. G., & Cook, L. (2017)). The alternative hypothesis (H1 and Ha) denotes that a statement between the variables is expected to be true (Travers, J. C., Cook, B. G., & Cook, L. (2017)).

As the data for the research were collected using interviews, online questionnaires, documentation, and observation to validate the research question, the statistical test involved 2 separate groups i.e. distributed agile software development team and co-located software development team. Based on the outcome of the data collection phase, a parametric hypothesis test was performed (Sedgwick, P. (2015)) to identify the difference in maturity assessment of distributed and co-located software development agile team.

RESEARCH FINDINGS

As the research intended to establish a direct relation between Distributed Agile software development and team performance, this section mentioned the findings observed in various literature and explored and enlisted the existing gaps and challenges associated with the dependent and independent variables i.e. Distributed Agile Software Development and Team Performance thus, helping the author in answering the research question.

Figure 9. Success factor for Agile Software Development

Dimension	Main Success Factors	Sub Success Factors
Organizational	Corporate Culture	Support from top management
		Team Environment
People	User Involvement	Handling commercial pressures
		Stakeholder politics
	Team Capability	Effective project management skills
		Ability to handle the project's complexity
		Decision time
		Effective communication and feedback
Process	Project Management Process	The minimum change in requirement
		Simplicity in process
		Good reporting of project status
	Project Definition Process	Risk management
		Time allocation
		Accurate estimates of project resources
	Active Testing	Code review
Project	Clear Objectives and Goals	Project type
		Project nature
	Realistic schedule	
	Realistic budget	Team distribution
		Team size
	Clear Requirements and Specifications	
Technical	Selecting the Proper Agile Method	Configuring the necessary tools and infrastructure
	Using Advanced Technology	Familiarity with technology

Figure 10. Impact of Distance (Carmel, E., & Agarwal, R. (2001))

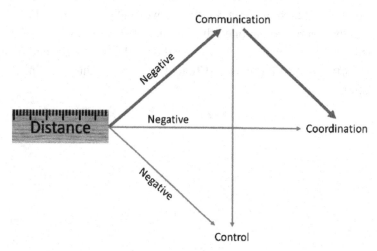

Distributed Agile Software Development

Success Factor with Agile Software Development

The Critical Success Factor (CSF) approached for identifying and measuring an organization's performance was first developed by (Rockart (1979)) and later on refined and became well- established (Chow, T., Cao, D., (1981)). CSF was defined by (Bullen and Rockart (1981)) as "the limited number of areas in which satisfactory results will ensure successful competitive performance for the individual, department, or organization."

Figure 9 depicted the list of all the success factors for agile software development along with its classification into 5 dimensions; organizational, process, project, people, and technical (RamadanDarwish, N., & M. Rizk, N. (2015)).

Practice: *It can thus be concluded that to have success in agile software development, the success factors associated with the 5 dimensions must then be addressed by the agile teams, and the management.*

Challenges with Distributed Agile Software Development

Figure 10 depicted the impact of distance (Carmel, E., & Agarwal, R. (2001)) and its co-relation between communication, control, and coordination.

Here Coordination was the challenge of "managing interdependencies, uncertainties and equivocality, conflicts, technology representations, and their interrelations" (Wiredu, G. O. (2006)). The control aspect was split into two categories: formal and informal control (Moore, J. E., Williams, C. K., & Sumner, M. (2012)). The self-organization of agile teams taps into the intrinsic motivation of team members and was a form of informal control. Communication, the coupling factor between coordination and control, was the exchange of information between a sender and a receiver to reach a mutual understanding. Figure 10 showed that distance directly impeded coordination and control as well as indirectly through its restraining impact on communication, while the bold arrows between distance-communication-coordination, "represented the main challenge of distributed software development".

On the other hand, Software Engineering Research and Practice, (2015) enlisted 13 challenges associated with the Distributed Agile software development. The same had also been referred to in various other journals (Hildenbrand, T., Geisser, M., Kude, T., Bruch, D., & Acker, T. (2008), Paasivaara, M., Durasiewicz, S., & Lassenius, C. (2008)).

Documentation

Agile teams did not give importance to the documentation (Shrivastava and Date, (2010)). This might affect the distributed teams as they would miss some details about the project and hence their understanding of the project would suffer.

Pair Programming

One of the agile development methods used pair programming in which two members of the teamwork on the same code side by side. This approach has been very challenging in a distributed environment

(Baheti, P., Gehringer, E., & Stotts, D. (2002)). Hence distributed groups might have to find some other similar methodology.

Different Work Hours

A distributed team working in distinct continents had different time zones and thus their working hours didn't match. Hence their working hours needed to be aligned so that they could communicate with each other. This helped to avoid rework and provided clarity to the project.

Communication

Reduced communication had more effects in the case of distributed teams (Coordination Theory, Malone and Crowston, (1994), Smite et al., (2010)). Most agile practices like test-driven development could be educated by providing one-on-one training. Many problems in distributed agile development were related to communication like unable to understand the customer, the system architecture, or system design. These had to be solved by participating in discussions or solving the problem manually.

Figure 11 illustrated communication channel complexity (Abilla, (2006)) i.e the team communication channel was dependent on team members in an agile team, thus the complexity was higher for large teams as compared to smaller teams. This complexity was denoted by the notation (N x (N - 1) / 2), where 'N' represented the number of team members.

Figure 11. Communication Channel Complexity (Abilla, (2006))

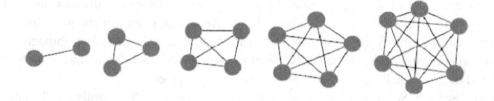

Knowledge Transition

Knowledge transition for the distributed team was highly dependent on various factors such as work hours, communication channels, documentation, etc (Espinosa, J. A., Slaughter, S. A., Kraut, R. E., & Herbsleb, J. D. (2007)). As the related factors were already an existing challenge, it made knowledge transition also a bit difficult for the teams working in different geographical locations.

Cultural Differences

Cultural issues could cause misunderstandings between team members. Several recent studies (Olson, J. S., & Olson, G. M. (2003), Abraham, L. R. (2009), Krishna, S., Sahay, S., & Walsham, G. (2004), Herbsleb, J., & Moitra, D. (2001)) had explored the cultural differences and measures to manage them in distributed teams. Some of the literature explicitly described the challenges when interacting with Indian culture. This was very relevant for this research as the selected Bank comprised of Indian IT resources.

Team Cohesion

As agile development focused on regular collaboration on all phases of the software project (Teasley, S., Covi, L., Krishnan, M., & Olson, J. (2002)), it got worse when implementing distributed agile software development (Espinosa, J. A., Slaughter, S. A., Kraut, R. E., & Herbsleb, J. D. (2007)). In the case of distributed development, members at distinctive locales were more averse to observe themselves as a major aspect of the same group when contrasted with co-placed members. The absence of togetherness, accompanied by a common view of goals, was an issue in that situation.

People vs. Process-Oriented Approaches

In co-located development, the process was people-oriented and informal methods (Casey, V., & Richardson, I. (2006)) were used to establish the control whereas distributed development needed the control to be achieved by formal methods (Shrivastava and Date, (2010)).

Knowledge Management

Knowledge management was crucial for success in co-located software development (Richardson et al., 2009). During the development process the experience of team members, decisions, methods, and skills must be gathered through knowledge sharing. However, in the distributed agile software development model, knowledge sharing was difficult due to the lack of face-to-face communication between team members (Boden and Avram, (2009), Holz and Maurer, (2003)).

Language Barriers

The language problem arose when the teams were non-collocated and hence, they were not able to understand each other due to their different languages (Shrivastava and Date, (2010), Kajko-Mattsson et al. (2010)).

Role of Specialist

Typical software organizations always had people with specialized knowledge like business analysts, testers and user interface specialists, etc. Their knowledge was expected to be utilized in the project when needed. The agile methodology didn't had any formal mechanism to request such expertise. Even if the specialists were brought, they might have faced problems similar to a new team member about gaining an understanding of the project requirements (Banerjee, U., Narasimhan, E., & Kanakalata, N. (2011)).

Figure 12. Teamwork Quality (Hoegl, M., & Gemuenden, H. G. (2001))

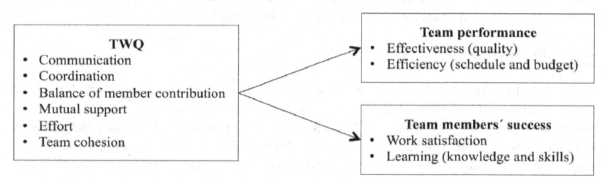

Developer Fear of Skill Deficiency Exposure

Some developers feared that agile processes could bring forward their deficiencies. Onsite customers, stand-up meetings, use of storyboards, and whiteboards brought the shortcoming of developers in front of the whole team because agile methodology involved constant communication and collaboration. Besides, continuous integration and automated testing meant that developers couldn't hide poor, low-quality code. Exposing the weaknesses of developers could lead to counterproductive (Conboy, K., Coyle, S., Wang, X., & Pikkarainen, M. (2011)).

Recruitment Challenges

It was difficult for agile companies to find the right people due to a lack of agile-specific recruitment policies. There were only a few universities or colleges that incorporated agile methods and skills into their programs. Moreover, degree programs tend to rely upon either technical or business skills but rarely involve both (Conboy, K., Coyle, S., Wang, X., & Pikkarainen, M. (2011)).

Practice: *It can thus be concluded that to have a better distributed agile software development, at one end, the 13 challenges must be acknowledged and worked upon by the organization along with the success factors associated with the 5 dimensions of agile software development to attain success in distributed agile software development.*

TEAM PERFORMANCE

A team could be deðned as a social system of three or more people, which is embedded in an organization, whose members perceived themselves as such and are perceived as members by others(identity), and who collaborated on a common task (team-work) (Alderfer (1987), Hackman (1987), Wiendieck (1992), Guzzo and Shea (1992)). They are usually organized hierarchically and sometimes dispersed geographically, they must integrate, synthesize, and share information; and they needed to coordinate and cooperate as task demands shift throughout a performance episode to accomplish their mission (Salas et al., (2008)).

Teamwork on the other hand was deðned as the interdependent components of performance required to effectively coordinated the performance of multiple individuals (Salas et al., (2008)).

Figure 13. Team performance (Dingsoyr, T., Faegri, T. E., Dyba, T., et al. (2016))

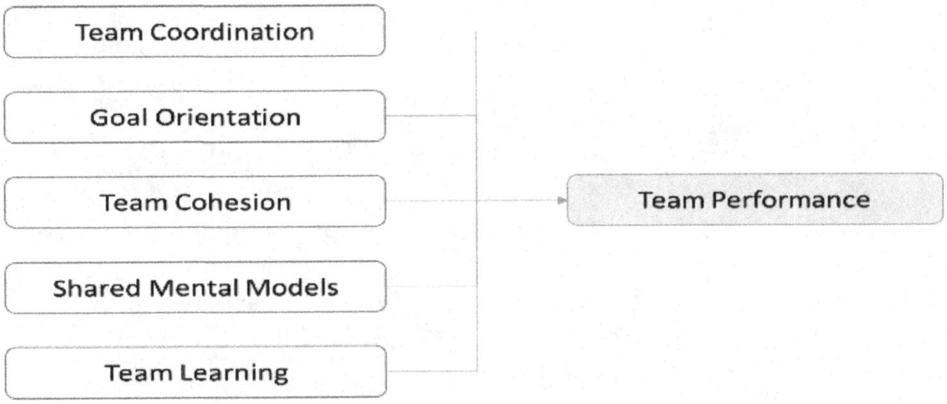

Team performance is conceptualized as a multilevel process (and not a product) arising as team members engage in managing their individual and team-level task work and teamwork processes (Kozlowski & Klein, (2000)). Team performance and team effectiveness (Singh, A. K., & Muncherji, N. (2007)) were often used synonymously in the literature; sometimes team performance was part of team effectiveness (Cohen and Bailey (1997)), and sometimes team effectiveness was part of team performance (Hoegl, M., & Gemuenden, H. G. (2001)). Most of the models of team performance (or team effectiveness) originated from management science and psychology (Salas et al., (2008)).

Team Performance Factors

Many team performances models and teamwork frameworks described teamwork quality (TWQ) and its relation to team performance in general (Hoegl, M., & Gemuenden, H. G. (2001), Mathieu et al. (2008), Cohen and Bailey (1997) and Rasmussen and Jeppesen (2006)). Teamwork quality is a measure for the quality of collaboration in teams and consisted of six facets: communication, coordination, the balance of member contributions, mutual support, effort, and cohesion. Figure 12 represented the six facets associated with teamwork quality and its positive influence and impact on team performance and personal success (Hoegl, M., & Gemuenden, H. G. (2001)).

Figure 14. Teamwork Quality vs Team Performance

Team Work Quality	Team Performance	Remarks
Communication	Shared Mental Models	Mostly matched
Coordination	Team Coordination	Matched
Balance of Member contribution	Team Learning	Partially matched
Mutual Support	Team Learning	Partially matched
Effort		Implicit characteristic
Team Cohesion	Team Cohesion	Matched
	Goal Orientation	No match

Figure 15. Tuckman's FSNP Model (Tuckman (1965))

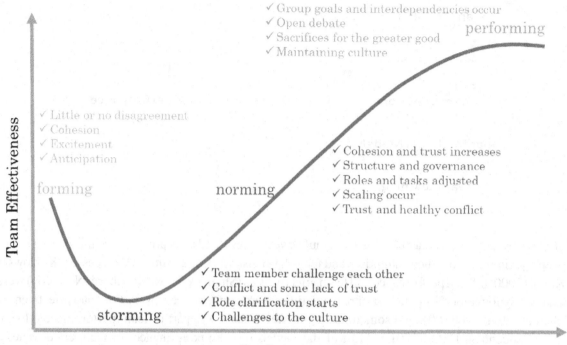

Another literature (Dingsoyr, T., Faegri, T. E., Dyba, T., Haugset, B., & Lindsjorn, Y. (2016)), showed the team performance as an accumulation of five factors as depicted in Figure 13.

As Figure 12 depicted a positive relation with team performance and Figure 13 represented the factors associated with team performance, Figure 14 represented the common factors which, as per literature contributed to deciding team performance.

It was very interesting to see that at one end team performance stressed on the goal orientation, whereas, teamwork quality didn't cover that factor to uncover the team objective which was also very critical for the team and its success (McGrath, (1984), Bunderson & Sutcliffe, (2003)).

Practice: *It can thus be concluded that there was a very strong correlation between teamwork quality and team performance, and to address and evaluate team performance all the factors must have been taken into account.*

Team Performance Model

There are various known team performance/team effectiveness models i.e. GRPI Model (Plovnick, Fry, and Rubin (1977)), Katzenbach and Smith Model (Katzenbach and Smith, (1993)), The T7 Model (Lombardo & Eichinger, (1995)), The LaFasto and Larson's Five Dynamics of Team Work (LaFasto & Larson, (2001)), The Hackman Model (Hackman (2002)), The Lencioni Model (Lencioni, P. M. (2010)), The Salas, Dickinson, Converse and Tannenbaum Model (Salas, E., Dickinson, T.L., Converse, S.A. and Tannenbaum, S.I. (1992)) and Tuckman's FSNP Model (Tuckman (1965)) etc.

Figure 16. Phases of Tuckman's FSNP Model (Tuckman (1965))

	Forming	**Storming**	**Norming**	**Performing**
General Observations	Uncertainty about roles, looking outside for guidance.	Growing confidence in the team, rejecting outside authority	Concern about being different, wanting to be part of the team	The concern with getting the job done
Content Issues	Some attempt to define the job to be done	Team members resist the task demands.	There is an open exchange of views about the team's problems	Resources are allocated efficiently; processes are in place to ensure that the final objective is achieved
Process Issues	Team members look outside for guidance and direction.	Team members deny the task and look for the reasons not to do it.	The team starts to set up the procedures to deal with the task.	The team can solve problems.
Feelings Issues	People felt anxious and are unsure of their roles. Most look to a leader or coordinator for guidance	People still felt uncertain and try to express their individuality. Concerns arise about the team hierarchy	People ignored individual differences and team members are more accepting of one another	People share a common focus, communicate effectively and become more efficient and flexible as a result

For depicting the phases of team performance, which was one of the key elements within the research, Tuckman's FSNP Model (Tuckman (1965)) was chosen and has been depicted in Figure 15 due to underlying reasons

1. It described the development sequence in a phase-wise approach.

Figure 17. Team performance curve model (Katzenbach and Smith, (1993))

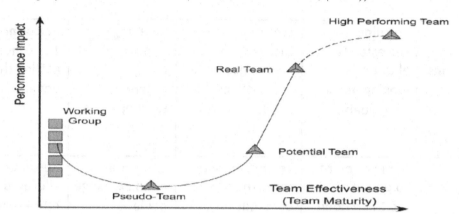

Figure 18. CHAOS 2020 Report on the success rate of mature teams

Maturity Level	Successful	Challenged	Failed
Highly Mature	66%	26%	8%
Mature	46%	41%	13%
Moderately Mature	21%	58%	21%
Not Mature	11%	54%	35%

Chart 1.5: Resolution is OnTime, OnBudget, with satisfied customer. The results are based on 50,000 projects in the 2020 CHAOS database.

2. Many models such as "GRPI Model" (Plovnick, Fry, and Rubin (1977)), "The T7 Model" (Lombardo & Eichinger, (1995)) and "The LaFasto and Larson's Five Dynamics of Team Work Model" (LaFasto & Larson, (2001)), focused on processes for better team performance, they didn't explicitly mentioned stage-wise approach to attain a particular team performance. These were good when attaining a better team performance on each stage of Tuckman's FSNP Model. Thus, by using Tuckman's FSNP Model, we could still avail of the benefits of other models.
3. There were a lot of similarity[9] between Katzenbach and Smith's Model (Katzenbach and Smith, (1993)) and Tuckman's FSNP Model (Tuckman (1965)). The "Team performance curve" of the Katzenbach and Smith Model depicted an identical representation of Tuckman's FSNP Model when the team became a high-performance team.

Tuckman suggested that the life cycle of a team involves four stages. At each stage, the dynamics of the team changed dramatically from periods of inefficiency and uneasiness through to a period of high performance. These changes have been summarised in Figure 16 illustrated below

Even though this model had its own set of limitations (Bonebright, D. A. (2010)) and other models might be a better choice for helping a team attain better team performance. The illustration was meant to depict the stages within existing team performance models.

Practice: *Thus, as per Tuckman, peak team performance can also be defined as a team that has undergone all phases of the Tuckman FSNP Model and resides at the performing stage.*

Assessment of the Frameworks

A framework generally provided a skeleton abstraction of a solution to several problems that had some similarities. A framework is demonstrated using an exemplary set of maturity models (ECIS Proceedings (2011)) and will generally outline the steps or phases that must be followed in implementing a solution without getting into the details of what activities are done in each phase (Mnkandla, E. (2009)).

Since the Software Engineering Institute had launched the Capability Maturity Model (Paulk, M. (2002)) almost twenty years ago, hundreds of maturity models had been proposed by researchers and practitioners across multiple application domains.

As the research aimed to identify the probable impact of distributed agile software development on team performance, this was concluded and tested with the help of maturity models (Poeppelbuss, J., Niehaves, B., Simons, A., & Becker, J. (2011)). The assessment was carried out using the data enlisting the factors contributing to the success of distributed agile software development and its associated challenges, and the factors contributing to team performance. These were further mapped with the data collected from one the leading bank within The Netherlands.

Team Performance vs Team Maturity

Katzenbach and Smith conceptually defined the relationship between team performance and team maturity in The Wisdom of Teams[10]: Creating the High-Performance Organization (1993). Figure 17 depicted the Katzenbach and Smith's "Team performance curve" model (Katzenbach and Smith, (1993)). This model showed the correlation between team maturity and team performance.

The importance of team maturity had also been introduced in CHAOS 2020 report. Figure 18 depicted the tabular representation of the success ratio concerning team maturity. It could be seen in the table below that a highly mature team has a success rate of 66% and a failure rate of only 8%. On the other hand, an immature team garners a success rate of only 11% and fails at 35% of their projects.

Practice: *Thus, as per Katzenbach and Smith's "Team performance curve" model, team maturity was directly proportional to team performance or in other words, if team maturity was high, team performance would also be high. This was further supported by the CHAOS 2020 report which indicated that highly mature teams would result in a better success rate in software development within the organization. Hence to analyze team maturity, the usage of the maturity model became evident and essential.*

MATURITY MODELS

Maturity could be regarded "as a measure to evaluate the capabilities of an organization regarding a certain discipline" (Rosemann & De Bruin (2005)). Wikipedia[11] stated the Maturity Model as "Maturity is a measurement of the ability of an organization for continuous improvement in a particular discipline. The higher the maturity, the higher will be the chances that incidents or errors will lead to improvements either in the quality or in the use of the resources of the discipline as implemented by the organization."

Figure 19. Squad Health Check (Rose, D. (2018))

Maturity models are conceptual models that outline anticipated, typical, logical, and desired evolution path towards maturity (Becker & Knackstedt & Poeppelbuss (2009)). Maturity models could be understood as reference models (Herbsleb & Zubrow & Goldenson& Hayes & Paulk (1997)) and could be used to assess the as-is situation of an organization, to derive and prioritize improvement measures in terms of evolutionary stages (Iversen et al. (1999), Gottschalk (2009), Kazanjian & Drazin (1989)). These distinctive stages provided a roadmap for improvement where each stage was superior to the previous stage (Mehta, Oswald, Mehta (2007), Subba Rao, Metts, Mora Monge (2003)). This advancement on the evolution path meant a step by step progression on team performance (Katzenbach and Smith, (1993)).

For the research, 80+ models[12] were evaluated and analyzed by going through its literature, assessment methods, pros and cons, target audience, type of framework, and awareness within the organization with the help of relevant success stories. These models were then compared with the bank's domain culture and their way of working, where the case study research was executed and final selection for appropriate models were concluded.

Figure 20. Agile Maturity Curve

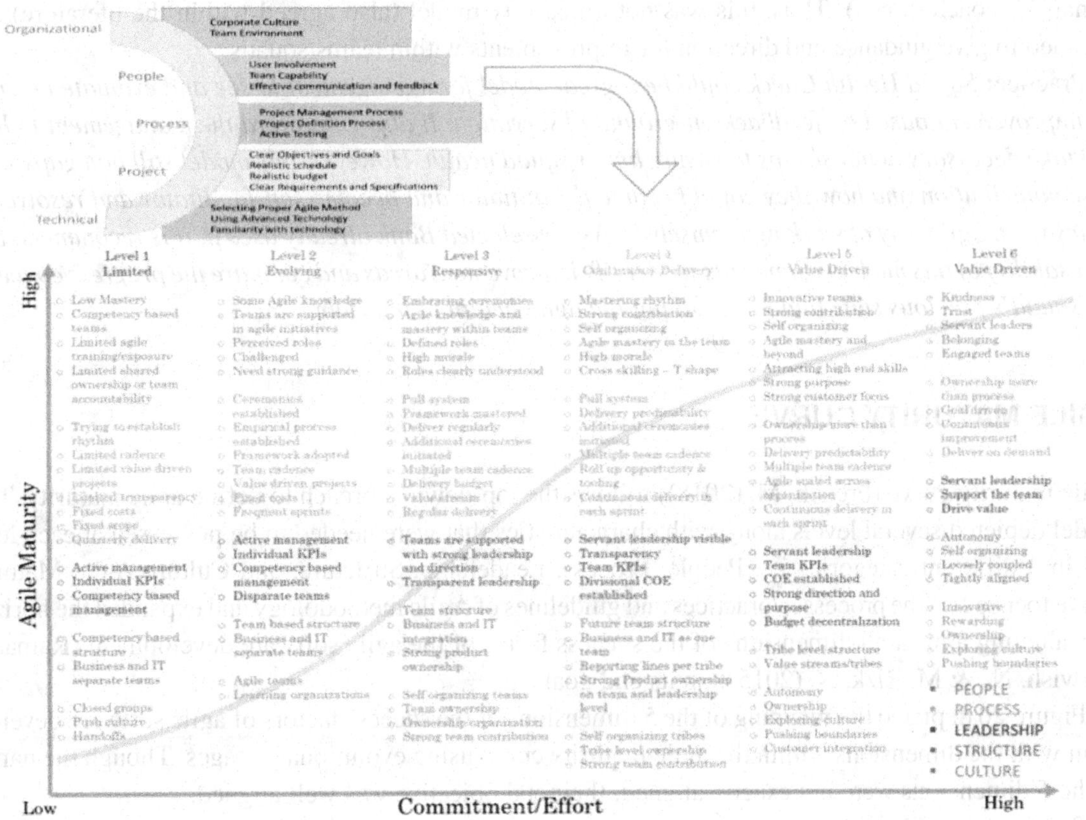

SQUAD HEALTH CHECK

Spotify[13], which was the initiator of Squad Health Check states "maturity models" as a model which involved in some sort of progression through different levels and was already been seen via the Agile Maturity Model. On the other hand, Squad Health Check model is a type of model which is usually benign and is for example intended for, managers or coaches in larger organizations who want to get a sense of focus areas for improvements, and help teams become more effective (Rose, D. (2018))

Depicted in Figure 19 are the 11 criteria on which the Squad Health Check is performed by the organization, with the status being Green, Yellow, and Red. Brief information about its interpretation is mentioned below

- **Green** doesn't mean perfect. It just means the squad is happy with this and see no major need for improvement right now.
- **Yellow** means some important problems need addressing, but it's not a disaster.
- **Red** means this needs to be improved.

This model was currently being used by the management to determine the health of their squads based on feedback from different teams within the bank used in this research. Squad Health Check helped the organization by providing health checks for Squad as well as people who were supporting it (especially managers, coaches, etc.). Thus, this was not a maturity model (also agreed within the literature) and intended to give guidance and direction for improvements within teams/squads.

Practice: *Squad Health Check could be a great model for management to see and evaluate the area for improvement based on feedback/on-ground observation. It also empowerd the management to lead and take necessary actions/steps to ensure better squad health. However, this model still had gaps w.r.t. team contribution and how they could become the initiator and become self-facilitator and resolve the issues w.r.t. agile way of working themselves. As the selected Bank already uses it, it is recommend that the usability of this model is done only to identify improvement areas and compare the progress concerning squad's previous state and for attaining the agile health of the team.*

AGILE MATURITY CURVE

Agile maturity curve (French, M. (2018)) follows the top-down approach to define agile maturity. This model depicted several levels along with characteristics that were needed to be possessed at each level and in 5 different categories i.e. People, Process, Leadership, Structure, and Culture Agile Maturity Curve focused on the processes/practices and guidelines of Agile methodology and expanded the horizon to evaluating almost all dimensions of the success factor of the agile software development (Ramadan Darwish, N., & M. Rizk, N. (2015)) on its end goal.

Figure 20 depicted the mapping of the 5 dimensions of the success factors of agile software development with the dimensions within the agile maturity curve using evolutionary stages. Though the names of the 5 dimensions were not exactly aligned, their end objective was well aligned.

Practice: *Agile Maturity Curve could help organizations to focus on all aspects which were vital to growing as an agile organization as it directly incorporates the success factors associated with agile software development.*

Despite the great value of this Maturity Model, it didn't integrate all the challenges faced within distributed agile software development (Software Engineering Research and Practice, (2015)) within its stages efficiently and clearly. Thus, this model could help in deriving the maturity of a team but might not help determine the root cause to enhance the team maturity and also expected the team to have an agile mindset and culture before their evaluation.

Agile Fluency Model

Agile Fluency Model[14], developed by Diana Larsen and James Shore in 2012 and substantially updated in 2018, is a framework to help teams understand their current position and to help them develop an individual road map and emphasizes on agile team's progress as they develop new capabilities. Agile teams pass through four distinct zones of fluency as they learn (fluency evolves). Diana Larsen defines fluency as things that you do automatically without thinking. Each zone brought specific benefits:

Focusing teams produced business value (agile fundamentals). The team thinks and plans in terms of the benefits their sponsors, customers, and users would see from their software.

Delivering teams delivered on the market cadence (agile sustainability). The team could release their latest work, at minimal risk and cost, whenever the business desiresd

Optimizing teams led their market (innovative business agility, agile promise). The team understood what their market wanted, what the business needed, and how to meet those needs.

Strengthening teams made their organizations stronger (possible future of agile). The team understood its role in the larger organizational system and actively works to make that system more successful.

The model depicted in Figure 21 gave a clear illustration based on

- Behaviors of teams at each level.
- How much time an organization must take to get matured at each level?
- What all investments must be taken by team/management to fulfill the requirements of each level?
- Supporting tools/methodology which can be utilized by the team for better productivity.

One of the articles "The Making of an Expert" (Harvard Business Review, (2006)), clearly emphasized the importance of becoming an expert associating it with time and effort. The below phrase from the article summarized the context in a very simple way

"It takes time to become an expert. Even the most gifted performers need a minimum of ten years of intense training before they win international competitions."

Figure 21. Agile Fluency Model

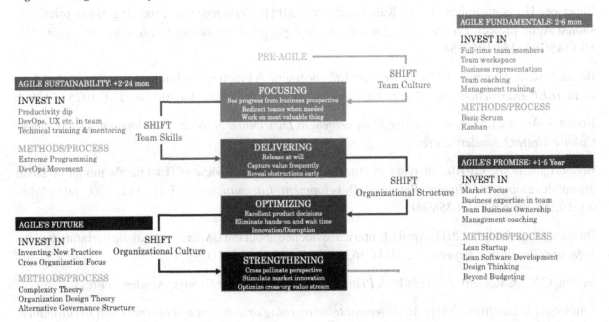

Practice: *Agile Fluency Model is a mature model that defined a roadmap with an immense focus on gaining maturity in each zone/level, thereby developing new capabilities and helping the organization grow to become a profound Agile Organization.*

Based on its literature, simplicity and the usability of the maturity models, Agile Maturity Curve and Agile Fluency Model were used for the research to evaluate the maturity and propose the findings and conclusions.

REFERENCES

Abilla. (2006). *People factors in agile software development and project management.* Academic Press.

Ågerfalk, P., & Fitzgerald, B. (2006). Introduction. *Communications of the ACM, 49*(10), 26–34. doi:10.1145/1164394.1164416

Ågerfalk, P. J., Fitzgerald, B., Holmström Olsson, H., & Ó Conchúir, E. (n.d.). Benefits of global software development: The known and unknown. *Making Globally Distributed Software Development a Success Story,* 1-9.

Allen, T. J., & Henn, G. (2006). *The Organization and Architecture of Innovation: Managing the Flow of Technology* (1st ed.). Butterworth-Heinemann.

Ambler, S. (2006). *The Agile Unified Process (AUP).* http://www.ambysoft.com/unifiedprocess/agileUP.html

Anderson, A., Beattie, R., & Beck, K. (1998). Chrysler Goes to Extremes. *Distributed Computing,* 24–28.

Baheti, P., Gehringer, E., & Stotts, D. (2002). Exploring the efficacy of distributed pair programming. *Extreme Programming and Agile Methods — XP/Agile Universe 2002,* 208-220. doi:10.1007/3-540-45672-4_20

Banerjee, U., Narasimhan, E., & Kanakalata, N. (2011). Experience of executing fixed price offshored Agile project. *Proceedings of the 4th India Software Engineering Conference on - ISEC '11.* 10.1145/1953355.1953364

Basili, V. R., & Turner, A. J. (1975). Iterative Enhancement: A Practical Technique for Software Development. *IEEE Transactions on Software Engineering, SE-1*(4), 390–396. doi:10.1109/TSE.1975.6312870

Boden & Avram. (2009). *Knowledge Management in Distributed Software Development Teams - Does Culture Matter?* Academic Press.

Bonebright, D. A. (2010). 40 years of storming: A historical review of Tuckman's model of small group development. *Human Resource Development International, 13*(1), 111–120. https://doi.org/10.1080/13678861003589099

Budgen, D., & Andy, J. (2011, April). Empirical evidence about the UML: A systematic literature review. *Software, Practice & Experience, 41*(4), 363–392. doi:10.1002pe.1009

Bullen, C.V., & Rockart, J.F. (1981). *A Primer on Critical Success Factors.* Academic Press.

Bunderson & Sutcliffe. (2003). *Management team learning orientation and business unit performance.* Academic Press.

Carmel, E., & Agarwal, R. (2001). Tactical approaches for alleviating distance in global software development. *IEEE Software, 18*(2), 22–29. doi:10.1109/52.914734

Casey, V. (2010). Imparting the importance of culture to global software development. *ACM Inroads, 1*(3), 51–57. doi:10.1145/1835428.1835443

Casey, V., & Richardson, I. (2006). Project management within virtual software teams. *2006 IEEE International Conference on Global Software Engineering (ICGSE'06).* 10.1109/ICGSE.2006.261214

Chow, T., & Cao, D. (1981). *A Survey Study of Critical Success Factors in Agile Software Projects.* Academic Press.

Coad, P., de Luca, J., & Lefebvre, E. (1999). *Java Modeling Color with Uml: Enterprise Components and Process with Cdrom.* Prentice Hall.

Cockburn, A. (2004). *Crystal Clear a Human-Powered Methodology for Small Teams.* Addison-Wesley.

Cohen, D., Lindvall, M., & Costa, P. (2004). An Introduction to Agile Methods. *Advances in Computers, 62*, 1–66. doi:10.1016/S0065-2458(03)62001-2

Cohen, S.G., & Bailey, D.E. (1997). What Makes Teams Work: Group Effectiveness Research from the Shop Floor to the Executive Suite. *Journal of Management, 23*, 239-290. doi:10.1177/014920639702300303

Conboy, K., Coyle, S., Wang, X., & Pikkarainen, M. (2011). *People over Process: Key Challenges in Agile Development.* Academic Press.

Dingsoyr, T., Faegri, T. E., Dyba, T., Haugset, B., & Lindsjorn, Y. (2016). Team Performance Software Development: Research Results versus Agile Principles. *IEEE Software, 33*(4), 106–110. doi:10.1109/MS.2016.100

Distributed Agile Development. (2015). A Survey of Challenges and Solutions. *Int'l Conf. Software Eng. Research and Practice.*

Eisele, P. (2015). The predictive validity of the team diagnostic survey. *Team Performance Management, 21*(5/6), 293–306. doi:10.1108/TPM-08-2014-0049

Eisele, P. (2015). The predictive validity of the team diagnostic survey. *Team Performance Management: An International Journal, 21*(5/6), 293–306.

Emmert-Streib, F., & Dehmer, M. (2019). Understanding statistical hypothesis testing: The logic of statistical inference. *Machine Learning and Knowledge Extraction, 1*(3), 945–961.

Espinosa, J. A., Slaughter, S. A., Kraut, R. E., & Herbsleb, J. D. (2007). Team knowledge and coordination in geographically distributed software development. *Journal of Management Information Systems, 24*(1), 135–169. doi:10.2753/MIS0742-1222240104

Herbsleb, J., & Moitra, D. (2001). Global software development. *IEEE Software, 18*(2), 16–20. doi:10.1109/52.914732

Herbsleb, J. D. (2007). Global software engineering: The future of socio-technical coordination. *Future of Software Engineering (FOSE'07).*

Hevner. (2004). *Design Science in Information Systems Research.* Academic Press.

Highsmith, J. (1997). Messy, Exciting, and Anxiety-Ridden: Adaptive Software Development. *American Programmer, 10*, 23–29.

Hoegl, M., & Gemuenden, H. G. (2001). Teamwork quality and the success of innovative projects: A theoretical concept and empirical evidence. *Organization Science, 12*(4), 435–449. doi:10.1287/orsc.12.4.435.10635

Holz & Maurer. (2003). *Knowledge management for distributed agile processes: models, techniques, and infrastructure.* Academic Press.

Joshi, A., Kale, S., Chandel, S., & Pal, D. (2015). Article. *Current Journal of Applied Science and Technology.*

Kajko-Mattsson. (2010). *Problems in Agile Trenches.* Academic Press.

Katzenbach & Smith. (1993). *The Wisdom of Teams: Creating the High-Performance Organization.* Academic Press.

Kim, T. K. (2015). T test as a parametric statistic. *Korean Journal of Anesthesiology, 68*(6), 540.

Kozlowski & Klein. (2000). *A multilevel approach to theory and research in organizations: Contextual, temporal, and emergent processes.* Academic Press.

Krishna, S., Sahay, S., & Walsham, G. (2004). Managing cross-cultural issues in global software outsourcing. *Communications of the ACM, 47*(4), 62–66. doi:10.1145/975817.975818

Ladas, C. (2009). *Scrumban-Essays on Kanban Systems for Lean Software Development.* Modus Cooperandi Press.

Lyoko, G., Phiri, J., & Phiri, A. (2016). Article. *International Journal of Future Computer and Communication.*

Malone & Crowston. (1994). *The Interdisciplinary Study of Coordination.* Academic Press.

Manikandan, S. (2011). Measures of central tendency: The mean. *Journal of Pharmacology & Pharmacotherapeutics, 2*(2), 140.

Mnkandla, E. (2009). About software engineering frameworks and methodologies. *AFRICON 2009.*

Moore, J. E., Williams, C. K., & Sumner, M. (2012). The role of informal control in PMO lite environments. *Proceedings of the 50th annual conference on Computers and People Research - SIGMIS-CPR '12.* 10.1145/2214091.2214101

Ó Conchúir, E., Holmström Olsson, H., Ågerfalk, P. J., & Fitzgerald, B. (2009). Benefits of global software development: Exploring the unexplored. *Software Process Improvement and Practice, 14*(4), 201–212. doi:10.1002pip.417

O'Leary, & Cummings. (2007). The spatial, temporal, and configurational characteristics of geographic dispersion in teams. *Management Information Systems Quarterly, 31*(3), 433. doi:10.2307/25148802

Olson, J. S., & Olson, G. M. (2003). Culture surprises in remote software development teams. *Queue, 1*(9), 52–59. doi:10.1145/966789.966804

Paasivaara, M., Durasiewicz, S., & Lassenius, C. (2008). Distributed Agile development: Using Scrum in a large project. *2008 IEEE International Conference on Global Software Engineering*. 10.1109/ICGSE.2008.38

Paulk, M. (2002). Capability maturity model for software. Encyclopedia of Software Engineering.

Peffers, K., Tuunanen, T., Rothenberger, M. A., & Chatterjee, S. (2007). A Design Science Research Methodology for Information Systems Research. *Journal of Management Systems, 24*(3), 45–77. doi:10.2753/MIS0742-1222240302

Poeppelbuss, J., Niehaves, B., Simons, A., & Becker, J. (2011). Maturity models in information systems research: Literature search and analysis. *Communications of the Association for Information Systems,* •••, 29.

Pöppelbuß, J., & Röglinger, M. (2011). What makes a useful maturity model? a framework of general design principles for maturity models and its demonstration in business process management. *ECIS 2011 Proceedings, 28.*

Poppendieck, M., & Poppendieck, T. (2003). *Lean Software Development: An Agile Toolkit*. Addison-Wesley.

RamadanDarwish, N., & Rizk, N. (2015). Multi-dimensional success factors of Agile software development projects. *International Journal of Computers and Applications, 118*(15), 23–30. doi:10.5120/20823-3453

RamadanDarwish, N., & Rizk, N. (2015). Multi-dimensional success factors of Agile software development projects. *International Journal of Computers and Applications, 118*(15), 23–30.

Rasmussen, T. H., & Jeppesen, H. J. (2006). Teamwork and Associated Psychological Factors: A Review. *Work and Stress, 20*(2), 105–128. doi:10.1080/02678370600920262

Recker, J. (2013). *Scientific Research in Information Systems*. Academic Press.

Recker, J. (2013). *Scientific Research in Information Systems*. Academic Press.

Richardson. (2009). *Knowledge Management and Competitive Advantage: Mediating Role of Innovation Capacity*. Academic Press.

Robson, C. (1993). *Real World Research. A Resource for Social Scientists and Practitioner Researchers*. Blackwell Publishers Inc.

Rockart, J. (1979). Chief Executives Define their own Data Needs. *Harvard Business Review, 52*(2), 81–93. PMID:10297607

Rose, D. (2018). *Enterprise agility for dummies*. John Wiley & Sons.

Salas, E., Cooke, N. J., & Rosen, M. A. (2008). On teams, teamwork, and team performance: Discoveries and developments. *Human Factors, 50*(3), 540–547. doi:10.1518/001872008X288457 PMID:18689065

Schwaber, K. (1996). *Controlled Chaos: Living on the Edge*. Academic Press.

Sedgwick, P. (2015). A comparison of parametric and non-parametric statistical tests. *BMJ, 350*(1), h2053-h2053.

Sedgwick, P. (2015). A comparison of parametric and non-parametric statistical tests. *BMJ (Clinical Research Ed.)*, *350*(1), h2053–h2053.

SEI – Software Engineering Institute. (2010). *CMMI for Development, version 1.3*. CMU/SEI-2010-TR-033

Shahin, M., Ali Babar, M., & Zhu, L. (2017). Continuous Integration, Delivery and Deployment: A Systematic Review on Approaches, Tools, Challenges and Practices. *IEEE Access: Practical Innovations, Open Solutions*, *5*, 3909–3943. doi:10.1109/ACCESS.2017.2685629

Shields & Rangarajan. (2013). *A Playbook for Research Methods: Integrating Conceptual Frameworks and Project Management*. Academic Press.

Shrivastava. (2010). Distributed Agile Software Development. *RE:view*.

Shrivastava, S. V., & Rathod, U. (2014). Risks in distributed Agile development: A review. *Procedia: Social and Behavioral Sciences*, *133*, 417–424. doi:10.1016/j.sbspro.2014.04.208

Singh, A. K., & Muncherji, N. (2007). Team effectiveness and its measurement. *Global Business Review*, *8*(1), 119–133. doi:10.1177/097215090600800108

Spriestersbach, A., Röhrig, B., Prel, J. D., Gerhold-Ay, A., & Blettner, M. (2009). *Descriptive statistics*. Deutsches Aerzteblatt Online.

Spriestersbach, A., Röhrig, B., Prel, J. D., Gerhold-Ay, A., & Blettner, M. (2009). *Descriptive statistics*. Deutsches Aerzteblatt Online.

Stadler, M., Vallon, R., Pazderka, M., & Grechenig, T. (2019). *Agile distributed software development in nine central European teams: Challenges, benefits and recommendations*. Academic Press.

Travers, J. C., Cook, B. G., & Cook, L. (2017). *Null Hypothesis Significance Testing and p Values*. Academic Press.

Tuckman, B. W. (1965). Developmental sequence in small groups. *Psychological Bulletin*, *63*(6), 384–399.

Tuffs, D., Stapleton, J., West, D., & Eason, Z. (1999). Inter-Operability of DSDM with the Rational Unified Process. *DSDM Consortium*, *1*, 1–29.

Vieira, E. T. (2017). *Introduction to real world statistics: with step-by-step SPSS instructions*. doi:10.4324/9781315233024

Williams, L. (2010). Agile software development methodologies and practices. *Advances in Computers*, 1–44.

Wiredu, G. O. (2006). A framework for the analysis of coordination in global software development. *Proceedings of the 2006 international workshop on Global software development for the practitioner - GSD '06*. 10.1145/1138506.1138516

ENDNOTES

1 See https://www.uantwerpen.be

2 See https://scholar.google.com

3 See www.benlinders.com

4 See www.benlinders.com

5 See https://en.wikipedia.org/wiki/Standard_deviation

6 See https://en.wikipedia.org/wiki/Variance

7 See https://en.wikipedia.org/wiki/Kurtosis

8 See https://en.wikipedia.org/wiki/Skewness

9 See https://www.praxisframework.org/en/library/katzenbach-and-smith

10 See Katzenbach, J. R. and Smith, D.K. (1993), The Wisdom of Teams: Creating the High-performance Organisation, Harvard Business School, Boston.

11 See https://en.wikipedia.org/wiki/Maturity_model

12 See https://www.benlinders.com/tools/agile-self-assessments/

13 See https://labs.spotify.com/2014/09/16/squad-health-check-model/

14 See https://www.agilefluency.org/

Chapter 6
Findings and Core Practices in the Domain of Agile Methodologies

Tapan Kumar
Cognizant, The Netherlands

ABSTRACT

This chapter analyzes the outcome of the research done on distributed agile teams working on digital platforms and software products. This chapter also aims to explore the gaps surrounding distributed agile and team performance which has been mentioned in the literature; hence, this was rarely co-related, and therefore, its probable impact was never explored in depth. In addition to this, the chapter highlights some key observations for the two selected maturity models and highlights its strength and weakness which seems to have been ignored by the organization while using the models. For this research, a large financial service company was examined as a case study. The chapter then concludes by drawing core practices to organizations having similar distributed agile software development model. In addition to this, the chapter also leaves open discussion points for the readers to think upon and question themselves on the way agile is being used within their organizations and how it contributes to working on digital systems and platforms.

INTRODUCTION

As the book "Handbook of research on Digital Platform Security Assurance" focuses on various dimensions associated with an organization ranging from small individual tech initiatives to complete business models with intertwined supply chains and "Platform" based business models. New ways of working, such as Agile, Distributed Agile and DevOps, are introduced, leading to new risks. These risks do not restrict themselves to the technology domain; new challenges arise by teams working together in a distributed manner to deliver high paced value at a higher pace by reducing the time to market.

This chapter analyzes the findings observed in the area of new way of working within Distributed Agile methodology and its impact on team performances thus having a direct impact on digital organiza-

DOI: 10.4018/978-1-7998-7367-9.ch006

tions and their time to market new products. In addition to evaluating the impact on team performance with the help of identifying gaps surrounding distributed agile team and team performance, the use of combined maturity models has also been explained as practices to effectively and efficiently evaluate the agile team performance of organizations.

As this book aims to follow the thread of the function of our business all the way to the basement of the individual organisation (engineering) working in an eco-system of platforms, way of working and people. It thus becomes very important for individual teams to evaluate their performance themselves by using appropriate tools correctly and effectively.

The chapter then concludes by drawing conclusions post data evaluation and highlights practices to organizations having similar distributed agile software development model. In addition to this, the chapter also leaves open discussion points for the readers to think upon and question themselves on the way Agile has been used within their organizations.

CASE STUDY RESEARCH METHOD

For the research to identify *the impact of distributed agile software development team on team performance*, single-case study with holistic approach research method was executed within the leading bank in The Netherlands.

DATA PREPARATION

As the earlier chapter "Research findings in the domain of Agile Methodologies" highlighted the characteristics about Distributed Agile Software Development and Team Performance, along with challenges and factors associated with distributed agile software development and team performance respectively, Figure 1 depicts the correlation between these independent and dependent variables (Recker, (2013)), i.e. distributed agile software development and team performance.

It can be seen that within literature there already exists gap concerning team performance factors (Dingsoyr, T., Faegri, T. E., Dyba, T., Haugset, B., & Lindsjorn, Y. (2016)) and the challenges identified within distributed agile software development methodology (Software Engineering Research and Practice, (2015)). All the challenges of distributed agile software development mentioned in Figure 1, which didn't find its co-related factor within team performance were directly or indirectly related with culture and its various aspects (Hall E. T., Beyond Culture, (1989), Wagner, (1995), Samovar, Porter, & McDa, (2009)).

This chapter tried to explore further this gap and see if this had any implication on team performance which was the objective of this research, and, tried to explore it using the data collected from the case study involving one of the Dutch National Bank.

Now that a gap had been identified using literature associated with distributed agile software development and team performance (Figure 1), it was also important to evaluate if the selected maturity assessment model i.e. Agile Maturity Curve could provide correct maturity of the team. This would thus give the data about team performance (Katzenbach and Smith, (1993)). Figure 2 tried to evaluate a few key questions and did a comparison between the selected maturity models for this research as per the author i.e. "Agile Maturity Curve" and "Agile Fluency Model".

Figure 1. Distributed Agile Software Development vs Team Performance
*Not addressed explicitly: the referred challenge is part of one of the factor's property
*Identified gap: This indicates that the challenge/factor is only present for one variable.

Challenges within Distributed Agile Software Development	Associated factor within Team Performance	Remarks
Documentation	Team Learning	Partially matched
Pair Programming	Team Learning	
Different Work Hours		Not addressed explicitly*
Communication	Team Coordination, Shared Mental Models	Partially matched
Knowledge Transition	Team Learning	
Cultural Differences		Identified gap*
Team Cohesion	Team Cohesion	
People vs. Process-Oriented Approaches		Identified gap*
Knowledge Management	Team Learning	
Language Barriers		Not addressed explicitly*
Role of Specialist	Team Learning	
Developer Fear of Skill Deficiency Exposure	Shared Mental Models	Partially matched
Recruitment Challenges	Team Learning	
	Goal Orientation	Identified gap*

Figure 2. Agile Maturity Curve vs Agile Fluency Model

Parameters	Agile Maturity Curve	Agile Fluency Model
Is the model reflecting all 5 dimensions of the success factors of agile software development?	YES	Partially
Can each dimension have a different maturity level at a given point in time?	YES	NO
Does the model focus on becoming an expert in each stage?	NO	YES
Does this model give guidance on the minimum timeframe needed in each phase?	NO	YES
Is the model easy to implement?	YES	YES
Can the basic fundamental principle of team performance model like Tuckman be integrated with the maturity model?	YES	YES

It can be seen within Figure 2 that both these models have their own set of strengths and weakness, it is also interesting to see that even though agile maturity curve covers several parameters of the success factor of agile software development, it also poses an ambiguity and uncertainty to organizations when these dimensions don't reflect the maturity of a team unanimously i.e. the maturity of one dimension is different from the other one, thus its consequence to the organizations is not known and can vary differently from one organization to the other. Also, this model expects a team to already possess an agile mindset and cultural shift before its evaluation as described in chapter" Research findings in the domain of Agile Methodologies". On the other hand, the agile fluency model mitigates these aspects in its stages.

As the intention was not to explore in depth the consequence of the gaps within agile maturity curve, which was not the objective of this research, this chapter tried to make use of agile maturity curve as the maturity model to evaluate the maturity of the team by using 5 categories which were also covered within the success factor of agile development model. As the research tries to explore if there is indeed an impact on team performance due to distributed agile software development model, using 1 maturity model was enough to derive the conclusion. It however doesn't state that using just 1 maturity model is the best approach an organization can take and the same would be concluded after the research question has been answered further in this chapter.

Data Evaluation

For the research, the data was collected using interviews (including online survey) and agile process documents along with observation as already described within chapter "Research findings in the domain of Agile Methodologies". Figure 3 and Figure 4 shows the maturity level against each categories with respect to the case study done in one of the leading bank in The Netherlands. The categories of the agile maturity curve had been used for the data collection method and using three data collection methods and plotting them with the parameters of the agile maturity curve.

For the research, the scoring had been done using the Mean Descriptive statistics (Manikandan, S. (2011)) for all the data collected as a result of the interview using the formula

$$\text{Interview Mean Score, } \bar{x} = \frac{\sum x}{n1}$$

Where "**x**" was the number of correct answers given during the interview and "**n1**" was the total count of the person who were part of the survey and interview process. The maturity level was set at a range of 0.165 difference based on the interview mean score (m_1).

In addition to the mean descriptive statistics used to derive the maturity level of data collection method comprising of interviews, for observation and documentation, Likert[1] scale was used to derive the maturity level (Joshi, A., Kale, S., Chandel, S., & Pal, D. (2015)), which was equivalent to the response set value and has been illustrated in Figure 3 and Figure 4. The Likert scale was determined as a result of interviews with the agile coaches, product owners, and delivery leads IT of the Bank used for the case study. For the interviews with team members, Likert 2 and 3 point scale had been used however for observation and documentation, Likert's 5 point scale had been used for calculation.

For the parameters comprising of multiple data collection methods, the average mean weight was used to calculate the maturity of the defined parameter using the formula

Average Maturity Mean Score, $\bar{y} = \dfrac{\sum y}{n2}$

Where "**y**" was the maturity value of each data collection method and "**n2**" was the number of data collection method used for the parameters of maturity assessment

The data shown in Figure 3 and Figure 4 showed an accumulated mean/average maturity of **2.6** based on the 5 categories as mentioned in the agile maturity curve for the case study. It was also important to notice that the categories didn't had a uniform distribution of the maturity level and showed a standard deviation (σ_x) of 0.89. Thus, it was difficult to infer if distributed agile software development did have any impact on team performance or not. Therefore, inferential statistics test needed to be conducted to get the answer to the main research question by comparing the statistics of the co-located software development team and distributed agile software, development teams.

Figure 5 showed overall scoring based on interviews and online surveys using a Likert 2,3 and 5-point scale (Joshi, A., Kale, S., Chandel, S., & Pal, D. (2015)), with respect to team maturity for the co-located software development team and distributed agile software development team. As the co-located team was of two variants i.e. co-located agile team with cultural diversity and co-located team without cultural diversity, Figure 5 showed the categorization of maturity score based on these variants only for representational purpose.

Along with the scoring, the table showed mean and variance using the formula

Mean value, $\bar{x} = \dfrac{\sum_{i=1}^{n} x_i}{n}$

Where, "x_i" was the total score of the team based on the online survey and interview and "n" was the number of samples taken for the calculation. Variance was also calculated using formula

Variance, $\sigma^2 = \dfrac{\sum_{i=1}^{n} \left(x_i - \bar{x}\right)^2}{n-1}$

Where, "x_i" was the value of each sample, "$\bar{x}$" was the mean value of individual teams and "n" was the total number of samples considered for executing the inferential statistical test. As the intention was to do an inferential statistical test using only two groups' i.e. co-located team and distributed agile team, the splitting of the co-located agile team was not done further (i.e. co-located agile team with cultural diversity and co-located team without cultural diversity) and was taken collectively to perform the inferential statistical test.

As the maturity values were normally distributed and Figure 5 could clearly show a difference in the average/mean for both the teams, thus, to get the answer to the research question, a parametric test (Sedgwick, P. (2015)) was performed by evaluating the maturity of co-located software development agile teams and distributed agile software development teams to determine team performance of each agile team.

Figure 3. Maturity assessment – 1 (Test Case result)

Parameters	Maturity Level	Data Collection Method	Interview Mean Score	Likert 5-point scale
People				
Adhering to all agile guidelines and practices	2	Interview, Observation	0.39	Frequency (2)
Agile training	3	Documentation	NA	Quality (3)
Role clarity	3	Documentation	NA	Familiarity (3)
Accountability to finish sprint	2	Interview	0.35	
Communication amongst individual	2	Interview	0.33	
Ceremonies to celebrate success	2	Observation, Documentation	NA	Frequency (2), Familiarity (2)
Individual Ownership on handling impediments	3	Interview, Observation	0.50	Frequency (3)
T-shaping	2	Documentation	NA	Quality (2)
Emphasis on code reviews and documentation	1	Interview	0.21	
Mean Average	**2**			
Process				
Fixed Budget	1	Interview	0.16	
Fixed scope	1	Interview	0.17	
Final product delivery after each sprint	2	Interview, Documentation	0.26	Quality (2)
Defined/documented quality assurance process	4	Interview, Documentation	0.66	Quality (3)
Team cadence	3	Interview	0.53	
The process to coordinate across multiple teams	2	Documentation	NA	Familiarity (2)
Discuss defects/incidents for knowledge sharing and constructive feedback	2	Interview, Observation	0.32	Frequency (2)
Accuracy of requirements by business	2	Interview, Documentation	0.17	Quality (2)
Mean Average	**2**			

Figure 4. Maturity assessment – 2 (Test Case result)

Leadership				
Providing support to agile teams	4	Observation, Documentation	NA	Frequency (4), Quality (3)
Transparency in the agile process and objectives	4	Observation, Documentation	NA	Frequency (4), Quality (4)
Establishing KPIs for agile teams	3	Documentation	NA	Quality (3)
Involvement of product owner with an Agile team	3	Interview	0.47	
Management involvement with the team	4	Interview	0.66	
Supporting teams by providing the necessary infrastructure needed for agile transformation	3	Interview	0.56	
Mean Average	**4**			
Structure				
Business and IT part of one team	4	Documentation	NA	Familiarity (4)
Product owner having strong business knowledge	3	Interview	0.50	
Static team members within the scrum team	1	Interview	0.04	
Involvement of COEs within teams/domain	4	Documentation	NA	Familiarity (4)
A governance framework to check team health/status	4	Interview	0.66	
Documented team hierarchy	4	Documentation	NA	Familiarity (4)
Mean Average	**3**			
Culture				
Social cohesion	2	Observation	NA	Frequency (2)
Team coordination	3	Interview	0.53	
Task cohesion	4	Interview	0.73	
Sharing the responsibility of the team member's task	3	Observation	NA	Frequency (3)
Awareness of business strategy of the tribe	2	Interview	0.33	
Enthusiasm on team building activities	2	Interview	0.30	
Ability to challenge the business on requirements	1	Interview	0.13	
Process of integrating new team members	1	Observation, Documentation	NA	Frequency (2), Familiarity (1)
Process of rewarding team members/team	2	Observation	NA	Frequency (2)
Mean Average	**2**			

Figure 5. Agile Team's maturity scoring

Co-located Software Development Teams		Distributed Agile Software Development Teams
Without Cultural Diversity	**With Cultural Diversity**	
73	44	34
79	48	37
	48	39
	54	43
	57	45
	58	45
	59	45
	61	46
	69	46
		47
		48
		49
Total Samples (n1) = 11 Average/Mean ($\bar{x}_1$) = 59.09 Variance (σ_1^2) = 119.69		Total Samples (n2) = 12 Average/Mean ($\bar{x}_2$) = 43.67 Variance (σ_2^2) = 21.33

Parametric Statistical Hypothesis Test

A statistical hypothesis test (Emmert-Streib, F., & Dehmer, M. (2019)) is a method of statistical inference[2]. For performing the Parametric Statistical Hypothesis Test, two statistical data sets need to be compared and the necessary hypothesis (null hypothesis (H_0) and alternate hypothesis (H_a)) will have be validated and tested. Thus, the hypothesis (H_0 and H_a) that were validated as part of the research were,

H_0: The maturity of a distributed agile team and co-located agile team were identical, thus, a distributed agile team did not have any impact on team performance.

H_a: The maturity of the co-located agile team were differing from the maturity of a distributed agile team, thus, the distributed agile team did had an impact on team performance.

To do the hypothesis test, the probability (P-value) was calculated to conclude the outcome. The "**P**" value (or the calculated probability) is the probability of the event occurring by chance if the null hypothesis is true. The "**P**" value is numerical between 0 and 1 and is interpreted in deciding whether

Figure 6. P-value interpretation (Emmert-Streib, F., & Dehmer, M. (2019))

P-Value	Result	Null Hypothesis (H_0)
< 0.01	Result is highly significant	Reject (null hypothesis) H_0
≥ 0.01 but < 0.05	Result is significant	Reject (null hypothesis) H_0
≥ 0.05	The result is not significant	Don't Reject (null hypothesis) H_0

to reject or retain the null hypothesis. Figure 6 depicts the correlation of this "**P**" value concerning the null hypothesis and alternate hypothesis.

As the null hypothesis and alternate hypothesis had been defined, the next step was to figure out the type of parametric test that would be suitable for the research. Due to the nature of data as depicted in Figure 5, which comprised of the maturity assessment of the teams within the Bank in two distinct categories i.e. one category comprising of the maturity assessment of co-located software development agile teams and other category comprising of the maturity assessment of distributed agile software development teams, the hypothesis test using "Student's Two-Sample Unpaired t-test assuming unequal variance[3]" was performed to draw conclusion.

Figure 7 depicted the derivation for the Student's Two-Sample Unpaired t-test assuming unequal variance using the maturity scores defined in Figure 5. Now that the hypothesis test had been concluded, the probable impact of a distributed agile team on team performance, the two-tail test result was considered for deciding which hypothesis holds true.

Figure 7. Student's Two-Sample Unpaired t-test assuming unequal variance

Sno.	Steps	Formula	Result
1	Calculate t-statistic (t_{stat})	$t_{stat} = \dfrac{(\bar{x}1 - \bar{x}2)}{\sqrt{\dfrac{\sigma_1^2}{n1} + \dfrac{\sigma_2^2}{n2}}}$ $\bar{x}1$: Mean of Co-located agile team $\bar{x}2$: Mean of the Distributed agile team n1: A sample size of the Co-located agile team n2: A sample size of the Distributed agile team σ_1^2 : Variance of the Co-located agile team σ_2^2 : Variance of the Distributed agile team	t_{stat} = 4.33
2	Calculate Degree of freedom (df)	$df = \dfrac{\left(\dfrac{\sigma_1^2}{n1} + \dfrac{\sigma_2^2}{n2}\right)^2}{\dfrac{(\sigma_1^2)^2}{(n1-1)} + \dfrac{(\sigma_2^2)^2}{(n2-1)}}$	df = 13
3	Fetch t-critical (t_{crit}) value from t-table[4]	Probability (α) = 0.05 df = 13 *Refer appendix 3	One-tail: t_{crit} = 1.77 Two-tail: t_{crit} = 2.16
4	Compare t_{stat} and t_{crit}		$t_{stat} > t_{crit}$
5	Derive p-value from t-table for df_{13}	p-value = value ($df_{13)}$ closest to t_{stat} *Refer appendix 3	One-tail: **p-value<0.0004** Two-tail: **p-value< 0.0008**

As P-value interpretation (Figure 6) depicted that if the "**p-value**" is less than "**0.01**", the result would be highly significant and that the null hypothesis (**H₀**) must be rejected. The same was also confirmed by looking at the comparison of "**t**$_{stat}$" and "**t**$_{crit}$", where the "**t**$_{stat}$" value was greater than "**t**$_{crit}$", re-confirming the conclusion drawn based on "**p-value**". Thus, the alternate hypothesis held true for the research, which was,

H$_a$: The maturity of the co-located agile team was different from the maturity of a distributed agile team, thus, the distributed agile team did had an impact on team performance.

CONCLUSION

As the research question intended to identify the impact of distributed agile software development on team performance, the "Student's Two-Sample Unpaired t-test assuming unequal variance", clearly provided below-mentioned conclusions,

- Based on the outcome from data evaluation, where the "Student's Two-Sample Unpaired t-test assuming unequal variance" proved that the alternative hypothesis held true with p-value way less than 0.01. Also, as the value of the mean/average was higher in the case of the co-located agile teams (*mean=58.90*), it proved that the maturity of the co-located agile team was better than that of a distributed agile team (*mean= 43.67*). The same was also confirmed with the help of the one-tail t-test result, where the "t$_{stat}$" value was greater than "t$_{crit}$". Hence team performance for the co-located agile teams was better than that of distributed agile teams. Thus, it strongly proved that there was a negative impact of distributed agile software development on team performance.
- The difference in maturity level between the co-located agile team and distributed agile could very well be justified by the observation as depicted in "Figure 1: Distributed Agile Software Development vs Team Performance".
  The table tried to identify the gaps between the challenges of distributed agile team and team performance factors, which were
  - Different work hours
  - Cultural Differences
  - People vs. Process-Oriented Approaches
  - Language Barriers
  - Goal Orientation

  These identified gaps seemed to be the decisive factor why such variations were observed between the distributed agile team and co-located agile team because all these gaps seemed to be more applicable and with higher impact for distributed agile teams when compared to the co-located agile teams.

- Assessing maturity using just a single maturity assessment model could give inconclusive results because of the non-uniform maturity level across various categories as observed within this research. Such inconclusive results might lead to wrong interpretation of maturity level of the team, thus leading to a perception challenge between agile team and their management as it was visible in the maturity assessment, were at one end leadership and structure scored a maturity level of 4, and on other hand, people, process and culture scored a maturity score of 2. This could lead to the organization not believing in current agile methodology implementation and could lead to making decisions that could impact them financially as well as could impact the morale of teams. Thus, the combination of the maturity assessment model along with the agile fluency model would help in minimizing this impact. The same had been illustrated in detail as practices within this chapter.

Practices

As the agile maturity curve was not able to have a uniform distribution of the maturity score throughout its category, using just one model for evaluating the team might not be helpful to the organization as mentioned in the conclusion section of this chapter.

- Thus, a recommended practice would be by making use of both agile maturity curve which gives strength to maturity assessment there by covering the 5 categories associated with an organization i.e. People, Process, Leadership, Structure and Culture. And agile fluency model, which could ensure that teams (including leadership), spent minimum time within one phase to become an expert thus adapting the agile way of working as a culture. By integrating both these powerful assessment tools/frameworks, the maturity assessment would have a uniformity thus helping the organization to focus on the areas which need attention.
- Each phase/level of the maturity model can also be sub-categorized by using the performance model like Tuckman. This approach might help the agile team in improving their maturity efficiently and can also act as a bridge when integrating the maturity model with the agile fluency model.
- Even though the maturity model has fixed categories, amending it by adding the gaps which were identified between the challenges of a distributed agile team and team performance factors, could help the team to act proactively in improving the team performance score/maturity score of the distributed agile software development team.

FURTHER RESEARCH

As the maturity assessment showcased a non-uniform distribution of the maturity score across its categories, it can be a viable option to execute a research to check the implication to organizations when they assess their team using just a single model, which showed similar characteristics i.e. multiple levels having varying maturity scores. Could this be a potential reason which is observed within the organization that tried to implement several methodologies in a quick period due to less confidence in the applied methodology?

In addition to this, as the research involved "Student's Two-Sample Unpaired t-test assuming unequal variance" for two samples i.e. co-located agile team and distributed agile team, however the data depicted in "Figure 5: Agile Team's maturity scoring", the co-located agile team were split for illustration purpose in two categories i.e. co-located team with cultural diversity and co-located team without cultural diversity. It could also be good research to find an impact of these two categories of a co-located agile team on team performance and associated factors that played a vital role in contributing to a better maturity score.

REFERENCES

Dingsoyr, T., Faegri, T. E., Dyba, T., Haugset, B., & Lindsjorn, Y. (2016). Team Performance Software Development: Research Results versus Agile Principles. *IEEE Software*, *33*(4), 106–110. doi:10.1109/MS.2016.100

Emmert-Streib, F., & Dehmer, M. (2019). Understanding statistical hypothesis testing: The logic of statistical inference. *Machine Learning and Knowledge Extraction*, *1*(3), 945–961. doi:10.3390/make1030054

Hall E. T. (1989). *Beyond Culture*. Academic Press.

Joshi, A., Kale, S., Chandel, S., & Pal, D. (2015). Article. *Current Journal of Applied Science and Technology*.

Katzenbach & Smith. (1993). *The Wisdom of Teams*. Creating the High-Performance Organization.

Manikandan, S. (2011). Measures of central tendency: The mean. *Journal of Pharmacology & Pharmacotherapeutics*, *2*(2), 140. doi:10.4103/0976-500X.81920 PMID:21772786

Recker, J. (2013). *Scientific Research in Information Systems*. Academic Press.

Samovar, P. (2009). *McDa*. Communication Between Cultures.

Sedgwick, P. (2015). A comparison of parametric and non-parametric statistical tests. *BMJ*, *350*(1), h2053-h2053.

ENDNOTES

1   See https://www.simplypsychology.org/likert-scale.html
2   See https://en.wikipedia.org/wiki/Statistical_inference
3   See https://en.wikipedia.org/wiki/Variance

Chapter 7
Problems of CI/CD and DevOps on Security Compliance

Yuri Bobbert
ON2IT BV, The Netherlands & Antwerp Management School, University of Antwerp, Belgium

Maria Chtepen
BNP Paribas Group, Belgium

ABSTRACT

In this chapter, the authors define the main problems when working on products in DevOps Teams and on CI/CD pipelines with regard to security and risk management. It focusses on the regulatory requirements and cyberthreats that have impact on organisations. Regulator requirements vary from industry and country. Working with multiple teams on products requires proper alignment in frameworks, controls, and architecture principles in order to be end-to-end protected throughout the connected platforms. This chapter examines the multiple compliance frameworks and architectural principles that can be applied to agile way of working and more precise to CICD pipelines. It defines the main problem statement and questions the authors wanted to answer. The authors looked with a lens of regulated industry since this industry suffers the most and therefore has the biggest benefit from this research project.

INTRODUCTION

CONCEPTS OF CI/CD AND (SEC)DEVOPS

The beginning of 2000 is a starting point of the creation of a novel software development methodology, known as Agile. The term officially appears for the first time in 2001 in *The Agile Manifesto* (Fowler, 2001), a publication by a group of software professionals, which covers the twelve principles of agile software development. These principles were the consequence of frustration that started to appear in industry a decade before, as the result of growing software complexity, increasing speed of technological evolution and the inability of the traditional Waterfall method to cope with the new reality. As stated

DOI: 10.4018/978-1-7998-7367-9.ch007

by Murray (Murray, 2015), the effect of Moore's Law is all around us. Not only computing power is doubled every 18 months, but chips get smaller, cell phones get more powerful and flat screen TV's get cheaper. If a software project takes 18 months to deliver, you've missed an entire Moore's Law cycle, because the business specifications were written 18 months ago.

Figure 1. Waterfall vs Agile software development method (Derksen, October 2018).

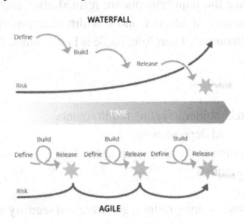

The **Waterfall approach** is considered as a traditional way of managing software development. It consists of consecutive phases leading to the development of a *complete* software product. The phases are summarized in Figure 1 (Derksen, October 2018):

- **Define**: collect software requirements
- **Build**: design the solution, code the software, test and debug the software
- **Release**: deliver the finished product and enter maintenance mode

The main disadvantages of this approach are (1) the need for complete detailed requirements to be known upfront, before the start of software development, and (2) the fact that the software product is only seen when the development is completed. It means that the project cannot take off before all the requirements are collected, and often the product is delivered after a long development process to only discover that the requirements were incomplete or misinterpreted.

Figure 2. DevOps toolchain (Gartner)

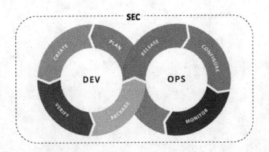

The **Agile approach** is the extreme opposite of the original sequential software development. It values early and continuous software delivery; changing requirements; close collaboration between business, customers and developers; and face-to-face communication as an alternative for extensive documentation. Figure 1 shows a graphical representation of a typical Agile software development cycle, where the project is split into short iterations. There is a deliverable and a feedback loop at the end of each iteration.

But also the Agile methodology has its drawbacks. For instance, detailed planning and cost estimation are major pain points. Since the requirements are refined after each iteration, it is very difficult to predict the scope and the outcome in advance and it is almost impossible to indicate how much it is going to cost to achieve a certain output. Therefore, Agile is better suited for projects with the following characteristics (Murray, 2015):

- Small and not too complex
- Constant availability of stakeholder to refine requirements
- Few complex impetrations and dependencies
- Leadership accepting planning and cost uncertainties
- Need for very fast time-to-market

At the same time, activities such as integrations, software and security architecture are often assumed to require a full overview and rigorous up-front planning to guarantee robustness and fewer security holes. The lack of certainty regarding these aspects can, in my experience, disqualify Agile for many organisations.

Figure 3. New work vs. unplanned work of High / Low DevOps performers (Forsgren, 2018).

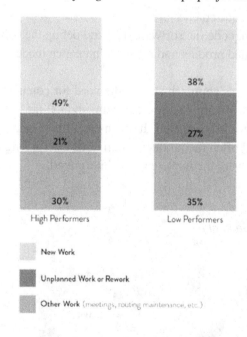

High Performers — New Work 49%, Unplanned Work or Rework 21%, Other Work 30%

Low Performers — New Work 38%, Unplanned Work or Rework 27%, Other Work 35%

New Work

Unplanned Work or Rework

Other Work (meetings, routing maintenance, etc.)

Scientific literature describes a number of approaches to address the inherent difficulty of integrating security in Agile software development. (Chivers, 2005) argue that security can grow organically within an agile project by using **incremental security architecture** that evolves with the code. On the one hand, unlike conventional security architectures, an incremental architecture remains true to agile principles by including only the essential features required for the current iteration – it does not try to predict future requirements. On the other hand, it preserves the link between local functions and system properties, and it provides a basis for an ongoing review of the system from the security prospective. Intel (Harkins, 2013) adopts the incremental security architecture approach by identifying four capabilities, or *cornerstones,* to achieve higher agility and dynamic architecture: *trust calculation* handles user identity and access management by dynamically determining what type of access (if any) a user should be granted to a resource; *security zones* provide different levels of protection to different types of data, depending on its criticality; *balanced controls* fulfil the need for a combination of detective, corrective and preventive controls (e.g. firewalls); *user and data perimeters* allow to treat users and data as additional security perimeters and protect them accordingly.

By the end of 2000, the Agile paradigm has extended its reach into many related fields, including product management, operations, organisational culture and learning, as well as IT infrastructure (Betz, 2016). **DevOps** (Development and Operations) is a movement in the IT community that uses agile/lean techniques to add value by increasing collaboration between the development and operations staff. This can take place at any time in the development life cycle when creating, operating, or updating an IT service. By using agile/lean, it allows for IT services to be updated continuously so the speed of delivery can dramatically increase while stability is improved (Colavita, 2016).

The main driving force for DevOps' wide adoption, is the use of less resources for software development and maintenance, which is achieved through automation (CA Technologies, 2014) (Delphic, 2016). It is the core activity in the DevOps practices – **the continuous integration & delivery (CI/CD)** (Sharma, 2015) (Bucena, 2017). Figure 2 shows activities (toolchain), as defined by Gartner, with which a set of DevOps tools to aid in the delivery, development and management of applications through the system development lifecycle. The research by (Forsgren, 2018) has shown that continuous integration & delivery predicts lower levels of unplanned work and rework in a statistically significant way. It was found that the amount of time spent on new work and unplanned work or rework was significantly different between CICD high performing companies and low performers. The differences are shown in Figure 3.

Also with the introduction of DevOps concerns remained how DevOps impacts security aspects of the developed software. In fact, security is among the major concerns that limit the adoption of DevOps processes (Mohan V. O., 2016) (CA Technologies, 2014). This triggers the coining of the terms **SecDevOps** or **DevSecOps**. Both refer to incorporating security practices in the DevOps processes by promoting collaboration between the development teams, the operations teams and the security teams.

A number of security practices for DevOps are described in literature. (Farroha, 2014) (Schneider, 2015) refer to the automation practices for integration into SecDevOps, such as automating testing to detect noncompliance, tracking compliance breaches through automated reporting of violations, continuous monitoring and maintenance of a service catalogue with tested and certified services, security scanning, configuration automation, etc. (Storms, 2015) believes that DevOps fails to include security throughout the process and leaves it to the end. The security problems due to DevOps that Storms lists include the high pace of deployments, the unclear access restrictions, and the lack of audit and control points. Finally, (Mattetti, 2015) and (Bass, 2015) approach security in DevOps through, respectively, platform and application hardening, while (Farroha, 2014) and (Storms, 2015) identified a set of tools that could be integrated into DevOps to support security, monitoring, and logging.

When looking at different solutions proposed in literature to address security concerns in the agile software development, the question arises: how can we determine which controls and mechanisms are the most suitable and effective for a particular organisation?

KEY TERMS

Before proceeding with the discussion of the subject, it is important to agree on the terms and definitions used. Since there are multiple interpretations of the same term in the literature, for consistency reasons, definitions from the Department of Defense (DoD) Enterprise DevSecOps Reference Design report (Lam, 2019) are used, if not indicated otherwise. The reason for this choice is that the DoD report is used for the validation of the findings of this research in the later chapters of this document.

Problem Definition

While rapid technological development forces companies to increase the speed of delivery through the adoption of methods like Agile Software Development and DevOps, security compliance seems to be an obstacle to deploying these methods in an effective and automated way (Plant, 2019). Companies are often required to achieve compliance with standards and laws that intend to reduce risks and create a traceable development process. Compliance is, however, often seen as an obstacle to employing DevOps because of required tests and controls that do not seem to fit into an automated process. Highly regulated environments usually demand segregation of duty, separated work groups and strict confidentiality as well as security measures. This contrasts DevOps where communication, collaboration and automation are central (Yasar, 2017).

One of the main problems that characterizes this misfit is the merging of development and operations in DevOps. Developers are assigned operational responsibilities such as debugging running production systems, but traditional compliance controls restrict access to production environments for developers (Michener, 2016). Multiple scholars therefore advocate for a hybrid environment in which the DevOps process is integrated into the specific environment as much as possible but stays restricted by applicable regulations (Yasar, 2017). Also (Mohan V. O., 2018) reveals that the main security concerns for DevOps automation are: separation of roles, enforcement of access controls, manual security tests, audit, security guidelines, management of security issues, and participation of the security team. The major recommended best practices for a transformation of current processes to SecDevOps are: good documentation and logging, strong collaboration and communication, automation of processes, and enforcement of separation of roles.

DevOps and automation often go together with the use of Cloud infrastructure. A major part of the risk inherent to any cloud scenario is the security posture of the cloud provider. Most reputable cloud providers offer a variety of controls attestation documents, such as the SSAE 16 SOC 2, ISO 27001 and ISO 27002 reports, or a report on the Cloud Security Alliance Cloud Controls Matrix (CCM) (Shackleford, D., SANS, 2016). Security teams should review this documentation carefully when choosing a cloud provider to decide whether cloud deployment is compliant with the requirements of their organisation.

Table 1.

Term	Definition
Access Control	Means to ensure that access to assets is authorized and restricted based on business and security requirements (ISO/IEC 27000:2014(E)).
Advanced Persistent Threat (APT)	An adversary that possesses sophisticated levels of expertise and significant resources which allow it to create opportunities to achieve its objectives by using multiple attack vectors (e.g., cyber, physical, and deception). These objectives typically include establishing and extending footholds within the information technology infrastructure of the targeted organisations for purposes of exfiltrating information, undermining or impeding critical aspects of a mission, program, or organisation; or positioning itself to carry out these objectives in the future. The advanced persistent threat: (i) pursues its objectives repeatedly over an extended period of time; (ii) adapts to defenders' efforts to resist it; and (iii) is determined to maintain the level of interaction needed to execute its objectives (NIST.SP.800-53r4).
Agile	A software development approach that is opposite to the original sequential software development. It values early and continuous software delivery; changing requirements; close collaboration between business, customers and developers; and face-to-face communication as an alternative for extensive documentation. Agile project is split into short iterations. There is a deliverable and a feedback loop at the end of each iteration (Murray, 2015).
Audit Log	A chronological record of information system activities, including records of system accesses and operations performed in a given period (NIST.SP.800-53r4).
Audit Record	An individual entry in an audit log related to an audited event (NIST.SP.800-53r4).
Authentication	Verifying the identity of a user, process, or device, often as a prerequisite to allowing access to resources in an information system (NIST.SP.800-53r4).
Availability	Ensuring timely and reliable access to and use of information (NIST.SP.800-53r4).
Baseline Configuration	A documented set of specifications for an information system, or a configuration item within a system, that has been formally reviewed and agreed on at a given point in time, and which can be changed only through change control procedures (NIST.SP.800-53r4).
Change Management	Change Management seeks to minimize the risk associated with Changes, where ITIL defines a Change as "the addition, modification of removal of anything that could have an effect on IT services". This includes Changes to the IT infrastructure, processes, documents, supplier interfaces, etc. (ITIL 4).
Capability	Capability is a measure of the ability of an entity (department, organization, person, system) to achieve its objectives, especially in relation to its overall mission (www.businessdictionary.com/definition/capability.html).
CI/CD Pipeline	The set of tools and the associated process workflows to achieve continuous integration and continuous delivery with build, test, security, and release delivery activities, which are steered by a CI/CD orchestrator and automated as much as practice allows.
CI/CD Pipeline Instance	A single process workflow and the tools to execute the workflow for a specific software language and application type. As much of the pipeline process is automated as is practicable.
Cloud computing / Cloud	The use of computing resources — servers, database management, data storage, networking, software applications, and special capabilities such as blockchain and artificial intelligence (AI) — over the internet, as opposed to owning and operating those resources yourself, on premises. Compared to traditional IT, cloud computing offers organisations a host of benefits: the cost-effectiveness of paying for only the resources you use; faster time to market for mission-critical applications and services; the ability to scale easily, affordably and — with the right cloud provider — globally; and much more. And many organisations are seeing additional benefits from combining public cloud services purchased from a cloud services provider with private cloud infrastructure they operate themselves to deliver sensitive applications or data to customers, partners and employees. There are 4 well-known types of Cloud services: • **Infrastructure as a Service (IaaS):** the original cloud computing service, which provides foundational computing resources — physical or virtual servers, operating system software, storage, networking infrastructure, data centre space — that you use over an internet connection on a pay-as-you-use basis. IaaS lets you rent physical IT infrastructure for building your own remote data centre on the cloud, instead of building a data centre on premises. • **Platform as a Service (PaaS):** provides a complete cloud-based platform for developing, running and managing applications without the cost, complexity and inflexibility of building and maintaining that platform on premises. The PaaS provider hosts everything — servers, networks, storage, operating system software, databases — at its data centre. Development teams can use all of it for a monthly fee based on usage, and can purchase more resources on demand, as needed. With PaaS you can deploy web and mobile applications to the cloud in minutes, and innovate faster and more cost-effectively in response to market opportunities and competitive threats. • **Serverless Computing:** is a hyper-efficient PaaS, differing from conventional PaaS in two important ways: o Serverless offloads all responsibility for infrastructure management tasks (scaling, scheduling, patching, provisioning) to the cloud provider, allowing developers to focus all their time and energy on code. o Serverless runs code only on demand — that is, when requested by the application, enabling the cloud customer to pay for compute resources only when they code is running. With serverless, you never pay for idle computing capacity. • **Software as a Service (SaaS):** application software that runs in the cloud, and which customers use via internet connection, usually in a web browser, typically for a monthly or annual fee. SaaS is still the most widely used form of cloud computing. SaaS lets start using software rapidly: just sign up and get to work. It lets you access your specific instance of the application and your data from any computer, and typically from any mobile device. If your computer or mobile device breaks, you don't lose your data (because it's all in the cloud). The software scales as needed. And the SaaS vendor applies fixes and updates without any effort on your part. (https://www.ibm.com/cloud/learn/cloud-computing).
Code	Software instructions for a computer, written in a programming language. These instructions may be I the form of either human-readable source code, or machine code, which is source code that has been compiled into machine executable instructions.
Common Secure Configuration	A recognized standardized and established benchmark that stipulates specific secure configuration settings for a given information technology platform (NIST.SP.800-53r4).
Configuration Item	An aggregation of information system components that is designated for configuration management and treated as a single entity in the configuration management process (NIST.SP.800-53r4).
Configuration Management	A collection of activities focused on establishing and maintaining the integrity of information technology products and information systems, through control of processes for initializing, changing, and monitoring the configurations of those products and systems throughout the system development life cycle (NIST.SP.800-53r4).
Configuration Settings	Set of parameters that can be changed in hardware, software, or firmware that affect the security posture and/or functionality of the information system (NIST.SP.800-53r4).
Containers	A standard unit of software that packages up code and all its dependencies, down to, but not including the Operating System (OS). It is a lightweight, standalone, executable package of software that includes everything needed to run an application except the OS: code, runtime, system tools, system libraries and settings. Several containers can run in the same OS without conflicting with one another. Since they run on the OS, no hypervisor (virtualization) is necessary (though the OS itself may be running on a hypervisor). Containers are much smaller than a VM, typically by a factor of 1,000 (MB vs GB), partly because they don't need to include the OS. Using containers allows denser packing of applications than VMs. Unlike VMs, containers are portable between clouds or between cloud and on-premise servers. This helps alleviate Cloud Service Provider (CSP) lock-in, though an application may still be locked-in to a CSP, if it uses CSP-specific services. Containers also start much faster than a VM (seconds vs. minutes), partly because the OS doesn't need to boot.

continued on following page

Table 1. Continued

Term	Definition
Control Objective	A Control Objective is an assessment object that defines the risk categories for a Process or Sub-Process. Control Objectives define the compliance categories that the Controls are intended to mitigate. Control Objectives can be classified into categories such as Compliance, Financial Reporting, Strategic, Operations, or Unknown. After a Control Objective is identified, the Risks belonging to that Control Objective can then be defined. In most cases, each Control Objective has one Risk that is associated with it. However, it might also have more than one Risk. For example, a financial services company employs traders that are aware of the required ethical standards. The HR department sets up a control objective called 'Personnel'. A risk that is associated with the Control Objective is, "Employees engage in business dealings that conflict with the company objectives for ethical and fair trading." (IBM Knowledge Center, https://www.ibm.com/support/knowledgecenter/SSFUEU_7.3.0/com.ibm.swg.ba.cognos.op_app_help.7.3.0.doc/c_about_ctrl_obj.html).
Cyber Attack	An attack, via cyberspace, targeting an enterprise's use of cyberspace for the purpose of disrupting, disabling, destroying, or maliciously controlling a computing environment/infrastructure; or destroying the integrity of the data or stealing controlled information (NIST.SP.800-53r4).
Cyber Security	The ability to protect or defend the use of cyberspace from cyber-attacks (NIST.SP.800-53r4).
DevOps	DevOps is an approach which streamlines interdependencies between development and operations through a set of protocols and tools. DevOps facilitates an enhanced degree of agility and responsiveness through continuous integration, continuous delivery, and continuous feedback loops between Development teams and Operation teams (McCarthy M. A., 2014).
Enterprise Architecture	A strategic information asset base, which defines the mission; the information necessary to perform the mission; the technologies necessary to perform the mission; and the transitional processes for implementing new technologies in response to changing mission needs; and includes a baseline architecture; a target architecture; and a sequencing plan (NIST.SP.800-53r4).
Event	Any observable occurrence in an information system (NIST.SP.800-53r4).
External Information System Service Provider	A provider of external information system services to an organisation through a variety of consumer-producer relationships including but not limited to: joint ventures; business partnerships; outsourcing arrangements (i.e., through contracts, interagency agreements, lines of business arrangements); licensing agreements; and/or supply chain exchanges (NIST.SP.800-53r4).
External Network	A network not controlled by the organisation (NIST.SP.800-53r4).
Hardware	The physical components of an information system (NIST.SP.800-53r4).
Impact	The effect on organisational operations, organisational assets, individuals, other organisations, or the Nation (including the national security interests of the United States) of a loss of confidentiality, integrity, or availability of information or an information system (NIST.SP.800-53r4).
Incident	An occurrence that actually or potentially jeopardizes the confidentiality, integrity, or availability of an information system or the information the system processes, stores, or transmits or that constitutes a violation or imminent threat of violation of security policies, security procedures, or acceptable use policies (NIST.SP.800-53r4).
Information	Any communication or representation of knowledge such as facts, data, or opinions in any medium or form, including textual, numerical, graphic, cartographic, narrative, or audio-visual (NIST.SP.800-53r4).
Information Leakage	The intentional or unintentional release of information to an untrusted environment (NIST.SP.800-53r4).
Information Security	The protection of information and information systems from unauthorized access, use, disclosure, disruption, modification, or destruction in order to provide confidentiality, integrity, and availability (NIST.SP.800-53r4).
Information Security Architecture	An embedded, integral part of the enterprise architecture that describes the structure and behaviour for an enterprise's security processes, information security systems, personnel and organisational subunits, showing their alignment with the enterprise's mission and strategic plans (NIST.SP.800-53r4).
Information Security Risk	The risk to organisational operations (including mission, functions, image, reputation), organisational assets, individuals, other organisations, and the Nation due to the potential for unauthorized access, use, disclosure, disruption, modification, or destruction of information and/or information systems (NIST.SP.800-53r4).
Information Technology	Any equipment or interconnected system or subsystem of equipment that is used in the automatic acquisition, storage, manipulation, management, movement, control, display, switching, interchange, transmission, or reception of data or information by the executive agency. For purposes of the preceding sentence, equipment is used by an executive agency if the equipment is used by the executive agency directly or is used by a contractor under a contract with the executive agency which: (i) requires the use of such equipment; or (ii) requires the use, to a significant extent, of such equipment in the performance of a service or the furnishing of a product. The term information technology includes computers, ancillary equipment, software, firmware, and similar procedures, services (including support services), and related resources (NIST.SP.800-53r4).
Integrity	Guarding against improper information modification or destruction, and includes ensuring information non-repudiation and authenticity (NIST.SP.800-53r4).
Internal Network	A network where: (i) the establishment, maintenance, and provisioning of security controls are under the direct control of organisational employees or contractors; or (ii) cryptographic encapsulation or similar security technology implemented between organisation-controlled endpoints, provides the same effect (at least with regard to confidentiality and integrity). An internal network is typically organisation-owned, yet may be organisation-controlled while not being organisation-owned (NIST.SP.800-53r4).
Local Access	Access to an organisational information system by a user (or process acting on behalf of a user) communicating through a direct connection without the use of a network (NIST.SP.800-53r4).
Network	Information system(s) implemented with a collection of interconnected components. Such components may include routers, hubs, cabling, telecommunications controllers, key distribution centres, and technical control devices (NIST.SP.800-53r4).
Organisation	An entity of any size, complexity, or positioning within an organisational structure (e.g., a federal agency or, as appropriate, any of its operational elements) (NIST.SP.800-53r4).
Penetration Testing	A test methodology in which assessors, typically working under specific constraints, attempt to circumvent or defeat the security features of an information system (NIST.SP.800-53r4).
Privileged Account	An information system account with authorizations of a privileged user (NIST.SP.800-53r4).
Privileged User	A user that is authorized (and therefore, trusted) to perform security-relevant functions that ordinary users are not authorized to perform (NIST.SP.800-53r4).
Risk	A measure of the extent to which an entity is threatened by a potential circumstance or event, and typically a function of: (i) the adverse impacts that would arise if the circumstance or event occurs; and (ii) the likelihood of occurrence. Information system-related security risks are those risks that arise from the loss of confidentiality, integrity, or availability of information or information systems and reflect the potential adverse impacts to organisational operations (including mission, functions, image, or reputation), organisational assets, individuals, other organisations, and the Nation (NIST.SP.800-53r4).
Risk Assessment	The process of identifying risks to organisational operations (including mission, functions, image, reputation), organisational assets, individuals, other organisations, and the Nation, resulting from the operation of an information system. Part of risk management, incorporates threat and vulnerability analyses, and considers mitigations provided by security controls planned or in place. Synonymous with risk analysis (NIST.SP.800-53r4).
Risk Management	The program and supporting processes to manage information security risk to organisational operations (including mission, functions, image, reputation), organisational assets, individuals, other organisations, and the Nation, and includes: (i) establishing the context for risk-related activities; (ii) assessing risk; (iii) responding to risk once determined; and (iv) monitoring risk over time (NIST.SP.800-53r4).

continued on following page

Table 1. Continued

Term	Definition
Risk Mitigation	Prioritizing, evaluating, and implementing the appropriate risk reducing controls/countermeasures recommended from the risk management process (NIST.SP.800-53r4).
Role-Based Access Control	Access control based on user roles (i.e., a collection of access authorizations a user receives based on an explicit or implicit assumption of a given role). Role permissions may be inherited through a role hierarchy and typically reflect the permissions needed to perform defined functions within an organisation. A given role may apply to a single individual or to several individuals (NIST.SP.800-53r4).
SecDevOps / DevSecOps	DevSecOps strives to automate core security tasks by embedding security controls and processes into the DevOps workflow. DevSecOps originally focused primarily on automating code security and testing, but now it also encompasses more operations-centric controls. Security can benefit from automation by incorporating logging and event monitoring, configuration and patch management, user and privilege management, and vulnerability assessment into DevOps processes. In addition, DevSecOps provides security practitioners with the ability to script and monitor security controls at a much larger and more dynamic scale than traditional in-house data centres (Shackleford, D., SANS, 2016).
Security	A condition that results from the establishment and maintenance of protective measures that enable an enterprise to perform its mission or critical functions despite risks posed by threats to its use of information systems. Protective measures may involve a combination of deterrence, avoidance, prevention, detection, recovery, and correction that should form part of the enterprise's risk management approach (NIST.SP.800-53r4).
Security Control	Can be effectively used to protect information and information systems from traditional and advanced persistent threats in varied operational, environmental, and technical scenarios. The controls can be used to demonstrate compliance with a variety of governmental, organisational, or institutional security requirements. Organisations have the responsibility to select the appropriate security controls, to implement the controls correctly, and to demonstrate the effectiveness of the controls in satisfying established security requirements. The security controls facilitate the development of assessment methods and procedures that can be used to demonstrate control effectiveness in a consistent/repeatable manner—thus contributing to the organisation's confidence that security requirements continue to be satisfied on an ongoing basis. In addition, security controls can be used in developing overlays for specialized information systems, information technologies, environments of operation, or communities of interest (NIST.SP.800-53r4).
Security Objective	Confidentiality, integrity, or availability (NIST.SP.800-53r4).
Security Policy	A set of criteria for the provision of security services (NIST.SP.800-53r4).
Security Requirement	A requirement levied on an information system or an organisation that is derived from applicable laws, Executive Orders, directives, policies, standards, instructions, regulations, procedures, and/or mission/business needs to ensure the confidentiality, integrity, and availability of information that is being processed, stored, or transmitted (NIST.SP.800-53r4).
Software	Computer programs and associated data that may be dynamically written or modified during execution (NIST.SP.800-53r4).
Threat	Any circumstance or event with the potential to adversely impact organisational operations (including mission, functions, image, or reputation), organisational assets, individuals, other organisations, or the Nation through an information system via unauthorized access, destruction, disclosure, modification of information, and/or denial of service (NIST.SP.800-53r4).
Threat Assessment	Formal description and evaluation of threat to an information system (NIST.SP.800-53r4).
User	Individual, or (system) process acting on behalf of an individual, authorized to access an information system (NIST.SP.800-53r4).
Vulnerability	Weakness in an information system, system security procedures, internal controls, or implementation that could be exploited or triggered by a threat source (NIST.SP.800-53r4).
Vulnerability Assessment	Systematic examination of an information system or product to determine the adequacy of security measures, identify security deficiencies, provide data from which to predict the effectiveness of proposed security measures, and confirm the adequacy of such measures after implementation (NIST.SP.800-53r4).
Waterfall	Is considered as a traditional way of managing software development. It consists of consecutive phases leading to the development of a *complete* software product: • **Define**: collect software requirements; • **Build**: design the solution, code the software, test and debug the software; • **Release**: deliver the finished product and enter maintenance mode (Derksen, October 2018).

While the examples mentioned above show compliance as an obstacle to deploying an efficient and automated DevOps process, (Laukkarinen T. K., May 2017) use the example of medical device and health software IEC/ISO standards to show that DevOps can in certain cases also be used as a helpful tool to ensure compliance (Plant, 2019). They found that DevOps was beneficial for implementing most requirements. For example, clause 5.8.6 of IEC 62304 for medical device software requires that the procedure and environment of the software creation has to be documented. In DevOps, this can easily be done with development tools such as the project management tool JIRA, source code repositories such as GIT and automation software such as Jenkins. Furthermore, using invariable Docker containers allows for a repeatable installation and release process which is required by clause 5.8.8. However, the authors also identified three obstacles that slow down the CI and CD procedures. Firstly, software units have to be verified, which means that Continuous Integration can only happen after all units have passed unit testing. Secondly, all tasks and activities such as unfinished documentation have to be completed before the release of a software unit. Lastly, Continuous Deployment through remote updating to the customer is not possible with IEC 82304-1 because the responsibility has to be transferred explicitly to the customer when taking the software into use.

(Laukkarinen (Laukkarinen T. K., 2018) concluded in a follow-up paper on DevOps in regulated software environments that tighter integration between development tools, requirements management, version control and the deployment pipeline would aid the creation of regulatory compliance development practices. However, the authors also note that regulations and accompanied standards could be improved to better relate regulations with DevOps practices. On the other hand, (Farroha, 2014) suggests that to integrate compliance and security throughout a deployed cloud application, automated testing for non-compliance and policy needs to be leveraged. Additionally, the enterprise should pre-build, test and certify services in a Services Catalog. To ensure compliance and security rules are adhered to, continuous monitoring/alerting and automation to detect and mitigate critical issues must be implemented. Finally, to track breaches in compliance, the application will need to do the following: a) automatic reporting for compliance violations; b) terminate access when exceeding a threshold; c) initiate alarms when new policy is not accepted.

The above discussion leads to the following problem statement:

Whether DevOps is a benefit to achieving security compliance or compliance is an obstacle to realizing DevOps heavily depends on the regulations applicable to the organisation at hand. In order to ensure compliance, the DevOps process in some cases needs to be slowed down or put on halt until other tasks are completed.

Scope

Not every organisation has to comply with regulatory requirements and also the impact of compliancy varies between companies. Many companies are driven not by regulatory compliance but by varying internal and customer demand for secure solutions. A number of interviews with representatives from companies that do not have regulatory compliance obligations[1], have shown that the approach of DevOps may differ from the approach of companies in heavily regulated industries. They may prioritise productivity and quality, facilitated by the use of DevOps and continuous deployment (Savor, 2016). At the same time, regulated companies should take into account the preservation of their operational licenses, while trying to achieve the same level of productivity and quality. This difference results in a significant impact on the way in which DevOps, and Agile in general, is integrated into the organisation.

Regulatory compliance of the organisation describes the efforts the organisation is taking to comply with relevant lows, policies and regulations (Lin, 2016). Due to the increasing complexity and number of regulations, organisations are often referring to the use of consolidated and approved sets of compliance controls described in compliancy frameworks and standards (Silveira, 2012). This approach allows to ensure that all the necessary governance requirements are met without duplicate efforts.

Applicable compliancy frameworks vary among countries and industries, with examples such as PCI-DSS for financial industry, FISMA for U.S. federal agencies, HACCP for food and beverage industry, and HIPAA in healthcare. Other commonly applicable references are COBIT framework and ISO and NIST standards.

Since it is impossible to cover the complete spectrum of organisations and their compliancy frameworks in a single study, the scope is limited to large-size European companies (more than 1000 employees) located in Belgium and the Netherlands. Their approach for addressing the regulatory compliancy in combination with DevOps is evaluated.

As a reference, providing the set of applicable compliancy controls, two standards are selected for evaluation within this research project: ISO/IEC 27001 and NIST.SP.800-53r4. ISO/IEC 27001 is an Information Security Standard and is a part of ISO/IEC 27000 family of standards. It specifies an Information Security Management System (ISMS). ISMS is a framework of policies and procedures that includes all legal, physical and technical controls involved in an organisation's information risk management processes. The use of ISO 27001 is widely spread in the European Union. Similar to ISO (International Organisation for Standardization), also NIST (National Institute of Standards and Technology) define industry-leading approach to Information Security Management. However, NIST 800-53 is more security control driven, more technical and less risk focused than ISO 27001. The standard is mainly used in the USA, but also regularly applied as a reference in European companies, which is facilitated by many synergies between both standards and predefined mappings of controls. All companies, considered in scope of this study are relying on ISO/IEC 27001, NIST.SP.800-53r4 or both to implement the ISMS.

Research Question

The main goal of this research is to investigate the impact of DevOps on the compliance with the ISO/IEC 27001 and NIST.SP.800-53r4 standards in heavily regulated large organisations. This goal will be achieved by addressing the following research questions:

- **Research Question 1:** What are the major security compliance controls that are impacted by DevOps adoption?
- **Research Question 2:** Which "sacrifices" have to be made in DevOps implantation in order to preserve compliance?
- **Research Question 3:** How can DevOps assist in assuring compliance?
- **Research Question 4:** What are the best practices facilitating the implementation of SecDevOps (i.e. the integration of security into DevOps)?

The deliverable of this research is an **SecDevOps Capability Artifact**, which consists of the following components:

- A list of governance and security control objectives, as defined by ISO/IEC 27001 and NIST.SP.800-53r4, which are generally impacted when DevOps capabilities are roll-out by an organisation
- A mapping between respective ISO/IEC 27001 and NIST.SP.800-53r4 control objectives, from the perspective of DevOps implementation
- DevOps control objectives, with a link to the corresponding ISO and NIST objectives, and an indication of the impact. The impact of DevOps control objectives on ISO and NIST objectives may be twofold: 1) it creates an **Opportunity** to achieve compliance assurance in a more effective or efficient way; 2) it creates a **Risk** of sacrificing a security control objective in favour of flexibility and speed
- A list of (Sec)DevOps controls that have proven to be effective in combining the agility of the DevOps paradigm with the security compliance assurance.

According to an online Business Dictionary[1], Capability is a measure of the ability of an entity (department, organization, person, system) to achieve its objectives, especially in relation to its overall mission. The SecDevOps Capability Artifact defined in this study, will, therefore, increase company's (Sec)DevOps capability in the context of regulatory compliance with ISO/IEC 27001 and NIST.SP.800-53r4 standards.

RESEARCH APPROACH

Research Methodology

The methodology selected to answer the above-mentioned research questions is known in the scientific literature as **Design Science Research** (DSR) (Hevner A. M., 2004). The primary differentiator of DSR, compared to other research methods, is that its focus lies with the design and investigation of artifacts in a specific context, making use of the existing knowledge base. According to Hevner, the main purpose of design science research is achieving knowledge and understanding of a problem domain by building and application of a designed artifact.

Figure 4. Design Science Research (DSR) Cycle

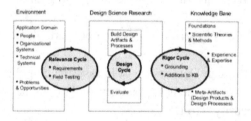

DSR consists of three inherent research cycles, as shown in Figure 4 (Hevner A., 2007):

- The **Relevance Cycle** bridges the contextual environment of the research project with the design science activities.
- The **Rigor Cycle** connects the design science activities with the knowledge base of scientific foundations, experience, and expertise that informs the research project.
- The central **Design Cycle** iterates between the core activities of building and evaluating the design artifacts and processes of the research.

In this research project a single artifact will be designed, a so-called **SecDevOps Capability Artifact.** A process for designing an artifact is following the approach described by (Johannesson, 2014), which can be found in Figure 4. Due to the limited timeline foreseen for this project, this research is limited to the Problem Space of the design (green rectangle in Figure 5): focussing on the literature field study, expert interviews and comparative analysis. The remaining steps of the design are left open for further study.

Figure 5. A Design Science Research approach to developing artifact requirements (green square indicates the research space).

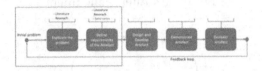

More precisely, the DSR research cycles will be approached as follows:

- The **Relevance Cycle** is to be addressed through expert interviews. This methodology bridges the existing knowledge gap by making the problem explicit based on the theoretical constructs and concepts, derived from the scientific literature. These concepts are then validated through semi-structured interviews with the domain experts.
- The **Rigor Cycle** will be addressed by a detailed literature study w.r.t. existing recommendations for the implementation of (Sec)DevOps capabilities within regulated environments, as well as earlier performed evaluations and research. It provides the scientific ground and ensures the research innovativeness.
- The **Design Cycle**, evaluating the proposed artifact, will be out of scope for this research project.

The main artifact developed in the scope of this study can be described as follows:

A mapping between ISO/IEC 27001 and NIST.SP.800-53 control objectives and the corresponding DevOps controls objectives & controls. This mapping allows large regulated organisations to design and implement their DevOps strategy and practices, while finding the optimal balance between the security compliance and the speed & flexibility of DevOps.

To start the exploration into the main capabilities for an effective & efficient DevOps strategy, in line with the control objectives of the selected standards, a combination of **literature** and **exploratory research** was used. A literature study is conducted to gather knowledge about the domain of the topic of interest and knowledge about relevant theories and research methods that can be applied to develop new knowledge (Recker, 2013). Exploratory research, on the other hand, is used to investigate the problem which is not clearly defined. It is conducted to have a better understanding of the existing problem, but will not provide conclusive results. For such a research, a researcher starts with a general idea and uses this research as a medium to identify issues, that can be the focus for future research. An important aspect of this approach is that the researcher should be willing to change the research direction subject to the revelation of new data or insight. Such a research is usually carried out when the problem is at a preliminary stage. It is often referred to as **grounded theory approach** or interpretive research as it used to answer questions like what, why and how.

Grounded Theory (Recker, 2013) is a type of qualitative research that relies on inductive generation (building) of theory based on ('grounded in') qualitative data systematically collected and analysed about a phenomenon, such as existing DevOps implementations within relevant organisations. The grounded theory approach essentially attempts to explore for, and develop, generalized formulations about the basic features of a phenomenon while simultaneously grounding the account in empirical observations or data. One of the key advantages – and challenges – of the grounded theory approach is that it is applicable to research domains that are new or emergent and may yet lack substantive theory.

Figure 6. Enterprise Strategic Alignment (Stackpole, 2010).

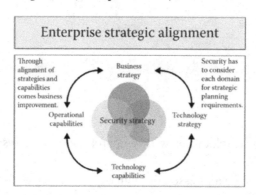

The theory follows two primarily principles:

- The process of theory building is highly iterative, during which theory and data are constantly compared, so-called **constant comparative analysis**. This step is achieved by deriving theoretical statements from the conducted literature study and comparing them against the practical examples and knowledge from the field via expert interviews.
- The theory is built upon theoretical sampling as a process of data collection and analysis, which is driven by concepts that emerge from the study and appear to be of relevance to the nascent theory. In the case of this research, the results collected from the previous step are validated through additional literature research, to confirm the established theory.

Theoretical Foundation

Software is nowadays a critical asset of almost every organisation, but producing great applications and services within a competitive timeline requires modern development and delivery processes.

Fundamental to this challenge is building trust, managing risk and exceeding the expectations of the customers for security and privacy with the business, online, via apps and in the data centres. It is becoming imperative to weave security into every step of the development process, from design, through coding, to release and operation, which is supported by the concepts such as SecDevOps. Research by (CA Technologies, 2014) confirmed that organisations which see effective security as an enabler of increased business performance significantly over-perform their mainstream peers. This manifests itself in the form of superior metrics and outcomes in relation to software delivery. It's probably also no coincidence that these security-minded organisations are seeing 40 percent higher revenue growth and 50 percent higher profit growth than their mainstream contemporaries.

As was demonstrated in the previous sections, many research efforts are dedicated to determining the right set of security measures in the context of the Agile development. These measures are not identical for each type of organisation and depend on varying factors, of which the strategy is the most important one (Stackpole, 2010). Stackpole points to the fact that many organisations try to shortcut the analysis phase and end up failing to include business drivers, business unit direction or big-picture input into their planning cycles. When not enough time has been spent gathering big-picture, the likelihood increases that the organisation will be more reactive to the environment than proactively helping shape the environment. In marketing jargon, this would be called market-shaping activities instead of market-reacting activities. Market-shaping activities involve the identification of the drivers shaping demand, a survey of what existing products and services might be to meet that demand, which in turn helps identify gaps in the market and the development of a strategy for market-shaping activities. A similar approach can be used to plan a proactive security strategy. First, gather the information needed to identify the issues affecting organisational security (now and into the future), and then compare existing and future requirements to your current capabilities to identify gaps in security functionality. Next, build a strategic plan to ðll those gaps. Figure 6 charts some of the basic domains within an enterprise that a security group must consider as it develops strategy.

A typical **security strategy** is a plan to mitigate risks while complying with legal, statutory, contractual, and internally developed requirements. But a security strategy resides inside an organisational strategy that may have very different drivers than a security strategic plan. In order for the two strategic plans to align and work well together, there must be a clear understanding of both plans and clear links between them. When designing the enterprise-wide, as well as the security strategy, the following input should be considered (Stackpole, 2010):

- Environmental scan (e.g. industry & competitor analysis, marketing research, technology trends, etc.)
- Regulations & legal environment
- Industry standards
- Customer base
- Organisational Culture
- Business Drivers

Defining the strategy means understanding the strategic risks, both internal and external. Security risks have a number of touchpoints with general risk management frameworks (see Figure 7), as outlined in CFO Forum Practitioner's Guide (CFO Forum):

- **Strategy definition**: assures alignment between the organisation's overall aspirations and the high-level security plan; identifies any associated risks or uncertainty that may arise; defines risk capacity & appetite.
- **Strategy implementation**: as with any other business or functional area, a set of objectives and plans to support the security strategy needs to be defined; standard risk management processes, including a detailed risk assessment, need to be implemented, followed by an appropriate risk response (treat; tolerate; terminate; transfer), establishment of a risk appetite and key risk indicators.
- **Strategy monitoring**: the strategy implementation progress should be monitored, together with the key risks and escalation of any breaches of risk appetite.

In the ideal world after the Strategy, Risks and the corresponding Threats are known, it would be possible to answer the question: "Which mechanism and controls shall be applied within Agile Software Development & Operations to make sure a software system is completely secure?". In reality, it is impossible to answer this question since the definition of *completely secure* means that all the possible threats of a software system are identified, which is impossible to guarantee. However, if an **Information**

Security Architecture, Frameworks and Standards are used, an assurance can be provided that security has been sufficiently implemented, taking into account all the *known* threats (Derksen, October 2018).

Figure 7. Enterprise Security Architecture domain (Vael, 2019).

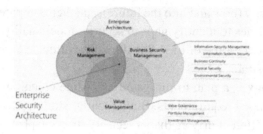

Software products can be extremely diverse and threats to a software system depend on numerous criteria: type of software, how it is built, the environment in which it is used and how it is used. Once the threats are defined, it becomes possible to verify if the software system is sufficiently protected against these threats. This protection is realized by a secure design, by implementing mitigation measures, by coding securely, security testing, etc. The use of Information Security Architecture and corresponding Frameworks and Standards aims to provide a secure software development process, by embedding them into secure organisational cultures, systems and behaviour, which are in line with the organisation's strategy. A more formal definition of Information Security Architecture is provided by Vael (Vael, 2019):

The practice of applying a comprehensive & rigorous method describing current and/or future structure & behaviour for an organisation's security processes, information security systems, personnel & organisational sub-units, so they align with the organisation's core goals & strategic direction.

Table 2.

COBIT® 5 (ISACA)
COBIT is an extensive business framework for governance and management of IT throughout the enterprise. COBIT 5 builds and expands on COBIT 4.1 by integrating other major frameworks, standards and resources, including ISACA's Val IT and Risk IT, Information Technology Infrastructure Library (ITIL®) and related standards from the International Organisation for Standardization (ISO). The following are the major objectives of the framework (COBIT5 website): • Maintain high-quality information to support business decisions • Achieve strategic goals through the effective and innovative use of IT • Achieve operational excellence through reliable, efficient application of technology • Maintain IT-related risk at an acceptable level • Optimize the cost of IT services and technology • Support compliance with relevant laws, regulations, contractual agreements and policies

Advantages	Disadvantages
• The breadth of the framework allows to link different aspects of governance, management & security together providing a holistic view of the whole enterprise • Consistence with main security techniques such as ISO 27005, 27001 and 31000, COSO ERM, NIST	• Selection of relevant components for COBIT roll-out, requires good preparatory work and knowledge of the enterprise • Not designed to managed standalone in-depth Information Security method and therefore it should be integrated with specific security standard

Table 3.

NIST Cybersecurity Framework
The National Institute of Standards and Framework's Cybersecurity Framework (CSF) was published in February 2014 in response to Presidential Executive Order 13636, "Improving Critical Infrastructure Cybersecurity," which called for a standardized security framework for critical infrastructure in the United States (Cipher Website). The NIST CSF is recognized by many as a resource to help improve the security operations and governance for public and private organisations. It provides a guideline for transforming the organisational security posture and risk management from a reactive to proactive approach. The framework is organized into five core Functions also known as the Framework Core. The functions are organized concurrently with one another to represent a security lifecycle. Each function is essential to a well-operating security posture and successful management of cybersecurity risk. Definitions for each Function are as follows: • Identify: Develop the organisational understanding to manage cybersecurity risk to systems, assets, data, and capabilities. • Protect: Develop and implement the appropriate safeguards to ensure delivery of critical infrastructure services. • Detect: Develop and implement the appropriate activities to identify the occurrence of a security event. • Respond: Develop and implement the appropriate activities when facing a detected security event. • Recover: Develop and implement the appropriate activities for resilience and to restore any capabilities or services that were impaired due to a security event

Advantages	Disadvantages
• Available for free from the NIST website are detailed checklists, mathematical formulas and other materials • Frequently assessed and restructured as new technologies/regulations arisen • May be used in conjunction with other NIST frameworks from NIST and/or other parties	• Heavily US-focused supporting documentation • Prescriptive and not easy to adopt • Finding appropriate support to effectively implement the methodology may be difficult outside the US

Table 4.

ISO / IEC 27002
A popular, internationally-recognised, standard developed by the International Organisation for Standardization (ISO) and the International Electrotechnical Commission (IEC). The standard is designed for organisations to use as a reference for selecting controls within the process of implementing an Information Security Management System (ISMS) based on ISO/IEC 27001 or as a guidance document for organisations implementing commonly accepted information security controls. This standard is also intended for use in developing industry- and organisation-specific information security management guidelines. It defines controls that should be established, monitored, reviewed and improved, where necessary, to ensure that the specific security and business objectives of the organisation are met. An ISMS such as that specified in ISO/IEC 27001 takes a holistic, coordinated view of the organisation's information security risks in order to implement a comprehensive suite of information security controls under the overall framework of a coherent management system. Many information systems have not been designed to be secure in the sense of ISO/IEC 2700 and this standard. The security that can be achieved through technical means is limited and should be supported by appropriate management and procedures. Identifying which controls should be in place requires careful planning and attention to detail. A successful ISMS requires support by all employees in the organisation. It can also require participation from shareholders, suppliers and other external parties. This standard helps to define the requirements for the different participants and stakeholders.

Advantages	Disadvantages
• Extensive set of best practices to build a comprehensive IT security program • Widely used by multinational corporations and companies that do not have to comply with US federal regulations • Less complex and easier to implement than NIST SP 800-53 • Contains security controls mapping with NIST SP 800-53	• Is not freely available and has to be purchased from ISO/IEC • Is a subset of NIST SP 800-53 in terms of defined security controls

Table 5.

NIST Special Publication 800-53
This Special Publication is published by the National Institute of Standards and Technology (NIST). NIST is responsible for developing information security standards and guidelines, including minimum requirements for federal information systems. This publication provides a catalogue of security and privacy controls for federal information systems and organisations and a process for selecting controls to protect organisational operations (including mission, functions, image, and reputation), organisational assets, individuals, other organisations, and the Nation from a diverse set of threats including hostile cyber-attacks, natural disasters, structural failures, and human errors. The controls are customizable and implemented as part of an organisation-wide process that manages information security and privacy risk. The controls address a diverse set of security and privacy requirements across the federal government and critical infrastructure, derived from legislation, Executive Orders, policies, directives, regulations, standards, and/or mission/business needs. The publication also describes how to develop specialized sets of controls, or overlays, tailored for specific types of missions/business functions, technologies, or environments of operation. Finally, the catalogue of security controls addresses security from both a functionality perspective (the strength of security functions and mechanisms provided) and an assurance perspective (the measures of confidence in the implemented security capability). Addressing both security functionality and security assurance ensures that information technology products and the information systems built from those products using sound systems and security engineering principles are sufficiently trustworthy. The goal of the publication is to achieve more secure information systems and effective risk management within the US federal government by: • Facilitating a more consistent, comparable, and repeatable approach for selecting and specifying security controls for information systems and organisations; • Providing a stable, yet flexible catalogue of security controls to meet current information protection needs and the demands of future protection needs based on changing threats, requirements, and technologies; • Providing a recommendation for security controls for information systems categorized in accordance with FIPS Publication 199, Standards for Security Categorization of Federal Information and Information Systems; • Creating a foundation for the development of assessment methods and procedures for determining security control effectiveness; and • Improving communication among organisations by providing a common lexicon that supports discussion of risk management concepts.

Advantages	Disadvantages
• Available for free from the NIST website • Frequently assessed and restructured as new technologies/regulations arisen • May be used in conjunction with other NIST frameworks from NIST and/or other parties • Contains security controls mapping with ISO 27002 standard, which is broadly applied in Europe • Contains more practical and detailed security controls than ISO 27002	• Heavily US-focused supporting documentation • Prescriptive and not easy to adopt • Finding appropriate support to effectively implement the methodology may be difficult outside the US

As shown in Figure 7 (Vael, 2019), Security Architecture is a sub-domain of Enterprise Architecture, Risk Management, Value Management and Business Security Management. One of the main advantages of the use of the security architecture is that it facilitates communication and collaboration with all stakeholders within an organisation through the use of a common framework. There exist a number of Security Frameworks and Standards, widely applied throughout the industry. Four of them are considered during this study and an overview is provided in Tables 2 through 5.

The examples above are just some of the many available frameworks. For a more complete list of cybersecurity & privacy related statutory, regulatory and industry frameworks see Annex A. Overview of Major Cybersecurity & privacy-related frameworks. There are many others generic or specific ones, such as PCI DSS, CSC / TOP 20, SCF (Secure Controls Framework), etc. However, the selection made gives an overview of high-level versus more detailed widely-used frameworks and standards.

It is important to mention that the selection of a cybersecurity framework is in the first place a business decision and less a technical decision. The choice must be driven by a fundamental understanding of what the organisation needs to comply with from a statutory, regulatory and contractual perspective, since that understanding establishes the minimum set of requirements necessary to (1) not be considered negligent with reasonable expectations for security & privacy; (2) comply with applicable laws, regulations and contracts; and (3) implement the proper controls to secure systems, applications and processes from reasonable threats (Compliance Forge Website, sd).

For the purpose of this study, the four considered frameworks / standards (COBIT 5, NIST Cybersecurity Framework, NIST SP 800-53 and ISO 27002) are reviewed from the point of view of the content each one is offering. The content had to provide sufficient security and privacy controls "out of the box" in order to avoid the need for further detailing and interpreting while mapping the DevOps controls. From that prospective, COBIT and NIST Cybersecurity Framework lacked the sufficient coverage to be considered a comprehensive cybersecurity framework, giving the preference to NIST SP 800-53 and ISO 27002. These considerations are extensively discussed with the promotor of this research project, prof. Yuri Bobbert, leading to the decision of limiting this study to the two above-mentioned and widely-used standards, NIST SP 800-53 and ISO 27002. While ISO 27002 is a subset of NIST SP 800-53, where fourteen ISO 27002 sections of security controls fit within twenty-six families of NIST SP 800-53, both are widely-used and suitable for large regulated companies. Therefore, both standards are considered for comparison and mapping with (Sec)DevOps controls by this research.

REFERENCES

Aljundi, M. (2018). *Tools and Practices to Enhance DevOps Core Values* (Master's Thesis). Lappeenranta University of Technology, School of Business and Management.

Bass, L. H. (2015). Securing a Deployment Pipeline. *Proceedings of the Third International Workshop on Release Engineering*.

Betz, C. O. (2016). The Impact of Digital Transformation, Agile, and DevOps on Future IT Curricula. Boston, MA: SIGITE'16.

Bucena, I. K. (2017). Simplifying the DevOps Adoption Process. *BIR Workshop*.

Callanan, M. S. (2016). DevOps: Making It Easy to Do the Right Thing. *IEEE Software*, *33*(3), 53–59.

Chivers, H. P. (2005). *Agile Security Using an Incremental Security Architecture* (Vol. 3556). Extreme Programming and Agile Processes in Software Engineering. doi:10.1007/11499053_7

Clager, J. R. (2016). Mitigating an Oxymoron: Compliance in a DevOps Environments. *2016 IEEE 40th Annual Computer Software and Applications Conference (COMPSAC)*.

Colavita, F. (2016). DevOps Movement of Enterprise Agile Breakdown Silos, Create Collaboration, Increase Quality, and Application Speed. *Proceedings of 4th International Conference in Software Engineering for Defence Applications Advances in Intelligent Systems and Computing*. 10.1007/978-3-319-27896-4_17

Compliance Forge Website. (n.d.). *Which framework is right for my business? NIST Cybersecurity Framework vs. ISO 27002 vs. NIST 800-53 vs. Secure Controls Framework*. Retrieved from https://www.complianceforge.com/faq/nist-800-53-vs-iso-27002-vs-nist-csf.html

Delphic, G. (2016). *2015 Annual State of DevOps*. Academic Press.

Derksen, B. N. (2018). *Agile Secure Software Lifecycle Management Secure by Agile Design.* Leiden: Secure Software Alliance.

Farroha, B. F. (2014). A Framework for Managing Mission Needs, Compliance, and Trust in the DevOps Environment. *Proceedings of the 2014 IEEE Military Communications Conference.* 10.1109/MILCOM.2014.54

Forsgren, N. H. (2018). *Accelerate: The Science of Lean Software and Devops: Building and Scaling High Performing Technology Organizations.* IT Revolution Press.

Forum, C. F. O. (n.d.). *Understanding and managing the IT risk landscape: A practitioner's guide.* https://www.thecroforum.org/2018/12/20/understanding-and-managing-the-it-risk-landscape-a-practitioners-guide/

Fowler, M. H. (2001). *The Agile Manifesto.* Academic Press.

Gill, A. L. (2017). DevOps for Information Management Systems. *VINE Journal of Information and Knowledge Management Systems.*

Harkins, M. (2013). A New Security Architecture to Improve Business Agility. In Managing Risk and Information Security. doi:10.1007/978-1-4302-5114-9_7

Hevner, A. (2007). A Three Cycle View of Design Science Research. Proceedings of Scandinavian Journal of Information Systems.

Hevner, A. M., March, Park, & Ram. (2004). Design science in information systems research. 75–105. *Management Information Systems Quarterly, 28*(1), 75. doi:10.2307/25148625

Hilton, M. N. (2017). *Trade-offs in continuous integration: assurance, security, and flexibility.* ESEC/FSE.

Johannesson, P. a. (2014). *An introduction to Design Science.* Springer. doi:10.1007/978-3-319-10632-8

Lam, T. C. (2019). *DoD Enterprise DevSecOps Reference Design.* Academic Press.

Laukkarinen, T. K. (2017). DevOps in regulated software development: Case medical devices. In *Proceedings of 2017 IEEE/ACM 39th International Con-ference on Software Engineering: New Ideas and Emerging Technologies Results Track (ICSE-NIER).* IEEE.

Laukkarinen, T. K., Kuusinen, K., & Mikkonen, T. (2018). Regulated software meets DevOps. *Information and Software Technology, 97,* 176–178. doi:10.1016/j.infsof.2018.01.011

Lin, T. (2016). *Compliance, Technology, and Modern Finance.* Temple University Legal Studies Research Paper No. 2017-06.

Mattetti, M. S.-P. (2015). Securing the infrastructure and the workloads of Linux containers. *Proceedings of the 2015 IEEE Conference on Communications and Network Security (CNS).* 10.1109/CNS.2015.7346869

McCarthy, M. A. (2014). A Compliance Aware Software Defined Infrastructure. *Proceedings of IEEE International Conference on Services Computing.*

McCarthy, M. H. (2015). Composable DevOps: Automated Ontology Based DevOps Maturity Analysis. *IEEE International Conference on Services Computing.*

Mohan, V. O. (2016). SecDevOps: Is It a Marteking Buzzword? *Proceedings of the 11th International Conference on Availability, Reliability and Security (ARES).*

Mohan, V. O. (2018). BP: Security Concerns and Best Practices for Automation of Software Deployment Processes: An Industrial Case Study. Proceedings of 2018 IEEE Cybersecurity Development (SecDev).

Murray, A. (2015). *The Complete Software Project Manager: Mastering Technology from Planning to Launch and Beyond – Agile, Waterfall, and the Key to Modern Project Management.* Academic Press.

Myrbakken, H. C.-P. R. (2017). Software Process Improvement and Capability Determination. In DevSec-Ops: A Multivocal Literature Review (pp. 17-29). Cham, Springer: Communications in Computer and Information Science.

Pastrana, M. (2019). Ensuring Compliance with Sprint Requirements in SCRUM: Preventive Quality Assurance in SCRUM. In Advances in Computer Communication and Computational Sciences (pp. 33-45). Springer.

Plant, O. (2019). *DevOps Under Control: Development of a Framework for Achieving Internal Control and Effectively Managing Risks in a DevOps Environment* (MSc thesis). University of Twente.

Rahman, A. W. (2016). Mitigating an Oxymoron: Compliance in a DevOps Environments. In *Proceedings of the Symposium and Bootcamp on the Science of Security (HotSos '16)* (pp. 109–111). New York, NY: Association for Computing Machinery.

Ravichandran, A. T. (2016). *DevOps for Digital Leaders*. Apress.

Recker, J. (2013). *Scientific Research in information Systems: A Beginner's Guide*. Springer. doi:10.1007/978-3-642-30048-6

Rindell, K. H. (2016). Case Study of Security Development in an Agile Environment: Building Identity Management for a Government Agency. *11th International Conference on Availability, Reliability and Security (ARES).*

Savor, T. D. (2016). Continuous Deployment at Facebook and OANDA. *Proceedings of 2016 IEEE/ACM 38th International Conference on Software Engineering Companion (ICSE-C).*

Schneider, C. (2015). Security DevOps - staying secure in agile projects. *Proceedings of OWASP AppSec Europe.*

Shackleford, D. (2016). A DevSecOps Playbook. SANS Institute.

Shahin, M. B. (2016). The Intersection of Continuous Deployment and Architecting Process: Practitioners' Perspectives. *10th ACM/IEEE International Symposium on Empirical Software Engineering and Measurement (ESEM).*

Shahin, M. B. (2017). Continuous Integration, Delivery and Deployment: A Systematic Review on Approaches, Tools, Challenges and Practices. *IEEE Access: Practical Innovations, Open Solutions, 5,* 3909–3943.

Shahin, M. B. (2017). Beyond Continuous Delivery: An Empirical Investigation of Continuous Deployment Challenges. *2017 ACM/IEEE International Symposium on Empirical Software Engineering and Measurement (ESEM).*

Sharma, S. C. (2015). DevOps for Dummies. John Wiley & Sons.

Silveira, P. R. (2012). Aiding Compliance Governance in Service-Based Business Processes. In Handbook of research on Service-Oriented Systems and Non-Functional Properties (pp. 524–548). Academic Press.

Stackpole, B. O. (2010). *Security Strategy: From Requirements to Reality.* Auerbach Publications. doi:10.1201/EBK1439827338

Stolt, S. N. (2013). Continuous Delivery? Easy! Just Change Everything (Well, Maybe It Is Not That Easy). 2013 Agile Conference.

Storms, A. (2015). How security can be the next force multiplier in devops. *Proceedings of RSA Conference.*

CA Technologies. (2014). *Devops: The Worst-Kept Secret to Winning in the Application Economy.* Author.

Vael, M. (2019, February 22). *Enterprise Information Security Architecture: How to design rock-solid security directly into information systems.* Antwerp, Belgium: Academic Press.

Vaishnavi, M. O. (2016). SecDevOps: Is It a Marketing Buzzword? - Mapping Research on Security in DevOps. *11th International Conference on Availability, Reliability and Security (ARES).*

Yasar, H. (2017). Implementing secure DevOps assessment for highly regulated environments. *Proceedings of the 12th International Conference on Availability, Reliability and Security - ARES '17.*

ENDNOTE

[1]   http://www.businessdictionary.com/definition/capability.html

APPENDIX

Overview of Major Cybersecurity & Privacy-Related Frameworks

Figure 8.

Geography	Icon	Source	Authoritative Source - Statutory / Regulatory / Contractual / Industry Framework	Version
Universal	AICPA	AICPA	Generally Accepted Privacy Principles (GAPP)	N/A
Universal	AICPA	AICPA	Service Organization Control - Trust Services Criteria (TSC) - SOC2	2016
Universal	AICPA	AICPA	Service Organization Control - Trust Services Criteria (TSC) - SOC2	2017
Universal	CIS	CIS	Critical Security Controls (CSC)	6.1
Universal	CIS	CIS	Critical Security Controls (CSC)	7.1
Universal	COSO	COSO	Committee of Sponsoring Organizations (COSO) 2013 Framework	2013
Universal	COSO	COSO	Committee of Sponsoring Organizations (COSO) 2017 Framework	2017
Universal	CSA	CSA	Cloud Controls Matrix (CCM)	3.0.1
Universal		EU	European Union Agency for Network and Information Security (ENISA)	2
Universal	ISACA	ISACA	Control Objectives for Information and Related Technologies (COBIT)	5
Universal	ISACA	ISACA	Control Objectives for Information and Related Technologies (COBIT)	2019
Universal	ISO	ISO	22301 - Security and resilience — Business continuity management systems — Requirements	2019
Universal	ISO	ISO	27001 - Information Security Management Systems (ISMS) - Requirements	2013
Universal	ISO	ISO	27002 - Code of Practice for Information Security Controls	2013

Figure 9.

Universal	ISO	ISO	27018 - Code of Practice for PI in Public Clouds Acting as PI Processors	2014
Universal	ISO	ISO	27701 - Security techniques - Extension to ISO/IEC 27001 and ISO/IEC 27002 for privacy information management — Requirements and guidelines	2019
Universal	ISO	ISO	29100 - Privacy Framework	2011
Universal	ISO	ISO	31000 - Risk Management	2009
Universal	ISO	ISO	31010 - Risk Assessment Techniques	2009
Universal	MPAA	MPAA	MPAA Content Security Best Practices Common Guidelines	4.04
Universal	NAIC	NAIC	Insurance Data Security Model Law (MDL-668)	N/A
Universal	NIST	NIST	SP 800-37 - Guide for Applying the RMF to Federal Information Systems rev1	1
Universal	NIST	NIST	SP 800-37 - Guide for Applying the RMF to Federal Information Systems rev2	2
Universal	NIST	NIST	SP 800-39 - Managing Information Security Risk	N/A
Universal	NIST	NIST	SP 800-53 - Security and Privacy Controls for Information Systems and Organizations	4
Universal	NIST	NIST	SP 800-53 - Security and Privacy Controls for Information Systems and Organizations	5 (draft)
Universal	NIST	NIST	SP 800-63B - Digital Identity Guidelines (partial mapping)	Jun-17
Universal	NIST	NIST	SP 800-160 - Systems Security Engineering	N/A
Universal	NIST	NIST	SP 800-171 - Protecting CUI in Nonfederal Systems and Organizations	1

Figure 10.

Universal	NIST	NIST	SP 800-171B - Protecting Controlled Unclassified Information in Nonfederal Systems and Organizations: Enhanced Security Requirements for Critical Programs and High Value Assets	draft
Universal	NIST	NIST	Cybersecurity Framework (CSF)	1.1 (Apr 19)
Universal	OWASP	OWASP	Top 10 Most Critical Web Application Security Risks	2017
Universal	PCI	PCI SSC	Payment Card Industry Data Security Standard (PCI DSS)	3.2
Universal	SWIFT	SWIFT	SWIFT Customer Security Controls Framework	2019
Universal	UL	UL	2900-1 - Software Cybersecurity for Network-Connectable Products	N/A
US		Federal	US DOJ / FBI - Criminal Justice Information Services (CJIS) Security Policy	5.5
US		Federal	US DOJ / FBI - Criminal Justice Information Services (CJIS) Security Policy	5.8
US		Federal	Children's Online Privacy Protection Act (COPPA)	N/A
US		Federal	Defense Federal Acquisition Regulation Supplement (DFARS) 252.204-7008	252.204-7008
US		Federal	Defense Federal Acquisition Regulation Supplement (DFARS) 252.204-7012	252.204-7012
US		Federal	Fair & Accurate Credit Transactions Act (FACTA) / Fair Credit Reporting Act (FCRA)	N/A
US		Federal	Family Educational Rights and Privacy Act (FERPA)	N/A
US		Federal	Federal Acquisition Regulation (FAR)	52.204-21

Figure 11.

US		Federal	Federal Financial Institutions Examination Council (FFIEC)	N/A
US		Federal	Federal Risk and Authorization Management Program (FedRAMP)	Moderate
US		Federal	Financial Industry Regulatory Authority (FINRA)	N/A
US		Federal	Food & Drug Administration (FDA)	21 CFR Part 11
US		Federal	Federal Trade Commission (FTC) Act	N/A
US		Federal	Gramm Leach Bliley Act (GLBA)	N/A
US		Federal	Health Industry Cybersecurity Practices (HICP) - Small / Medium / Large Practice	N/A
US		Federal	Health Insurance Portability and Accountability Act (HIPAA)	N/A
US		Federal	Internal Revenue Service (IRS) 1075	N/A
US		Federal	National Industrial Security Program Operating Manual (NISPOM)	N/A
US		Federal	North American Electric Reliability Corporation Critical Infrastructure Protection (NERC CIP)	N/A
US		Federal	Privacy Shield	N/A
US		Federal	Sarbanes Oxley Act (SOX)	N/A
US		Federal	Social Security Administration (SSA) Electronic Information Exchange Security Requirements	8
US		State	AK - Alaska Personal Information Protection Act (PIPA)	N/A

Figure 12.

US		State	CA - SB327	N/A
US		State	CA - SB1121 - California Consumer Privacy Act (CCPA)	N/A
US		State	CA - SB1386	N/A
US		State	MA - 201 CMR 17.00	N/A
US		State	NY - NY DFS 23NYCRR500	N/A
US		State	NV - SB220	N/A
US		State	OR - ORS 646A	N/A
US		State	SC - South Carolina Insurance Data Security Act	N/A
US		State	TX - BC521	N/A
US		State	TX - Cybersecurity Act	N/A
US		State	TX - 2019 - SB820	N/A
EMEA		EU	ePrivacy Directive	draft
EMEA		EU	General Data Protection Regulation (GDPR)	N/A
EMEA		EU	Second Payment Services Directive (PSD2)	N/A
EMEA		Austria	Federal Act concerning the Protection of Personal Data (DSG 2000)	N/A

Figure 13.

EMEA		Belgium	Act of 8 December 1992	N/A
EMEA		Czech Republic	Act No. 101/2000 on the Protection of Personal Data	N/A
EMEA		Denmark	Act on Processing of Personal Data (Act No. 429 of May 31, 2000)	N/A
EMEA		Finland	Personal Data Act (986/2000)	N/A
EMEA		France	78 17 / 2004 8021 - Information Technology, Data Files & Civil Liberty	N/A
EMEA		Germany	Cloud Computing Compliance Controls Catalogue (C5)	N/A
EMEA		Germany	Federal Data Protection Act	N/A
EMEA		Greece	Protection of Individuals with Regard to the Processing of Personal Data (2472/1997)	N/A
EMEA		Hungary	Informational Self-Determination and Freedom of Information (Act CXII of 2011)	N/A
EMEA		Ireland	Data Protection Act (2003)	N/A
EMEA		Israel	Cybersecurity Methodology for an Organization	1
EMEA		Israel	Protection of Privacy Law, 5741 – 1981	N/A
EMEA		Italy	Personal Data Protection Code	N/A
EMEA		Luxembourg	Protection of Personals with Regard to the Processing of Personal Data	N/A

Figure 14.

EMEA		Netherlands	Personal Data Protection Act	N/A
EMEA		Norway	Personal Data Act	N/A
EMEA		Poland	Act of 29 August 1997 on the Protection of Personal Data	N/A
EMEA		Portugal	Act on the Protection of Personal Data	N/A
EMEA		Russia	Federal Law of 27 July 2006 N 152-FZ	N/A
EMEA		Russia	Russian Labor Code	N/A
EMEA		Slovak Republic	Protection of Personal Data (122/2013)	N/A
EMEA		South Africa	Protection of Personal Information Act (POPIA)	N/A
EMEA		Spain	Royal Decree 1720/2007 (protection of personal data)	N/A
EMEA		Sweden	Personal Data Act	N/A
EMEA		Switzerland	Federal Act on Data Protection (FADP)	N/A
EMEA		Turkey	Regulation on Protection of Personal Data in Electronic Communications Sector	N/A
EMEA		UAE	Data Protection Law No. 1 of 2007	N/A
EMEA		United Kingdom	Data Protection Act	N/A
APAC		Australia	Privacy Act of 1998	N/A

Figure 15.

APAC		Australia	Australian Government Information Security Manual (ISM)	2017
APAC		China	Decision on Strengthening Network Information Protection	N/A
APAC		Hong Kong	Personal Data Ordinance	N/A
APAC		India	Information Technology Rules (Privacy Rules)	N/A
APAC		Indonesia	Government Regulation No. 82 of 2012	N/A
APAC		Japan	Act of the Protection of Personal Information	N/A
APAC		Malaysia	Personal Data Protection Act of 2010	N/A
APAC		New Zealand	Privacy Act of 1993	N/A
APAC		New Zealand	New Zealand Information Security Manual (NZISM)	N/A
APAC		Philippines	Data Privacy Act of 2012	N/A
APAC		Singapore	Personal Data Protection Act of 2012	N/A
APAC		Singapore	Monitory Authority of Singapore (MAS) Technology Risk Management (TRM) Guidelines	N/A
APAC		South Korea	Personal Information Protection Act	N/A
APAC		Taiwan	Personal Data Protection Act	N/A
Americas		Argentina	Protection of Personal Law No. 25,026	N/A

Figure 16.

Americas		Argentina	Protection of Personal Data - MEN-2018-147-APN-PTE	N/A
Americas		Bahamas	Data Protection Act	N/A
Americas		Brazil	General Data Protection Law (LGPD)	N/A
Americas		Canada	Personal Information Protection and Electronic Documents Act (PIPEDA)	N/A
Americas		Chile	Act 19628 - Protection of Personal Data	N/A
Americas		Colombia	Law 1581 of 2012	N/A
Americas		Costa Rica	Protection of the Person in the Processing of His Personal Data	N/A
Americas		Mexico	Federal Law on Protection of Personal Data held by Private Parties	N/A
Americas		Peru	Personal Data Protection Law	N/A
Americas		Uruguay	Law No. 18,331 - Protection of Personal Data and Action 'Habeas Data'	N/A

Chapter 8
Research Findings in the Domain of CI/CD and DevOps on Security Compliance

Yuri Bobbert
ON2IT BV, The Netherlands & Antwerp Management School, University of Antwerp, Belgium

Maria Chtepen
BNP Paribas Group, Belgium

ABSTRACT

This chapter studies the mapping of governance and security control objectives impacted by DevOps to the corresponding DevOps control objectives. These DevOps objectives introduce either an opportunity or a risk for the achievement of the security and governance control objectives. Finally, the artifact defines a list of SecDevOps controls that have proven to be effective in combining the agility of the DevOps paradigm with the security compliance assurance. The authors examine in collaboration with experts the multiple frameworks to be suitable. The authors define SecDevOps controls that have proven to be effective in combining the agility of the DevOps paradigm with the security compliance assurance. To design this artefact, four widely-used frameworks/standards (COBIT 5, NIST cybersecurity framework, NIST SP 800-53, and ISO 27002) were reviewed for sufficiently detailed security and privacy control objectives and controls. Based on these criteria, NIST SP 800-53 and ISO 27002 standards were selected for comparison and mapping with (Sec)DevOps controls in this research.

LITERATURE REVIEW

As mentioned in the Research Methodology chapter, this study is relying upon the methodology that is making use of the extensive literature study in order to provide a solid scientific basis.

Before initiating this study, the following open questions related to academic literature review are stated:

- How to assure sufficient coverage of the search terms?
- How to find relevant papers with no publicly available content?

DOI: 10.4018/978-1-7998-7367-9.ch008

Within this research we review current (Sec)DevOps and CI/CD practices and determine their control objectives that either pose a risk or introduce an opportunity to the compliance. For each control objectives controls are determined, which facilitate the achievement of the compliance with ISO 27002 and NIST SP 800-53.

The literature review requires a two-fold investigation:

1. Determine which (Sec)DevOps and CI/CD practices are currently in use;
2. Determine the controls objectives and controls applicable to (Sec)DevOps and CI/CD practices, and evaluate their impact on the compliance with ISO 27002 and NIST SP 800-53 security standards.

To answer the first question, this research refers to the parallel study currently ongoing within Antwerp Management School: "A framework for continuous security & security velocity metering in DevOps". This study reviews, among others, the recent academic literature to determine the most relevant Agile and DevOps practices applicable today. The selected list of practices serves as a basis for the investigation of the appropriate controls. This investigation brings us to the second question, which periodically appears in the recent academic literature but is not fully developed and answered by the current research efforts.

To answer the second question, literature review is conducted using the Google Scholar web search engine for academic literature, which indexes peer-reviewed academic journals, books, conference papers, etc. Since DevOps and CI/CD are rapidly evolving domains, the search was limited to recent publications, going back a maximum of 5 years, i.e. from 2014 to 2019.

The following search strings were used as an input in Google Scholar: "DevOps compliance", "DevSecOps compliance", "SecDevOps compliance", "Continuous Delivery compliance", "Continuous Deployment compliance", "Continuous integration compliance", "ISO 27001 DevOps", "ISO 27002 DevOps", "ISO 27001 Agile", "ISO 27002 Agile", "NIST SP DevOps". The search was performed on the Title and the Content of the publications. Only publications written in English were considered.

The first search strings provided a large number of potentially relevant results. The further the search progressed, the larger was the number of duplicate publications found, indicating the exhaustion of the search space. The total number of potentially relevant publications, based on their title and abstract, across all search strings amounts to 88 publications, of which 36 were maintained as being relevant for this research after thorough abstract review. Furthermore, the text of the 36 selected publication was examined in detail, including the forward search on the recent references. The forward search provided 13 additional relevant publications that were used as the foundation of this study.

Figure 1. Literature review search method

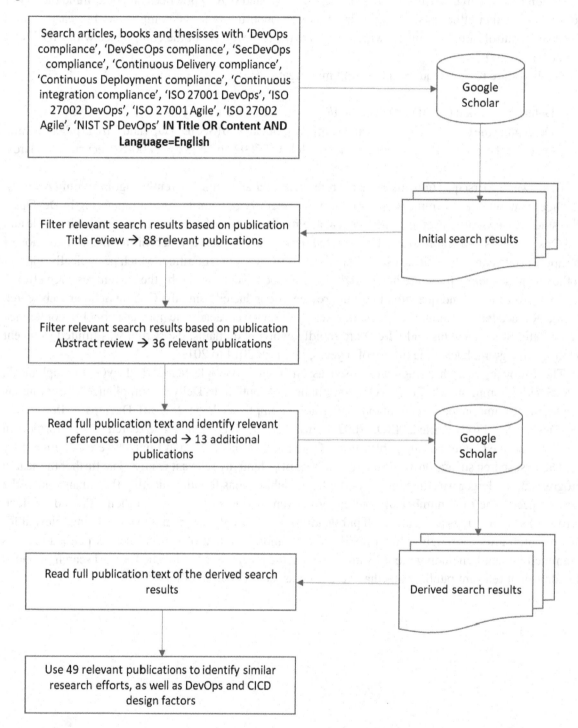

DEFINING THE ARTIFACT

ISO/IEC 27001 and NIST SP 800-53 Relevant Control Objectives and Controls

The creation of **SecDevOps Capability Artifact** starts with determining the ISO/IEC 27001 and NIST SP 800-53 control objectives and controls that are relevant for the adoption of (Sec)DevOps. In fact, not all control objectives are impacted by the choice of one or another development and operations methodology. Certain objectives, for example "Management direction for information security" (ISO/IEC 27001, section 5.1) or "Mobile devices and teleworking" (ISO/IEC 27001, section 6.2) are not sensitive to the choice between Waterfall or Agile development, since all the controls are still applicable in exactly the same way to both. While it is clear that policies for information security have to be defined/reviewed and secure teleworking should be foreseen in all circumstances, some controls are less obvious to review from the perspective of their impact on (Sec)DevOps. For instance, if we take a closer look at "Organisation of information security" (ISO/IEC 27001, section 6), we observe that the information security roles and responsibilities may undergo a major shift as a result of the new form of cohesion between Security, Development and Operations responsibilities triggered by SecDevOps. The same applies to the "Segregation in networks" (ISO/IEC 27001, section 13.1.3), since the segregation determines the ease with which Developers can access the traditionally Operational environments (i.e. Production) and vice versa.

The control objectives and controls of ISO/IEC 27001 and NIST SP 800-53 impacted by the introduction of (Sec)DevOps were determined as follows:

- **Step 1**: security control objectives and the corresponding controls are listed, which are mentioned by the scientific literature studies (see Literature Review) as relevant within this context
- **Step 2**: the results of Step 1 are reviewed by domain experts during the following individual interviews (see Subject matter expert interviews) and the original input is modified based on their experience and recommendations

Below is an overview of the outcome. Taking into the account the widespread use of ISO/IEC 27001 within the European Union, this standard is used as the core for structuring the SecDevOps Capability Artifact. However, it is extended with more extensive related controls of NIST SP 800-53. The mapping between the two standards is following Appendix H of SP 800-53 Revision 4. However, it is important to mention that the reference mapping is very exhaustive and the sections that have no direct link to (Sec) DevOps capabilities are left out, following the same logic as described above.

(Sec)DevOps Capability Artifact

The artifact described below shows the relationship between the ISO/IEC 27002 and NIST SP 800-53 control objectives and controls, which are impacted by the implementation of (Sec)DevOps within an organisation, and the corresponding (Sec)DevOps control objectives. In general, the security standard's control objectives are either impacted in a negative (Risk) or positive (Opportunity) way by (Sec)DevOps. However, the Risks created can often be mitigated by appropriate controls. Sometimes the introduction of a certain control may, however, have an impact on the control objectives of (Sec)DevOps as the

Table 1.

ISO/IEC 27001 control objectives impacted by (Sec)DevOps	ISO/IEC 27001 controls impacted by (Sec)DevOps	NIST SP 800-53 controls impacted by (Sec)DevOps
6.1 Internal organisation To establish a management framework to initiate and control the implementation and operation of information	**6.1.1 Information security roles and responsibilities** • Define and allocate all information security responsibilities to individuals competent in the area • Individuals with allocated information security responsibilities may delegate security tasks to others, but remain accountable and should verify that all tasks are correctly performed	**CM-1 CONFIGURATION MANAGEMENT POLICY AND PROCEDURES** **CM-9 CONFIGURATION MANAGEMENT PLAN** • The organisation assigns responsibility for developing the configuration management process to organisational personnel that are not directly involved in information system development. This separation of duties ensures that organisations establish and maintain a sufficient degree of independence between the information system development and integration and configuration management processes to facilitate quality control and more effective oversight
	6.1.2 Segregation of duties • Conflicting duties and areas of responsibility should be segregated to reduce opportunities for unauthorized or unintentional modification or misuse of the organisation's assets • Care should be taken that no single person can access, modify or use assets without authorization or detection • The initiation of an event should be separated from its authorization. The possibility of collusion should be considered in designing the controls	**AC-5 SEPARATION OF DUTIES** • Separation of duties includes, for example: (i) dividing mission functions and information system support functions among different individuals and/or roles; (ii) conducting information system support functions with different individuals (e.g., system management, programming, configuration management, quality assurance and testing, and network security); and (iii) ensuring security personnel administering access control functions do not also administer audit functions
	6.1.5 Information security in project management • Information security should be integrated into the organisation's project management method(s) to ensure that information security risks are identified and addressed as part of a project • The project management methods in use should require that: a) information security objectives are included in project objectives; b) an information security risk assessment is conducted at an early stage of the project to identify necessary controls; c) information security is part of all phases of the applied project methodology • Responsibilities for information security should be defined and allocated to specified roles defined in the project management methods	**SA-3 SYSTEM DEVELOPMENT LIFE CYCLE** • Manages the information system using [Assignment: organisation-defined system development life cycle] that incorporates information security considerations • Defines and documents information security roles and responsibilities throughout the system development life cycle • Identifies individuals having information security roles and responsibilities • Integrates the organisational information security risk management process into system development life cycle activities • Security awareness and training programs can help ensure that individuals having key security roles and responsibilities have the appropriate experience, skills, and expertise to conduct assigned system development life cycle activities • The effective integration of security requirements into enterprise architecture also helps to ensure that important security considerations are addressed early in the system development life cycle and that those considerations are directly related to the organisational mission/business processes
7.2 During employment To ensure that employees and contractors are aware of and fulfil their information security responsibilities	**7.2.2 Information security awareness, education and training** • All employees of the organisation and, where relevant, contractors should receive appropriate periodic awareness education and training and regular updates in organisational policies and procedures, as relevant for their job function	**AT-2 SECURITY AWARENESS TRAINING** **AT-3 ROLE-BASED SECURITY TRAINING** **PM-13 INFORMATION SECURITY WORKFORCE** • The organisation provides basic security awareness training to information system users (including managers, senior executives, and contractors): a. As part of initial training for new users; b. When required by information system changes; and c. [Assignment: organisation-defined frequency] thereafter • The organisation provides role-based security training to personnel with assigned security roles and responsibilities: a. Before authorizing access to the information system or performing assigned duties; b. When required by information system changes; and c. [Assignment: organisation-defined frequency] thereafter • The organisation establishes an information security workforce development and improvement program
9.2 User access management To ensure authorized user access and to prevent unauthorized access to systems and services	**9.2.3 Management of privileged access rights** • The allocation and use of privileged access rights should be restricted and controlled • Privileged access rights should be allocated to users on a need-to-use basis and on an event-by-event basis in line with the access control policy i.e. based on the minimum requirement for their functional roles	**AC-6 LEAST PRIVILEGE** • The organisation explicitly authorizes access to [Assignment: organisation-defined security functions (deployed in hardware, software, and firmware) and security-relevant information] • The organisation prohibits privileged access to the information system by non-organisational users • The information system audits the execution of privileged functions
9.4 System and application access control To prevent unauthorized access to systems and applications	**9.4.4 Use of privileged utility programs** • The use of utility programs that might be capable of overriding system and application controls should be restricted and tightly controlled • Limitation of the availability of utility programs, e.g. for the duration of an authorized change • Logging of all use of utility programs • Not making utility programs available to users who have access to applications on systems where segregation of duties is required	**AC-3 ACCESS ENFORCEMENT** **AC-6 LEAST PRIVILEGE** • The principle of least privilege is also applied to information system processes, ensuring that the processes operate at privilege levels no higher than necessary to accomplish required organisational missions/business functions • The information system enforces approved authorizations for logical access to information and system resources in accordance with applicable access control policies • The information system enforces a role-based access control policy over defined subjects and objects and controls access based upon [Assignment: organisation-defined roles and users authorized to assume such roles] • The organisation employs an audited override of automated access control mechanisms under [Assignment: organisation-defined conditions]
	9.4.5 Access control to program source code • Access to program source code should be restricted. For program source code, this can be achieved by controlled central storage of such code • An audit log should be maintained of all accesses to program source libraries • Maintenance and copying of program source libraries should be subject to strict change control procedures • If the program source code is intended to be published, additional controls to help getting assurance on its integrity (e.g. digital signature) should be considered	**CM-5 ACCESS RESTRICTIONS FOR CHANGE** • The organisation employs an audited override of automated access control mechanisms under [Assignment: organisation-defined conditions] • Organisations maintain records of access to ensure that configuration change control is implemented and to support after-the-fact actions should organisations discover any unauthorized changes. Access restrictions for change also include software libraries • The information system prevents the installation of [Assignment: organisation-defined software and firmware components] without verification that the component has been digitally signed using a certificate that is recognized and approved by the organisation • The organisation enforces dual authorization for implementing changes to [Assignment: organisation-defined information system components and system-level information] • The organisation limits privileges to change software resident within software libraries • The organisation: (a) Limits privileges to change information system components and system-related information within a production or operational environment; and (b) Reviews and re-evaluates privileges [Assignment: organisation-defined frequency]

continued on following page

Table 1. Continued

ISO/IEC 27001 control objectives impacted by (Sec) DevOps	ISO/IEC 27001 controls impacted by (Sec)DevOps	NIST SP 800-53 controls impacted by (Sec)DevOps
12.1 Operational procedures and responsibilities To ensure correct and secure operations of information processing facilities	**12.1.2 Change management** • Changes to the organisation, business processes, information processing facilities and systems that affect information security should be controlled • The following items should be considered: identification and recording of significant changes; planning and testing of changes; assessment of the potential impacts, including information security impacts, of such changes; formal approval procedure for proposed changes; verification that information security requirements have been met; fallback procedures, including procedures and responsibilities for aborting and recovering from unsuccessful changes and unforeseen events; provision of an emergency change process to enable quick and controlled implementation of changes needed to resolve an incident • Formal management responsibilities and procedures should be in place to ensure satisfactory control of all changes	**CM-3 CONFIGURATION CHANGE CONTROL** **CM-5 ACCESS RESTRICTIONS FOR CHANGE** **SA-10 DEVELOPER CONFIGURATION MANAGEMENT** • Configuration change control includes changes to baseline configurations for components and configuration items of information systems, changes to configuration settings for information technology products (e.g., operating systems, applications, firewalls, routers, and mobile devices), unscheduled/unauthorized changes, and changes to remediate vulnerabilities • Typical processes for managing configuration changes to information systems include, for example, Configuration Control Boards that approve proposed changes to systems • Auditing of changes includes activities before and after changes are made to organisational information systems and the auditing activities required to implement such changes • The organisation employs automated mechanisms to: (a) Document proposed changes to the information system; (b) Notify [Assignment: organized-defined approval authorities] of proposed changes to the information system and request change approval; (c) Highlight proposed changes to the information system that have not been approved or disapproved by [Assignment: organisation-defined time period]; (d) Prohibit changes to the information system until designated approvals are received; (e) Document all changes to the information system; and (f) Notify [Assignment: organisation-defined personnel] when approved changes to the information system are completed • The organisation employs automated mechanisms to implement changes to the current information system baseline and deploys the updated baseline across the installed base • The organisation requires an information security representative to be a member of the [Assignment: organisation-defined configuration change control element] • The information system implements [Assignment: organisation-defined security responses] automatically if baseline configurations are changed in an unauthorized manner • The organisation defines, documents, approves, and enforces physical and logical access restrictions associated with changes to the information system • The information system prevents the installation of [Assignment: organisation-defined software and firmware components] without verification that the component has been digitally signed using a certificate that is recognized and approved by the organisation • The organisation enforces dual authorization for implementing changes to [Assignment: organisation-defined information system components and system-level information] • The organisation requires the developer of the information system, system component, or information system service to: a. Perform configuration management during system, component, or service [Selection (one or more): design; development; implementation; operation]; b. Document, manage, and control the integrity of changes to [Assignment: organisation-defined configuration items under configuration management]; c. Implement only organisation-approved changes to the system, component, or service; d. Document approved changes to the system, component, or service and the potential security impacts of such changes; and e. Track security flaws and flaw resolution within the system, component, or service and report findings to [Assignment: organisation-defined personnel] • The organisation requires the developer of the information system, system component, or information system service to employ tools for comparing newly generated versions of security relevant hardware descriptions and software/firmware source and object code with previous versions • The organisation requires the developer of the information system, system component, or information system service to execute procedures for ensuring that security-relevant hardware, software, and firmware updates distributed to the organisation are exactly as specified by the master copies
	12.1.3 Capacity management • The use of resources should be monitored, tuned and projections made of future capacity requirements to ensure the required system performance • Detective controls should be put in place to indicate problems in due time	**AU-4 AUDIT STORAGE CAPACITY** • The organisation allocates audit record storage capacity in accordance with [Assignment: organisation-defined audit record storage requirements]
12.4 Logging and monitoring To record events and generate evidence	**12.4.1 Event logging** • Event logs recording user activities, exceptions, faults and information security events should be produced, kept and regularly reviewed • Event logging sets the foundation for automated monitoring systems which are capable of generating consolidated reports and alerts on system security	**AU-3 CONTENT OF AUDIT RECORDS** **AU-6 AUDIT REVIEW, ANALYSIS, AND REPORTING** **AU-12 AUDIT GENERATION** • The information system generates audit records containing information that establishes what type of event occurred, when the event occurred, where the event occurred, the source of the event, the outcome of the event, and the identity of any individuals or subjects associated with the event • The organisation employs automated mechanisms to integrate audit review, analysis, and reporting processes to support organisational processes for investigation and response to suspicious activities • The information system provides the capability to centrally review and analyse audit records from multiple components within the system • The information system: a. Provides audit record generation capability for the auditable events
12.6 Technical vulnerability management To prevent exploitation of technical vulnerabilities	**12.6.2 Restrictions on software installation** • Rules governing the installation of software by users should be established and implemented. The organisation should define and enforce strict policy on which types of software users may install • The principle of least privilege should be applied	**CM-11 USER-INSTALLED SOFTWARE** • The information system alerts [Assignment: organisation-defined personnel or roles] when the unauthorized installation of software is detected • The information system prohibits user installation of software without explicit privileged status
13.1 Network security management To ensure the protection of information in networks and its supporting information processing facilities	**13.1.3 Segregation in networks** • Groups of information services, users and information systems should be segregated on networks	**AC-4 INFORMATION FLOW ENFORCEMENT** • Transferring information between information systems representing different security domains with different security policies introduces risk that such transfers violate one or more domain security policies

continued on following page

Table 1. Continued

ISO/IEC 27001 control objectives impacted by (Sec)DevOps	ISO/IEC 27001 controls impacted by (Sec)DevOps	NIST SP 800-53 controls impacted by (Sec)DevOps
14.2 Security in development and support processes To ensure that information security is designed and implemented within the development lifecycle of information systems	**14.2.1 Secure development policy** • Rules for the development of software and systems should be established and applied to developments within the organisation • Within a secure development policy, the following aspects should be put under consideration: security of the development environment; guidance on the security in the software development lifecycle: security in the software development methodology; secure coding guidelines for each programming language used; security requirements in the design phase; security checkpoints within the project milestones; secure repositories; security in the version control; required application security knowledge; developers' capability of avoiding, finding and fixing vulnerabilities • Developers should be trained in their use and testing and code review should verify their use. If development is outsourced, the organisation should obtain assurance that the external party complies with these rules for secure development	**SA-15 DEVELOPMENT PROCESS, STANDARDS, AND TOOLS** **SA-17 DEVELOPER SECURITY ARCHITECTURE AND DESIGN** • The organisation requires the developer of the information system, system component, or information system service to: (a) Define quality metrics at the beginning of the development process; and (b) Provide evidence of meeting the quality metrics [Selection (one or more): [Assignment: organisation-defined frequency]; [Assignment: organisation-defined program review milestones]; upon delivery] • The organisation requires the developer of the information system, system component, or information system service to select and employ a security tracking tool for use during the development process • The organisation requires that developers perform threat modeling and a vulnerability analysis for the information system at [Assignment: organisation-defined breadth/depth] • The organisation requires the developer of the information system, system component, or information system service to reduce attack surfaces to [Assignment: organisation-defined thresholds] • The organisation requires the developer of the information system, system component, or information system service to implement an explicit process to continuously improve the development process • The organisation requires the developer of the information system or system component to archive the system or component to be released or delivered together with the corresponding evidence supporting the final security review • The organisation requires the developer of the information system, system component, or information system service to: (a) Produce, as an integral part of the development process, a formal policy model describing the [Assignment: organisation-defined elements of organisational security policy] to be enforced; and (b) Prove that the formal policy model is internally consistent and sufficient to enforce the defined elements of the organisational security policy when implemented
	14.2.2 System change control procedures • Changes to systems within the development lifecycle should be controlled by the use of formal change control procedures • Formal change control procedures should be documented and enforced to ensure the integrity of system, applications and products, from the early design stages through all subsequent maintenance efforts • This process should include a risk assessment, analysis of the impacts of changes and specification of security controls needed. This process should also ensure that existing security and control procedures are not compromised, that support programmers are given access only to those parts of the system necessary for their work and that formal agreement and approval for any change is obtained • Maintaining a version control for all software updates • Maintaining an audit trail of all change requests	**CM-3 CONFIGURATION CHANGE CONTROL** **SA-10 DEVELOPER CONFIGURATION MANAGEMENT** **SI-2 FLAW REMEDIATION** • The organisation employs automated mechanisms to: (a) Document proposed changes to the information system; (b) Notify [Assignment: organized-defined approval authorities] of proposed changes to the information system and request change approval; (c) Highlight proposed changes to the information system that have not been approved or disapproved by [Assignment: organisation-defined time period]; (d) Prohibit changes to the information system until designated approvals are received; (e) Document all changes to the information system; and (f) Notify [Assignment: organisation-defined personnel] when approved changes to the information system are completed • The organisation tests, validates, and documents changes to the information system before implementing the changes on the operational system • The organisation employs automated mechanisms to implement changes to the current information system baseline and deploys the updated baseline across the installed base • The organisation requires an information security representative to be a member of the [Assignment: organisation-defined configuration change control element] • The information system implements [Assignment: organisation-defined security responses] automatically if baseline configurations are changed in an unauthorized manner • The organisation requires the developer of the information system, system component, or information system service to: a. Perform configuration management during system, component, or service [Selection (one or more): design; development; implementation; operation]; b. Document, manage, and control the integrity of changes to [Assignment: organisation-defined configuration items under configuration management]; c. Implement only organisation-approved changes to the system, component, or service; d. Document approved changes to the system, component, or service and the potential security impacts of such changes; and e. Track security flaws and flaw resolution within the system, component, or service and report findings to [Assignment: organisation-defined personnel] • The organisation requires the developer of the information system, system component, or information system service to enable integrity verification of hardware components • The organisation requires the developer of the information system, system component, or information system service to execute procedures for ensuring that security-relevant hardware, software, and firmware updates distributed to the organisation are exactly as specified by the master copies • The organisation: a. Identifies, reports, and corrects information system flaws; b. Tests software and firmware updates related to flaw remediation for effectiveness and potential side effects before installation; c. Installs security-relevant software and firmware updates within [Assignment: organisation defined time period] of the release of the updates; and d. Incorporates flaw remediation into the organisational configuration management process
	14.2.7 Outsourced development • The organisation should supervise and monitor the activity of outsourced system development • Contractual requirements for secure design, coding and testing practices; provision of the approved threat model to the external developer; acceptance testing for the quality and accuracy of the deliverables; provision of evidence that security thresholds were used to establish minimum acceptable levels of security and privacy quality; provision of evidence that sufficient testing has been applied to guard against the absence of both intentional and unintentional malicious content upon delivery; provision of evidence that sufficient testing has been applied to guard against the present and known vulnerabilities; contractual right to audit development processes and controls; effective documentation of the build environment used to create deliverables; the organisation remains responsible for compliance with applicable laws and control efficiency verification	**SA-4 ACQUISITION PROCESS** **SA-12 SUPPLY CHAIN PROTECTION** • The organisation requires the developer of the information system, system component, or information system service to provide a description of the functional properties of the security controls to be employed • The organisation requires the developer of the information system, system component, or information system service to: a. Create and implement a security assessment plan; b. Perform [Selection (one or more): unit; integration; system; regression] testing/evaluation at [Assignment: organisation-defined depth and coverage]; c. Produce evidence of the execution of the security assessment plan and the results of the security testing/evaluation; d. Implement a verifiable flaw remediation process; and e. Correct flaws identified during security testing/evaluation • The organisation employs [Assignment: organisation-defined tailored acquisition strategies, contract tools, and procurement methods] for the purchase of the information system, system component, or information system service from suppliers • The organisation conducts a supplier review prior to entering into a contractual agreement to acquire the information system, system component, or information system service • The organisation conducts an assessment of the information system, system component, or information system service prior to selection, acceptance, or update • The organisation establishes inter-organisational agreements and procedures with entities involved in the supply chain for the information system, system component, or information system service

continued on following page

Table 1. Continued

ISO/IEC 27001 control objectives impacted by (Sec)DevOps	ISO/IEC 27001 controls impacted by (Sec)DevOps	NIST SP 800-53 controls impacted by (Sec)DevOps
		CA-7 CONTINUOUS MONITORING • The organisation develops a continuous monitoring strategy and implements a continuous monitoring program that includes: a. Establishment of [Assignment: organisation-defined metrics] to be monitored; b. Establishment of [Assignment: organisation-defined frequencies] for monitoring and [Assignment: organisation-defined frequencies] for assessments supporting such monitoring; c. Ongoing security control assessments in accordance with the organisational continuous monitoring strategy; d. Ongoing security status monitoring of organisation-defined metrics in accordance with the organisational continuous monitoring strategy; e. Correlation and analysis of security-related information generated by assessments and monitoring; f. Response actions to address results of the analysis of security-related information; and g. Reporting the security status of organisation and the information system to [Assignment: organisation-defined personnel or roles] [Assignment: organisation-defined frequency]
18.2 Information security reviews To ensure that information security is implemented and operated in accordance with the organisational policies and procedures	**18.2.2 Compliance with security policies and standards** **18.2.3 Technical compliance review** • Managers should regularly review the compliance of information processing and procedures within their area of responsibility with the appropriate security policies, standards and any other security requirements • Automatic measurement and reporting tools should be considered for efficient regular review • Results of reviews and corrective actions carried out by managers should be recorded and these records should be maintained • Information systems should be regularly reviewed for compliance with the organisation's information security policies and standards • Technical compliance should be reviewed preferably with the assistance of automated tools, which generate technical reports for subsequent interpretation by a technical specialist. Alternatively, manual reviews (supported by appropriate software tools, if necessary) by an experienced system engineer could be performed • If penetration tests or vulnerability assessments are used, caution should be exercised as such activities could lead to a compromise of the security of the system. Such tests should be planned, documented and repeatable • Any technical compliance review should only be carried out by competent, authorized persons or under the supervision of such persons	**CM-1 CONFIGURATION MANAGEMENT POLICY AND PROCEDURES** **CA-2 SECURITY ASSESSMENTS** • The organisation: a. Develops, documents, and disseminates to [Assignment: organisation-defined personnel or roles]: 1. A configuration management policy that addresses purpose, scope, roles, responsibilities, management commitment, coordination among organisational entities, and compliance; and 2. Procedures to facilitate the implementation of the configuration management policy and associated configuration management controls; and b. Reviews and updates the current: 1. Configuration management policy [Assignment: organisation-defined frequency]; and 2. Configuration management procedures [Assignment: organisation-defined frequency] • The organisation includes as part of security control assessments, [Assignment: organisation defined frequency], [Selection: announced; unannounced], [Selection (one or more): in-depth monitoring; vulnerability scanning; malicious user testing; insider threat assessment; performance/load testing; [Assignment: organisation-defined other forms of security assessment]]
		CM-2 BASELINE CONFIGURATION • The organisation develops, documents, and maintains under configuration control, a current baseline configuration of the information system • The organisation reviews and updates the baseline configuration of the information system: (a) [Assignment: organisation-defined frequency]; (b) When required due to [Assignment organisation-defined circumstances]; and (c) As an integral part of information system component installations and upgrades • The organisation employs automated mechanisms to maintain an up-to-date, complete, accurate, and readily available baseline configuration of the information system • The organisation retains [Assignment: organisation-defined previous versions of baseline configurations of the information system] to support rollback • The organisation maintains a baseline configuration for information system development and test environments that is managed separately from the operational baseline configuration
		SI-4 INFORMATION SYSTEM MONITORING • The organisation connects and configures individual intrusion detection tools into an information system-wide intrusion detection system • The organisation employs automated tools to support near real-time analysis of events. Supplemental Guidance: Automated tools include, for example, host-based, network-based, transport-based, or storage-based event monitoring tools or Security Information and Event Management (SIEM) technologies that provide real time analysis of alerts and/or notifications generated by organisational information systems • The organisation employs automated tools to integrate intrusion detection tools into access control and flow control mechanisms for rapid response to attacks by enabling reconfiguration of these mechanisms in support of attack isolation and elimination • The information system monitors inbound and outbound communications traffic [Assignment: organisation-defined frequency] for unusual or unauthorized activities or conditions • The information system alerts [Assignment: organisation-defined personnel or roles] when the following indications of compromise or potential compromise occur: [Assignment: organisation-defined compromise indicators] • The information system notifies [Assignment: organisation-defined incident response personnel (identified by name and/or by role)] of detected suspicious events and takes [Assignment: organisation-defined least-disruptive actions to terminate suspicious events] • The organisation employs automated mechanisms to alert security personnel of the following inappropriate or unusual activities with security implications: [Assignment: organisation-defined activities that trigger alerts] • The organisation correlates information from monitoring tools employed throughout the information system • The organisation correlates information from monitoring physical, cyber, and supply chain activities to achieve integrated, organisation-wide situational awareness

Figure 2. Process of creation of SecDevOps Capability Artifact

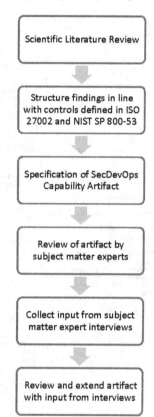

methodology. It means that each organisation should find the right balance between the Risk introduced and the speed and flexibility of development and deployment processes.

The (Sec)DevOps controls described in what follows are the result of the extensive scientific literature study (see Literature Review) and the review / suggestions by subject domain expert (see Subject matter expert interviews). They are specifically focussing on large regulated organisations, since these organisations are primarily dealing with compliancy issues and often follow the above-mentioned standards rigorously. The proposed controls provide a guidance on either how to reduce the Security Risk without significantly sacrificing the (Sec)DevOps speed gain, or how to optimally use (Sec)DevOps capacities to improve the Security Compliance.

EVALUATING ARTIFACT

Introduction

The **SecDevOps Capability Artifact** described in detail in the previous section is created using a stepwise approach. The initial design is derived from a detailed scientific literature review, combined with the study of the controls covered by the applicable standards and frameworks. The outcome artifact is validated through interviews with subject domain experts from diverse large regulated organisations

Table 2.

ISO/IEC 27001 controls impacted by (Sec) DevOps	NIST SP 800-53 controls impacted by (Sec) DevOps	(Sec)DevOps control objectives	(Sec)DevOps controls
6.1.1. Information security roles and responsibilities	CM-1 CONFIGURATION MANAGEMENT POLICY AND PROCEDURES CM-9 CONFIGURATION MANAGEMENT PLAN	**Opportunity.** Introduce new security-oriented roles & responsibilities (Rindell, 2016)	• **Part of the tasks from the Security department should be moved to the SecDevOps teams.** In order to achieve fast security feedback in each step of the development process, as opposed to the late reviews at fixed security gates, the Security department should steer the activities in a different way. Part of the security tasks should be executed by SecDevOps teams. • **Define the role of the Security Champion within the agile organisation.** This role combines the responsibility of ensuring the appropriate Security Gating with support for testing and security automation within the agile team. The Security Champion himself is supported by a Cyber Defence team for the execution of his tasks. It creates scalability within Cyber Defence teams, since it is not feasible to foresee a dedicated Cyber Defence professional within each team/squad. • **Work according to "5 amigo's" principle, stimulating collaboration between Business, Development, Testing, Security & Operations.** In fact, instead of the traditional "3 amigo's" Agile principle, where the work is examined from 3 different perspectives (Business, Development & Testing), we need to start speaking in terms of "5 amigo's", adding Security and Operations into the list. • **Security should be positioned at the front of the development pipeline.** Often Security does not have enough bandwidth and becomes reactive. Therefore, a merge between Security and Enterprise Architecture is a must. Security should steer the design and implementation process using SABSA principles: test planning, certificate management, judge & advocate principle, segregation of duties, etc. The role of Security as the police agent should be replaced by the effective design of secure architectures. • **Shift the Security role from operations to process and controls design.** Security is no longer fulfilling the operational role (it is automated), but it defines the process, the pipeline controls (e.g. definition of firewall ports), and share the best practices (champions). Once or a couple of times a year the process should be reviewed, and thus Security should be organising hackathons, red teaming and sharing of best practices in order to achieve an improvement. Security department should fulfil, in first place, a coaching role for the teams and facilitate the use of different tools.
6.1.2 Segregation of duties	AC-5 SEPARATION OF DUTIES	**Opportunity.** Information sharing between Development, Operations and Testing teams (McCarthy M. H., 2015) **Risk.** Allow developers to make decisions independently (Savor, 2016) **Risk.** Grant team members complete access to production in order to fulfil operational tasks (Plant, 2019) **Risk.** All team members must be able to know, understand, and modify the source code (Pastrana, 2019) **Risk.** Provide developers with the freedom and possibility to commit changes to production (Shahin M. B., 2017)	• **Automate the production deployment process so no person can execute the deployment without passing the automated controls first.** To reduce opportunities for unauthorized or unintentional modification or misuse of the organisation's assets it is suggested to automate the production deployment process. The same procedure should be used when deploying to non-production environments. • **Foresee full governance for sensitive phases in software development and deployment.** For sensitive phases such as production deployment, full governance may be required. To support these cases, systems should prevent mistakes by allowing only the permitted users to approve the execution of these phases. • **Code should be peer-reviewed.** In order to ensure that no single person has end-to end control of a process without a separate check point, code that is checked-in should always be peer-reviewed. The same can be achieved by encouraging developers to create merge requests and assigning those to a more senior developer who has to check the code and merge it. The reviewed code should be signed with personal cryptographic signatures of the developer and the reviewer. When the code moves through the deployment pipeline it should be automatically checked after every step of the process that both signatures are still valid and that the code has not been tampered with. For changes or products with major impact, the number of reviewers could be increased to three or four. However, the sense of responsibility for the review reduces with the increase in the number of reviewers. • **Make use of separate local and central code repositories to enforce the control of the release process.** Developer may be required to commits changes to a local repository, after passing all build and automated tests, and only after the review those changes would be committed to a central repository. • **A separate team can be foreseen for all products to release to production, while guaranteeing the segregation of duties.** To avoid the risk for CI/CD pipeline manipulation by developers, a separate release team can be foreseen. Separate team, however, may create a lot of fringing with the development teams and has to be enforced. Normally, it should not pose an issue: a release pipeline for a product does not change much, if designed in a stable way.

continued on following page

Table 2. Continued

ISO/IEC 27001 controls impacted by (Sec) DevOps	NIST SP 800-53 controls impacted by (Sec) DevOps	(Sec)DevOps control objectives	(Sec)DevOps controls
6.1.5 Information security in project management 14.2.1 Secure development policy	SA-3 SYSTEM DEVELOPMENT LIFE CYCLE SA-15 DEVELOPMENT PROCESS, STANDARDS, AND TOOLS SA-17 DEVELOPER SECURITY ARCHITECTURE AND DESIGN	**Opportunity**. Use collaboration among teams to collect data about the process to identify weaknesses, assess the performance of the teams, and check compliance with standards (Vaishnavi, 2016) **Opportunity**. Develop expertise and processes to best discover, protect against, and find solutions to threats and risks, ahead of time (Myrbakken H., 2017) **Opportunity**. Avoid custom development processes since it brings complexity, increase maintenance costs, installation costs, etc. (Hilton, 2017) **Opportunity**. Build systems that are designed to be deployed easily into multiple environments, can detect and tolerate failures, and can have various components of the system updated independently (Forsgren, 2018) **Opportunity**. Automate security controls, where the controls can be deployed and managed without manual interference (Myrbakken H., 2017) **Opportunity**. Application release deployments must be fully automated end-to-end across development, test and production environments (McCarthy M. H., 2015) (Mohan V. O., 2016) (Stolt, 2013) (Gill, 2017) **Opportunity**. Development teams must use automated testing across the software development lifecycle (Plant, 2019) (McCarthy M. H., 2015) (Farroha, 2014) **Opportunity**. Organize development activities around a threat model which is maintained (Plant, 2019) **Opportunity**. Constantly assessing the security readiness across the software development lifecycle (Ravichandran, 2016) **Opportunity**. A risk management approach should be integrated into the process (Plant, 2019)	• **Well-defined policies regarding information exchange across teams should be in place to prevent security threats due to collaboration.** For instance, SecDevOps may facilitate data sharing and knowledge exchange between teams if sharing is happening in a safe way and all data is stored in a shared secure platform, such as O365. • **Risk assessments should be performed from the first planning stage and continuously before every iteration.** It is important as a way to prioritize risks, examine controls already in place and decide which are needed for going forward. • **Manage security controls as API or "Security/Compliance as a Code".** SecDevOps should facilitate preventive scanning during the development process. Scanning needs to be performed against a number of matrices with foreseen threats, vulnerabilities and bad programming habits (e.g. SAST, DAST). For example, code scanning (e.g. Sonar) maybe performed by the developers in Test and Quality Assurance environments, followed by security flow scanning (e.g. Fortify) performed by a security specialist / champion, and, finally, followed by IAST (Interactive Application Security Testing) as the final step, when the application is already running in Production. Furthermore, tools like NexusIQ can scan source code for vulnerabilities, which allows rapidly to limit the use of vulnerable tools to only DEV and TEST environments and to get them out of production. • **Design for failure.** The software design should be conducted in such a way that if a failure occurs, the user is able to continue working with minor intervention. Minimum workflow should be always supported. To achieve this there is a need for a new application architecture. For example, if the application is designed to be stateless, you can easily restart without losing data: only a single user session will be impacted and not all users simultaneously. For new applications, this design feature is easier to achieve, since you have a green field and a new technology. For traditional applications, on the other hand, it is much more complicated and the success strongly depends on the type of application and its original design. A huge difference is that old applications were built on top of the infrastructure that was always available and failures were addressed at the infrastructure level. In the cloud, you have to modify the application to deal with the failure. Not only the application architecture should be modified, but also the supporting tooling. To constantly implement changes in a product, the tooling in the pipeline should be continuously modified (e.g. key management). Changing tooling costs more energy, since these activities generally involve manual work. • **Standardize security controls and security assurance process.** For instance, the creation of architecture generic services, using Archimate, facilitates standardization and allows to externalize the security review (by third party). An overview of controls will largely take away the need to do individual control reviews (e.g. when buying external cloud services). It provides a complete view on the level of compliance, demonstrating the overall level of security. From the security control overview, we can move towards the security assurance through validation and certification (eventually, by accredited third parties). Architecture should be applied continuously and consistently, leading to continuous security assurance. • **Adopt convention over configuration to reduce the complexity of CI/CD system.** Developers should strive to keep their processes as simple as possible, to avoid adding unneeded complexity. For development tools (e.g. IDE) selection freedom of choice can be given to the developers and operational people. However, the pipeline should be the same and homogeneous for everyone. • **Use Service-oriented architectures and micro-services architecture to build systems that can easily be deployed in multiple environments.** Bounded contexts and APIs are a way to decouple large domains into smaller, more loosely coupled units. Virtualization can be used to test services or components in isolation. Service-oriented architectures are supposed to enable these outcomes, as should any true micro-services architecture. • **Make the engineer accountable for the code in production he created.** It reduces the time to resolution of system failures. • **Use automated activities to simplify and to improve software development.** These activities may include the automated code review, automated monitoring, automated testing, etc. • **Secure the pipeline by restricting the attack surface of the code base.** This involves restricting the reach of parts of the code base from critical parts. The approach suggests that untrustworthy components need to communicate via trustworthy components to reach sensitive parts of the pipeline. • **Security teams should be informed about the challenges faced by operators and developers, and vice versa.** It will improve the security processes they develop. • **Put the responsibility for handling the security considerations within the scope of the development team.** This is an advantage, because the development team itself knows best of the requirements, the architecture and the technologies used for a specific application component (i.e. a micro-service). • **Perform automatic patching.** If patching can be turned off, it may result into delays with patch management. Therefore, it is important to push patches automatically as much as possible. • **Let a Quality Assurance officer to support and monitor the DevOps process from a high-level view.** While the teams are mainly responsible themselves for the quality of their products, the QA officer checks general quality criteria such as whether the Definition of Done is correctly defined.
7.2.2 Information security awareness, education and training	AT-2 SECURITY AWARENESS TRAINING AT-3 ROLE-BASED SECURITY TRAINING PM-13 INFORMATION SECURITY WORKFORCE	**Opportunity**. Foresee security training for development team members (Rahman, 2016)	• **Train Security Champion.** Security Champion plays a central role in the successful implementation of SecDevOps and the question of training the right skills of the Security Champion becomes prominent. Specific, detailed, training is required for this role, including a "Survival Kit". • **Train Security Team.** Security team can be a police defining the rules or an advisor. The first way of working is against the agile principles and is not scalable. For instance, static code scanning should be a part of the work of an Agile team, a part of the tasks of a Security Champions. For the second way of working Security team needs new competences, in order to give advice on specific topics, such as coding guidelines. • **Train software developers.** Technical training is required for developers, to make sure they are aware of the best practices for the secure software development. Security people often do not understand development specifics. Visa versa, developers expect to receive a clear TO-DO list, while only translating OWASP will have 260 items, ISO 110 items, PCI 400 items, etc. So you need to create a real team with different profiles. For example, Agile Skills Framework can be used, which advises to create COP (Community Of Practice). Before giving full control to developers as a security team, they need to be aware of the consequences. They also should receive a hardened operational image for deployment, without errors. Developers should not be allowed to modify this image any more, but if they do the security team should be immediately informed via automatic monitoring and alerting.

continued on following page

Table 2. Continued

ISO/IEC 27001 controls impacted by (Sec) DevOps	NIST SP 800-53 controls impacted by (Sec) DevOps	(Sec)DevOps control objectives	(Sec)DevOps controls
9.2.3 Management of privileged access rights 9.4.4 Use of privileged utility programs 9.4.5 Access control to program source code	AC-6 LEAST PRIVILEGE AC-3 ACCESS ENFORCEMENT CM-5 ACCESS RESTRICTIONS FOR CHANGE	**Risk**. Search the right balance between security and compliance, and operational flexibility for developers and testers (Michener, 2016) **Risk**. Secure and manage access control to CI pipelines (Hilton, 2017) **Opportunity**. Information security should be considered by all developers (Hilton, 2017)	• **Any privileged accounts (such as root and the local administrator accounts) should be monitored very closely (or ideally be disabled completely).** For instance, a version control system can be adopted as a central repository that store the historical changes made to the code. Appropriate individuals must be notified of the event and associated report for review. All event reports must be stored securely and made available to internal or external reviewers and auditors, as appropriate. • **Use timed access rights by allowing developers to request timed passwords.** It grants developers restricted access, for example, to the production environment to perform a change, once it is authorized. The access is logged and a standard notification is sent to the security department. Tooling may help to simplify the task: RepoMan is a tool that uses activity history to reduce account privileges over a certain period of time. While this control seems like an effective compromise, it depends on the circumstances whether this is practical. When developers need access to the production twice per week, requesting timed passwords for this frequent access would be rather impractical and not add much security since there will be a regular opportunity to tamper with the production environment. • **Grant employees dynamic access rights per sprint by assigning them specific responsibilities and access rights every time.** • **It is often sufficient to give DevOps engineers two accounts for the different environments.** Developers often need administrator rights in the development environment and restricted user level rights in the production environment. Do not give to developers default privileges to access configuration tools in the pipeline. Only specific roles should be granted this access. • **Maintain a separate CI server for development done outside the company.** Maintain an internal CI server that operates behind a company firewall, as well as an external one. Internal one cannot be exposed due to confidentiality requirements, but external CI may be used to maintain a positive relationship with the external developer community. • **Provide fine -grained account management with different levels of access.** For instance, less trusted accounts should be allowed to view only build results, and trusted accounts to have full access to build results and management features in the CI system. • **All changes in production should preferably be executed by CI/CD pipeline.** No one in the organisation should be allowed to execute changes in products, except for the pipeline. In CI/CD, process changes could be deployed, after automatic or manual review. Only read-only access rights to production are allowed. When there is an incident, a manager could approve exceptionally a temporary access to the production environment. • **Organize release pipeline admins is a separate team.** For each environment a functional user and a password are required to start the deployment. Password in each tool should be encrypted. Therefore, the source code should be stored separately from work item management.
12.1.1 Documented operating procedures	CM-1 CONFIGURATION MANAGEMENT POLICY AND PROCEDURES SA-5 INFORMATION SYSTEM DOCUMENTATION	**Opportunity**. Facilitate documentation of procedures and environments of the software creation (Plant, 2019) **Opportunity**. Use automated deployment and testing outcomes as documentation (e.g. evidence for audits) (Mohan V. O., 2018)	• **Use tooling to generate documentation.** In DevOps, documentation of procedures and environment can easily be done with development tools such as the project management tool JIRA, source code repositories like GIT and automation software like Jenkins. Furthermore, using invariable Docker containers allow for a repeatable installation and release process. • **Use logging for documentation and traceability purposes.** Document repositories should log the details of process owners at each step of the deployment process, for example, to indicate who approved the last tests set. • **Documentation should not be shared through a medium which cannot be traced.** For example, data should not be shared through only the chat clients. All required information need to be stored in a central repository that can be accessed by all the stakeholders of the process. • **Differentiate between operational and non-operational documentation.** While the first category of documentation can be automated, the second category (e.g. test cases, architecture documentation, threat modelling) should still be written. It can be limited in volume and should be stored centrally, using repositories like Sharepoint or Wikipages.
12.1.2 Change management 14.2.2 System change control procedures	CM-3 CONFIGURATION CHANGE CONTROL CM-5 ACCESS RESTRICTIONS FOR CHANGE SA-10 DEVELOPER CONFIGURATION MANAGEMENT SI-2 FLAW REMEDIATION	**Risk**. Allow incremental changes directly to be made to components (Plant, 2019) (Michener, 2016) (Ravichandran, 2016) **Opportunity**. Delivering smaller changes more frequently (Plant, 2019) **Opportunity**. Before functional changes can be safely made the threat model must be updated and the new and existing components reviewed for processing of untrusted data and crossing trust boundaries. All threats meeting the security criteria much be fixed or mitigated before deployment can move to the new model (Michener, 2016) **Risk**. Use a lightweight change approval process (Forsgren, 2018) **Opportunity**. Implement automatic change management (Savor, 2016)	• **Integrate Change Management into the Release Management process.** Direct changes to components during the development process are difficult to achieve in a regulated environment. There is often no complete view on the dependencies. Thus there is no guarantee that that there is no security impact on the underlying components. However, there is no longer need for a periodically organized Change Advisory Board, but instead the Change Management process should be incorporated into the Release Management process. Change Management is becoming a slave of the Release Management. For instance, the change request process can be integrated into the release automation. • **Changes should only be applied to production using a process that forms part of a deployment pipeline.** This process should be preferably fully or semi-automated. That is, no changes should be able to be made to production unless they have been committed to version control, validated by the standard build and test process, and then deployed through an automated process triggered through a deployment pipeline. A deployment pipeline is, thus, particularly valuable in the context of safety-critical or highly regulated industries. In this pipeline, not all changes can go life without an approval. The expectation is that 65% of changes should be processed fully automatically, the rest should require a "check the box" approval. When introduced for the first time, the change should fall into the second category (gradual learning), but as it gets known, controlled and fine-tuned, it can move to the fully automated segment (gradual learning). • **Integrate automatic change checks into the deployment pipeline that halt the process if necessary.** Foresee automatic change notifications in the most critical parts. These controls can assess the impact of a change based on the dependencies of the system. If change is significant (e.g. changes to untrusted data processing), security specialists and architects have to be involved. Other possibilities are to use the results/metrics of a peer review or compliance tests run on every commit to decide whether to extend the change or to roll back. • **If CAB is in place, allow CAB leadership determining which DevOps-related changes do not require the normal process rigor and may bypass the traditional controls.** • **Allow teams communicate themselves with other teams if they are implementing a change that might impact other systems.** Changes that also affect other applications and teams have to be approved by another person from the affected team. • **Foresee mechanism for automatic software roll back if any code changes leave the software in a less than fully functioning state.** For example, companies might have two production environments (a "blue" and a "green" one) of which one is live. During a release, the software is deployed to the other production environment. In case of problems, a rapid rollback can be performed by just putting the first production environment live again. • **Instead of releasing individual changes, we should think more in terms of releasing a product as a whole.** Through the pipeline the following product types can be released: compiled programs, packages, containers of Virtual Machines.

continued on following page

Table 2. Continued

ISO/IEC 27001 controls impacted by (Sec) DevOps	NIST SP 800-53 controls impacted by (Sec) DevOps	(Sec)DevOps control objectives	(Sec)DevOps controls
12.1.3 Capacity management	AU-4 AUDIT STORAGE CAPACITY	**Opportunity**. Teams must have visibility into capacity in order to understand the impact on infrastructure when there are increases in workload, transactions, application upgrades, and hardware refresh (McCarthy M. H., 2015)	• **Operations tools should provide key insights into the performance of business transactions before the system goes live.** This enables development teams to quickly understand infrastructure dependencies, how functional changes impact business performance, and where refactoring is needed. • **Maintain a dynamic infrastructure environment.** In SecDevOps provisioning of new servers is facilitated by cloud capacity management. In a static environment the concept of SecDevOps is difficult to implement. Everything in the cloud is elastic and can upsize or downsize. You should not have too much capacity as well, since it comes with a significant cost. The environment should be periodically reviewed from the point of view of capacity: optimization versus service. You should also monitor that upscaling and downscaling does not happen too often.
12.4.1 Event logging	AU-3 CONTENT OF AUDIT RECORDS AU-6 AUDIT REVIEW, ANALYSIS, AND REPORTING AU-12 AUDIT GENERATION CA-7 CONTINUOUS MONITORING SI-4 INFORMATION SYSTEM MONITORING	**Opportunity**. Tools should not only monitor and automatically report incidents such as compliance breaches but should also continuously perform logging to create traceable processes and valid audit trails (Plant, 2019) **Opportunity**. Use logs to share technical information within an organisation (Aljundi, 2018) **Opportunity**. Use new logging & alerting techniques to better predict application performance and usage (Ravichandran, 2016) **Opportunity**. Tools should provide teams fast feedback on software code effectiveness and problem components (Ravichandran, 2016) **Opportunity**. Make logs readable to all stakeholders (Shahin M. B., 2016) **Opportunity**. Foresee monitoring infrastructure to quickly identify newly-deployed software that is misbehaving (Savor, 2016) **Opportunity**. Use strong monitoring as a mean to compensate for a potential lack of preventive controls (Plant, 2019)	• **Integrate process monitoring tools into the deployment pipeline.** It allows to minimize risk and create reliable reporting which can be used by auditors. In case of a problem or if compliance conditions are not fulfilled, these tools can halt the deployment process and alert the developers. Teams need to embrace active monitoring methods to build an understanding about issues before they affect customers. Part of this involves finding better ways to remove misleading alarms and false-positives. • **Define monitoring metrics.** Mean time to repair (MTTR) and mean time to restore service (MTRS) are more important to track than the mean time between failures (MTBF) in most cases because DevOps teams should focus on learning and moving on from mistakes instead of not making any. Other useful metrics are overall process quality, cost of development, cost of maintenance, accessibility, reliability, interoperability, and availability for audits. • **An effective, dynamic inventory must quickly and continuously discover and validate new assets, or changes in existing assets, as soon as they appear online.** • **Have a log-driven, log-specific architecture in the middle of the applications.** Software architectures of those applications should continuously collect operational data at the appropriate levels and more importantly make it easy to aggregate logs, convert them into appropriate format and make them searchable. • **Use quality assurance monitoring to check if the code is following the defined standard.** QA monitoring can be automated to reduce resources consumption and faster defects detection. Encourage DevOps teams to monitor because they also have to provide operational services for their applications. Without the feedback they receive from monitoring these systems it is more difficult for them to detect problems. • **Use tooling to process logs.** There are many different logging, which can be created in the cause of the years. For example, access logging is becoming very important for cybersecurity, but this logging generates a huge amount of data. New tooling, like Splunk, help to process that data. It allows you to create intelligence over the whole pile of loggings in the technology stack: pattern recognition (user login, network access, etc.), exceptional situation detection, etc.
12.6.2 Restrictions on software installation	CM-11 USER-INSTALLED SOFTWARE	**Risk**. Delegate tool choice to teams to contributes to software delivery performance (Forsgren, 2018)	• **In a contained environment the tooling choice can be delegated to the development team to improve the software delivery performance.** However, in a wide network, the choice should be limited to protect the eco system. Furthermore, the free choice is not suitable for the regulated environments. • **Define policies for free tooling download.** Developers are often interested in downloading components/libraries freely without going to security each time. In the past you had an approved list of components. Now the approach can be to allow the download of any component, but a set of policies needs to be applied: no known vulnerabilities, component may not be too old, there must be a large community behind the component, etc. No need for manual checks exist any longer, since there is tooling to control the compliance: Black Duck or SLM (Service Level Management), or Trend Micro (automation of security within CI/CD pipeline). • **Teams in a common technology stack and common domains should standardize on the same tooling and configuration of CI/CD.** Otherwise application development becomes very complicated.
13.1.3 Segregation in networks	AC-4 INFORMATION FLOW ENFORCEMENT	**Opportunity**. Implement network isolation and policy control (Shackleford, D., SANS, 2016)	• **Use micro-segmentation for network isolation and policy control in the cloud environment.** With this technique, each cloud instance adopts a "zero trust" policy model that allows for very granular network interaction controlled at the virtual machine network interface controller. This allows each cloud system to essentially take its network access control and interaction policy with it as it migrates through virtual and cloud environments, minimizing disruption and reliance on physical and hypervisor-based network technology, although software-defined networking and automation techniques can definitely play a role in micro segmentation operations. • **Development environment should be separated from Production.** This allows to download and use random products in the Test environment. • **Consider moving from standard segmentation towards more flexible networking zones for business production, infrastructure testing and application factory.** The latter is similar to business production, but for emergency release, where the whole factory has the same criticality as the business itself.
14.2.7 Outsourced development	SA-4 ACQUISITION PROCESS SA-12 SUPPLY CHAIN PROTECTION	**Opportunity**. Teams must have the visibility and insight you need to ensure reliable business service delivery for the applications and business services that have moved to third-party providers (McCarthy M. H., 2015)	• **Foresee service level agreements with hosting providers that include security.** • **With off-shore development, the 3rd party needs to be on the same platform, using the same tooling.** Just passing the code does not work. There should be time alignment as well, since there are more time dependencies. We should also be aware of vulnerabilities introduced by the third party. • **With cloud, we rely on the configuration of the supplier, who provides the assurance.** You have an auditor who controls the cloud provider: compliance & assurance. We are working with the cascading control frameworks: the more you reuse from cloud, the less controls you need to implement yourself. And cloud provider guarantees that all controls are correctly implemented.

continued on following page

Table 2. Continued

ISO/IEC 27001 controls impacted by (Sec) DevOps	NIST SP 800-53 controls impacted by (Sec) DevOps	(Sec)DevOps control objectives	(Sec)DevOps controls
18.2.2 Compliance with security policies and standards 18.2.3 Technical compliance review	CM-1 CONFIGURATION MANAGEMENT POLICY AND PROCEDURES CA-2 SECURITY ASSESSMENTS	**Opportunity.** Trace automatically compliance breaches and verify standards compliance (Mohan V. O., 2016) (Callanan, 2016) (Plant, 2019).	• Integrate into SecDevOps the following set of best practices: automating tests to detect noncompliance, tracking compliance breaches through automated reporting of violations, continuous monitoring and maintenance of a service catalogue with tested and certified services. • The compliance tests should run on every commit and during deployment to the acceptance test environments. Release candidates should be blocked if any of the standards criteria are not met. • Introduce item tracking from requirement to the final product as a standard practice in DevOps. Software items related to requirements should be traced over the complete version history and at every point of its lifecycle. In order to enable this, workflow tools, version histories and CI tools have to form automatic connections. In order to achieve further compliance, tools should include standard templates that comply with regulations. These tools should work hierarchically by linking requirements, subsequent items and their test items and reports to each other. Lastly, the tools should guide the developer to follow the regulated workflow.
	CM-2 BASELINE CONFIGURATION	**Opportunity.** Use automated configuration management tools (Plant, 2019) **Opportunity.** Systems should be able to establish the definition of dependencies between components, with automatic alerts when dependency conditions are not met (Ravichandran, 2016) **Opportunity.** It should be possible to provision environments and build, test, and deploy software in a fully automated fashion purely from information stored in version control (Forsgren, 2018) **Opportunity.** Put application code, system configurations, application configurations and scripts for automating build in a version control system (Forsgren, 2018) **Opportunity.** Integrate performance baselines into the continuous deployment pipeline (Shahin M. B., 2017)	• Always integrate version control into DevOps processes. This can be done by using version control systems such as Git or Subversion. • Start each new build from an up-to-date image housing the latest patched operating system and middleware. • Ensure that the state of production systems can be reproduced (with the exception of production data) in an automated fashion from information in version control. • Considered an approach that kept track of which versions were deployed and tested together, and then deploying those applications together as "snapshots" or "release sets". • Deployable images should be available, with each Operating System and the corresponding configurations that can be deployed automatically. Easily modifiable configuration templates should be available for reuse. The baseline can be defined and modified per application type.

from varying industries. The final validation is performed against the recommendations proposed in the recent Department of Defence "Enterprise DevSecOps Reference Design" report (Lam, 2019).

The process of deriving the artifact is summarized in

Subject Matter Expert Interviews

The review of almost 100 scientific articles provided a sufficiently detailed view on the (Sec)DevOps control objectives and on the security control objectives they are impacting. Also many articles suggested particular (Sec)DevOps controls to address the control objectives. However, depending on the author's background, the controls are defined from a specific prospective and sometimes contradict. For instance, controls defined for non-regulated environments, like Facebook, are often less strict on the aspects such as segregation of duties and change management. These controls are primarily focusing on the speed of the release process and the frequency with which new features can be deployed in production.

As was mentioned earlier, the scope of this study is limited to a specific company segment: large regulated organisations. Therefore a set of eight interviews was set up with employees of varying organisations in Belgium and The Netherlands that fall under this category. The goal is to validate the

Table 3.

SecDevOps Capability Artifact controls	DoD Enterprise DevSecOps Reference Design
Information security roles and responsibilities	
• Part of the tasks from the Security department should be moved to the SecDevOps teams • Define the role of the Security Champion within the agile organisation • Work according to "5 amigo's" principle, stimulating collaboration between Business, Development, Testing, Security & Operations • Security should be positioned at the front of the development pipeline • Shift the Security role from operations to process and controls design	• There are nine DevSecOps software lifecycle phases: plan, develop, build, test, release, deliver, deploy, operate and monitor. Security is embedded within each phase • Change the organisational culture to take a holistic view and share the responsibility of software development, security and operations • Break down organisational silos. Increase the team communication and collaboration in all phases of the software lifecycle
Segregation of duties	
• Automate the production deployment process so no person can execute the deployment without passing the automated controls first • Foresee full governance for sensitive phases in software development and deployment • Code should be peer-reviewed • Make use of separate local and central code repositories to enforce the control of the release process • A separate team can be foreseen for all products to release to production, while guaranteeing the segregation of duties	• Most of the processes should be automatable via tools and technologies
Secure development policy	
• Well-defined policies regarding information exchange across teams should be in place to prevent security threats due to collaboration • Risk assessments should be performed from the first planning stage and continuously before every iteration • Manage security controls as an API or "Security/Compliance as a Code" • Design for failure • Standardize security controls and security assurance process • Adopt convention over configuration to reduce the complexity of CI/CD system • Use Service- oriented architectures and micro-services architecture to build systems that can easily be deployed in multiple environments • Make the engineer accountable for the code in production he created • Use automated activities to simplify and to improve software development • Secure the pipeline by restricting the attack surface of the code base • Security teams should be informed about the challenges faced by operators and developers, and vice versa • Put the responsibility for handling the security considerations within the scope of the development team • Perform automatic patching • Let a Quality Assurance officer to support and monitor the DevOps process from a high-level view	• The "big bang" style delivery of the Waterfall process is replaced with small but more frequent deliveries, so that it is easier to change course as necessary. Each small delivery is accomplished through a fully automated process or semi-automated process with minimal human intervention to accelerate continuous integration and delivery • Build a culture of safety by sharing after-action reports on both positive and negative events across the entire organisation. Teams should use both success and failure as learning opportunities to improve the system design, harden the implementation, and enhance the incident response capability as part of the DevSecOps practice • A software system can start with a Continuous Build pipeline, which only automates the build process after the developer commits code. Over time, it can then progress to Continuous Integration, Continuous Delivery, Continuous Deployment, Continuous Operation, and finally Continuous Monitoring, to achieve the full closed loop of DevSecOps. A program could start with a suitable process and then grow progressively from there. The process improvement is frequent, and it responds to feedback to improve both the application and the process itself • Accept that change can be required at any time, and all options are available to achieve it. Fail fast, fail small, and fail forward. An example of failing forward is when a developer finds that a release does not work. Then instead of restoring the server to its pre-deployment state with the previous software, the developer's change should be discrete enough that they can fix it and address the issue through a newer release
Information security awareness, education and training	
• Train Security Champion • Train Security Team • Train software developers	• Train staff with DevSecOps concepts and new technologies. Gradually gain buy-in from all stakeholders
Management of privileged access rights	
• Any privileged accounts (such as root and the local administrator accounts) should be monitored very closely (or ideally be disabled completely) • Use timed access rights by allowing developers to request timed passwords • Grant employees dynamic access rights per sprint by assigning them specific responsibilities and access rights every time • It is often sufficient to give DevOps engineers two accounts for the different environments • Maintain a separate CI server for development done outside the company • Provide fine -grained account management with different levels of access • All changes in production should preferably be executed by CI/CD pipeline • Organize release pipeline admins as a separate team	• Offer open access across the organisation to view the activities occurring within the automated process and to view the auto-generated Artifacts of Record
Documented operating procedures	
• Use tooling to generate documentation • Use logging for documentation and traceability purposes • Documentation should not be shared through a medium which cannot be traced • Differentiate between operational and non-operational documentation	
Change management	

Table 3. Continued

SecDevOps Capability Artifact controls	DoD Enterprise DevSecOps Reference Design
• Integrate Change Management into the Release Management process • Changes should only be applied to production using a process that forms part of a deployment pipeline • Integrate automatic change checks into the deployment pipeline that halt the process if necessary • If CAB is in place, allow CAB leadership determining which DevOps-related changes do not require the normal process rigor and may bypass the traditional controls • Allow teams communicate themselves with other teams if they are implementing a change that might impact other systems • Foresee mechanism for automatic software roll back if any code changes leave the software in a less than fully functioning state • Instead of releasing individual changes, we should think more in terms of releasing a product as a whole	• Make many small, incremental changes instead of fewer large changes. The scope of smaller changes is more limited and thus easier to manage • The DevSecOps lifecycle is an iterative closed loop. Start small and build it up progressively to strive for continuous improvement. Set up human intervention at the control gates when necessary, depending on the maturity level of the process and the team's confidence level in the automation. Start with more human intervention and gradually decrease it as possible • AO (Authorizing Official) should consider automating the Authority to Operate (ATO) process as much as possible • The tags added to artifacts in the artifact repository help guarantee that the same set of artifacts move together along a pipeline • Push down or delegate responsibility to the lowest level: • Strategic: This is related to the Change Control Board (CCB) or Technical Review Board (TRB); it involves "Big Change" unstructured decisions. These infrequent and high-risk decisions have the potential to shape the strategy and mission of an organisation. • Operational: (Various Scrum) Cross-cutting, semi-structured decisions. In these frequent and high-risk decisions, a series of small, interconnected decisions are made by different groups as part of a collaborative, end-to-end decision process. • Tactical: (Global Enterprise Partners (GEP)/Product Owner/Developers Activities) Delegated, structured decisions. These frequent and low-risk decisions are effectively handled by an individual or working team, with limited input from others • DoD Centralized Artifact Repository (DCAR) holds the hardened VM images and hardened OCI compliant container images of: DevSecOps tools, container security tools, and common program platform components (e.g. COTS or open source products) that DoD program software teams can utilize as a baseline to facilitate the authorization process.
Capacity management	
• Operations tools should provide key insights into the performance of business transactions before the system goes live • Maintain a dynamic infrastructure environment	
Event logging, monitoring & reporting	
• Integrate process monitoring tools into the deployment pipeline • Define monitoring metrics • An effective, dynamic inventory must quickly and continuously discover and validate new assets, or changes in existing assets, as soon as they appear online • Have a log-driven, log-specific architecture in the middle of the applications • Use quality assurance monitoring to check if the code is following the defined standard • Use tooling to process logs	• Actionable security and quality assurance (QA) information, such as security alerts or QA reports, must be automatically available to the teams at each software lifecycle phase to make collaborative actions possible • Governance activities do not stop after ATO but continue throughout the software lifecycle, including operations and monitoring. DevSecOps can facilitate and automate many governance activities
Restrictions on software installation	
• In a contained environment the tooling choice can be delegated to the development team to improve the software delivery performance • Define policies for free tooling download • Teams in a common technology stack and common domains should standardize on the same tooling and configuration of CI/CD	
Segregation in networks	
• Use micro-segmentation for network isolation and policy control in the cloud environment • Development environment should be separated from Production • Consider moving from standard segmentation towards more flexible networking zones for business production, infrastructure testing and application factory	
Outsourced development	
• Foresee service level agreements with hosting providers that include security • With off-shore development, the 3rd party needs to be on the same platform, using the same tooling • With cloud, we rely on the configuration of the supplier, who provides the assurance	• The program should have a formal Service Level Agreement (SLA) with the underlying infrastructure provider about what services are included and what authorizations can be inherited. This affects the status of applicable assessment procedures and prepares the stage for inheritance into the operations environment and application
Compliance with security policies and standards	
• Integrate into SecDevOps the following set of best practices: automating tests to detect noncompliance, tracking compliance breaches through automated reporting of violations, continuous monitoring and maintenance of a service catalog with tested and certified services • The compliance tests should run on every commit and during deployment to the acceptance test environments • Introduce item tracking from requirement to the final product as a standard practice in DevOps	• Taking the same approach as Infrastructure as Code (IaC), security teams program security policies directly into configuration code, as well as implement security compliance checking and auditing as code, which are referred as Security as Code (SaC) • Technologies and tools play a key role in DevSecOps practice to shorten the software lifecycle and increase efficiency. They not only enable software production automation as part of a software factory, but also allow operations and security process orchestration
Configuration management	
• Always integrate version control into DevOps processes. This can be done by using version control systems such as Git or Subversion. • Start each new build from an up-to-date image housing the latest patched operating system and middleware. • Ensure that the state of production systems can be reproduced (with the exception of production data) in an automated fashion from information in version control. • Considered an approach that kept track of which versions were deployed and tested together, and then deploying those applications together as "snapshots" or "release sets". • Deployable images should be available, with each Operating System and the corresponding configurations that can be deployed automatically	• The instantiation of the DevSecOps environments can be orchestrated from configuration files instead of setting up one component at a time manually. The infrastructure configuration files, the DevSecOps tool configuration scripts, and the application run-time configuration scripts are referred to as Infrastructure as Code (IaC) • Both IaC and SaC are treated as software and go through the rigorous software development processes including design, development, version control, peer review, static analysis, and test

Figure 3. Major capabilities highlighted in SecDevOps Capability Artefact

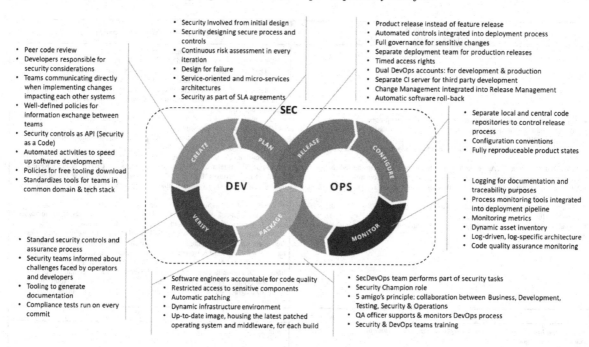

findings from the scientific literature study and to compare them against the practical experience of the interviewees. It allows to filter out the controls that are less suitable or inapplicable to the organisations considered in the scope of this research.

For confidentiality reasons, the names of the companies and the interviewees are not disclosed. However, all interviewees were selected based on the following criteria:

- Over ten years of experience in Digital / IT industry in at least one of the following roles: Security Architect / Responsible, Delivery/Release Manager, Information Technology Architect, Integration Engineer or similar roles;
- Minimum five years of experience in a large (more than 1,000 employees) organisation in a sector that has strict regulatory compliance requirements, such as banking, insurance, medical, etc.

The interviewees were selected to provide a mixture of security oriented and operational profiles, to highlight (Sec)DevOps compliance from different perspectives and to find the optimal set of controls, which allow to optimize the compliancy requirements as well as the operational needs for speed and flexibility.

All eight interviews were organized following the same scheme:

- An interviewee receives a draft version of the SecDevOps Capability Artifact, containing the selection of security control objectives and controls from ISO 27002 and NIST SP 800-53 impacted by (Sec)DevOps, (Sec)DevOps control objectives and the corresponding controls obtained from the literature review. The draft artifact is provided to the interviewee at least a month before the interview takes place and is accompanied with the necessary explanation.

- A semi-structured interview is organized with a duration varying from 1 to 4 hours, depending on the interviewee availability and the interviewee progress. The goal of the interview is, first of all, to obtain the feedback on the draft SecDevOps Capability Artifact and to refine the selection of the security control objectives and controls impacted by (Sec)DevOps. Secondly, the (Sec)DevOps controls proposed in the literature are discussed to determine if they are applicable and effective in regulated environments. Finally, the list of controls is extended, taking into account the interviewee practical experience with (Sec)DevOps.
- The following questions are covered in the indicated order and the input from the interview is recorded (recordings are available on demand):
- How are Information Security roles and responsibilities impacted by the introduction of SecDevOps? Is there impact at all?
- How to address the segregation of duties in SecDevOps environment where the boundaries between different roles are faded in favour of flexibility & speed?
- How does SecDevOps impact the software development life cycle?
- Which information security awareness, education and training is required to support the new way of working?
- How to control access to the source code in the SecDevOps environment which promotes sharing?
- Can we guarantee compliance while simplifying or completely removing the change approval process?
- Compliance requires a sufficient amount of documentation, while SecDevOps strives mainly towards speed and flexibility. What is the best way to address this challenge?
- Is there any impact of SecDevOps on capacity management?
- How are event logging & monitoring impacted by SecDevOps?
- Should the installation of software be restricted or not, in order to boost performance?
- How does SecDevOps assist in the management of technical vulnerabilities?
- What is the best way to segregate the network while preserving the flexibility?
- What is the impact in information security analysis and specification on SecDevOps?
- Is there a change in the role of Security within the Development Lifecycle due to SecDevOps?
- Does SecDevOps have influence on outsourcing and its compliance management?
- What is the impact of SecDevOps on compliance management?
- What is the impact of SecDevOps on configuration management?

After the interview, transcripts were created and are left out of this book but can be extracted from the author of this chapter.

Finally, the draft SecDevOps Capability Artifact was modified and extended with the input obtained from the subject matter experts.

Result comparison with DoD Enterprise DevSecOps Reference Design Report

The DoD Enterprise DevSecOps Reference Design has as a purpose to describe the DevSecOps lifecycle, supporting pillars, and DevSecOps ecosystem. It goes beyond the boundaries of this research by listing the tools and activities for the DevSecOps software factory and ecosystem; introducing the DoD enterprise DevSecOps container service that provides hardened DevSecOps tools and deployment templates to the program application DevSecOps team to select; and showcasing a sampling of software factory reference designs and application security operations. However, in what follows we can evaluate the findings of our research against the DevSecOps high-level implementation and operational guidance provided in DoD report.

The reason for choosing DoD Enterprise DevSecOps Reference Design as a means for the validation of SecDevOps Capability Artifact is that this recent report provides a practical reference for the implementation of the DevSecOps capability within a regulated organisation, in line with NIST Cybersecurity Framework, which is a high-level framework building upon specific controls and processes defined by NIST SP 800-53, COBIT 5 and ISO 27000 series.

The table below shows how the findings of this research, defined by SecDevOps Capability Artifact, are reflected and confirmed by the report.

REFERENCES

Aljundi, M. (2018). *Tools and Practices to Enhance DevOps Core Values* (Master's Thesis). Lappeenranta University of Technology, School of Business and Management.

Bass, L. H. (2015). Securing a Deployment Pipeline. *Proceedings of the Third International Workshop on Release Engineering*.

Betz, C. O. (2016). The Impact of Digital Transformation, Agile, and DevOps on Future IT Curricula. SIGITE'16.

Bucena, I. K. (2017). Simplifying the DevOps Adoption Process. *BIR Workshop*.

Callanan, M. S. (2016). DevOps: Making It Easy to Do the Right Thing. *IEEE Software*, *33*(3), 53–59.

Chivers, H. P. (2005). *Agile Security Using an Incremental Security Architecture* (Vol. 3556). Extreme Programming and Agile Processes in Software Engineering. doi:10.1007/11499053_7

Clager, J. R. (2016). Mitigating an Oxymoron: Compliance in a DevOps Environments. *2016 IEEE 40th Annual Computer Software and Applications Conference (COMPSAC)*.

Colavita, F. (2016). DevOps Movement of Enterprise Agile Breakdown Silos, Create Collaboration, Increase Quality, and Application Speed. *Proceedings of 4th International Conference in Software Engineering for Defence Applications Advances in Intelligent Systems and Computing*. 10.1007/978-3-319-27896-4_17

Compliance Forge Website. (n.d.). *Which framework is right for my business? NIST Cybersecurity Framework vs. ISO 27002 vs. NIST 800-53 vs. Secure Controls Framework*. Retrieved from https://www.complianceforge.com/faq/nist-800-53-vs-iso-27002-vs-nist-csf.html

Delphic, G. (2016). *2015 Annual State of DevOps*. Academic Press.

Derksen, B. N. (2018). *Agile Secure Software Lifecycle Management Secure by Agile Design*. Leiden: Secure Software Alliance.

Farroha, B. F. (2014). A Framework for Managing Mission Needs, Compliance, and Trust in the De- vOps Environment. *Proceedings of the 2014 IEEE Military Communications Conference.* 10.1109/ MILCOM.2014.54

Forsgren, N. H. (2018). *Accelerate: The Science of Lean Software and Devops: Building and Scaling High Performing Technology Organizations.* IT Revolution Press.

Forum, C. F. O. (n.d.). *Understanding and managing the IT risk landscape: A practitioner's guide.* https:// www.thecroforum.org/2018/12/20/understanding-and-managing-the-it-risk-landscape-a-practitioners- guide/

Fowler, M. H. (2001). *The Agile Manifesto.* Academic Press.

Gill, A. L. (2017). DevOps for Information Management Systems. *VINE Journal of Information and Knowledge Management Systems.*

Harkins, M. (2013). A New Security Architecture to Improve Business Agility. Managing Risk and Information Security. doi:10.1007/978-1-4302-5114-9_7

Hevner, A. (2007). A Three Cycle View of Design Science Research. Proceedings of Scandinavian Journal of Information Systems.

Hevner, A. M., March, Park, & Ram. (2004). Design science in information systems research. 75–105. *Management Information Systems Quarterly, 28*(1), 75. doi:10.2307/25148625

Hilton, M. N. (2017). *Trade-offs in continuous integration: assurance, security, and flexibility.* ESEC/FSE.

Johannesson, P. A. (2014). *An introduction to Design Science.* Springer. doi:10.1007/978-3-319-10632-8

Lam, T. C. (2019). *DoD Enterprise DevSecOps Reference Design.* Academic Press.

Laukkarinen, T. K. (2017). DevOps in regulated software development: Case medical devices. In *Pro- ceedings of 2017 IEEE/ACM 39th International Con-ference on Software Engineering: New Ideas and Emerging Technologies Results Track (ICSE-NIER).* IEEE.

Laukkarinen, T. K., Kuusinen, K., & Mikkonen, T. (2018). Regulated software meets DevOps. *Informa- tion and Software Technology, 97,* 176–178. doi:10.1016/j.infsof.2018.01.011

Lin, T. (2016). *Compliance, Technology, and Modern Finance.* Temple University Legal Studies Re- search Paper No. 2017-06.

Mattetti, M. S.-P. (2015). Securing the infrastructure and the workloads of Linux containers. *Proceedings of the 2015 IEEE Conference on Communications and Network Security (CNS).* 10.1109/CNS.2015.7346869

McCarthy, M. A. (2014). A Compliance Aware Software Defined Infrastructure. *Proceedings of IEEE International Conference on Services Computing.*

McCarthy, M. H. (2015). Composable DevOps: Automated Ontology Based DevOps Maturity Analysis. *IEEE International Conference on Services Computing.*

Michener, J. C. (2016). Mitigating an oxymoron: compliance in a DevOps environment. *2016 IEEE 40th Annual Computer Software and Applications Conference (COMPSAC).*

Mohan, V. O. (2016). SecDevOps: Is It a Marteking Buzzword? *Proceedings of the 11th International Conference on Availability, Reliability and Security (ARES).*

Mohan, V. O. (2018). BP: Security Concerns and Best Practices for Automation of Software Deployment Processes: An Industrial Case Study. Proceedings of 2018 IEEE Cybersecurity Development (SecDev).

Murray, A. (2015). *The Complete Software Project Manager: Mastering Technology from Planning to Launch and Beyond – Agile, Waterfall, and the Key to Modern Project Management.* Academic Press.

Myrbakken, H. C.-P. R. (2017). Software Process Improvement and Capability Determination. In DevSec-Ops: A Multivocal Literature Review (pp. 17-29). Cham: Springer, Communications in Computer and Information Science.

Pastrana, M. (2019). Ensuring Compliance with Sprint Requirements in SCRUM: Preventive Quality Assurance in SCRUM. In Advances in Computer Communication and Computational Sciences (pp. 33-45). Springer.

Plant, O. (2019). *DevOps Under Control: Development of a Framework for Achieving Internal Control and Effectively Managing Risks in a DevOps Environment* (Master's thesis). University of Twente.

Rahman, A. W. (2016). In Proceedings of the Symposium and Bootcamp on the Science of Security (HotSos '16). In *Security practices in DevOps* (pp. 109–111). New York, NY: Association for Computing Machinery.

Ravichandran, A. T. (2016). *DevOps for Digital Leaders*. Apress.

Recker, J. (2013). *Scientific Research in information Systems: A Beginner's Guide*. Springer. doi:10.1007/978-3-642-30048-6

Rindell, K. H. (2016). Case Study of Security Development in an Agile Environment: Building Identity Management for a Government Agency. *11th International Conference on Availability, Reliability and Security (ARES).*

Savor, T. D. (2016). Continuous Deployment at Facebook and OANDA. *Proceedings of 2016 IEEE/ACM 38th International Conference on Software Engineering Companion (ICSE-C).*

Schneider, C. (2015). Security DevOps - staying secure in agile projects. *Proceedings of OWASP AppSec Europe.*

Shackleford, D. (2016). A DevSecOps Playbook. SANS Institute.

Shahin, M. B. (2016). The Intersection of Continuous Deployment and Architecting Process: Practitioners' Perspectives. *10th ACM/IEEE International Symposium on Empirical Software Engineering and Measurement (ESEM).*

Shahin, M. B. (2017). Continuous Integration, Delivery and Deployment: A Systematic Review on Approaches, Tools, Challenges and Practices. *IEEE Access: Practical Innovations, Open Solutions, 5,* 3909–3943.

Shahin, M. B. (2017). Beyond Continuous Delivery: An Empirical Investigation of Continuous Deployment Challenges. *2017 ACM/IEEE International Symposium on Empirical Software Engineering and Measurement (ESEM)*.

Sharma, S. C. (2015). DevOps for Dummies. John Wiley & Sons.

Silveira, P. R. (2012). Aiding Compliance Governance in Service-Based Business Processes. In Handbook of research on Service-Oriented Systems and Non-Functional Properties (pp. 524–548). Academic Press.

Stackpole, B. O. (2010). *Security Strategy: From Requirements to Reality*. Auerbach Publications. doi:10.1201/EBK1439827338

Stolt, S. N. (2013). Continuous Delivery? Easy! Just Change Everything (Well, Maybe It Is Not That Easy). 2013 Agile Conference.

Storms, A. (2015). How security can be the next force multiplier in devops. *Proceedings of RSA Conference*.

CA Technologies. (2014). *Devops: The Worst-Kept Secret to Winning in the Application Economy*. Author.

Vael, M. (2019, February 22). *Enterprise Information Security Architecture: How to design rock-solid security directly into information systems*. Antwerp, Belgium: Academic Press.

Vaishnavi, M. O. (2016). SecDevOps: Is It a Marketing Buzzword? - Mapping Research on Security in DevOps. *11th International Conference on Availability, Reliability and Security (ARES)*.

Yasar, H. (2017). Implementing secure DevOps assessment for highly regulated environments. *Proceedings of the 12th International Conference on Availability, Reliability and Security - ARES '17*.

Chapter 9
Findings and Core Practices in the Domain of CI/CD and DevOps on Security Compliance

Yuri Bobbert
ON2IT BV, The Netherlands & Antwerp Management School, University of Antwerp, Belgium

Maria Chtepen
BNP Paribas Group, Belgium

ABSTRACT

In this chapter, the authors describe the findings and conclusions on "The SecDevOps Capability Artifact." It is validated by means of an extensive academic literature review and interviews with multiple domain experts and practitioners. An additional validation was performed by comparing the findings of this study with high-level implementation and operational guidance of the DoD enterprise DevSecOps reference design report. The report has as a purpose to describe the DevSecOps lifecycle and supporting pillars, in line with NIST cybersecurity framework, which is a high-level framework building upon specific controls and processes defined by NIST SP 800-53, COBIT 5, and ISO 27000 series. This chapter is concluded with a pragmatic set of core practices academics, and practitioners can use them to ensure security compliance in CI/CD pipelines that ultimately enable teams to work agile on digital platforms.

FINDINGS

This research investigates a challenge large regulated organisations are exposed to: how to increase the speed and quality of delivery, using DevOps, while remaining compliant with the applicable security standards and regulations? This question is particularly relevant since the speed and flexibility propagated by DevOps are often contradicting the core controls addressed by security standards: segregation of duties, change control, network segregation, etc. On the contrary, certain DevOps objectives are simplifying the implementation of security controls through development process automation, continuous monitoring, earlier integration of security requirements into design, etc.

DOI: 10.4018/978-1-7998-7367-9.ch009

In the previous chapters we investigated whether DevOps is at the end an opportunity or a risk to security compliance. The answer to this question definitely depends on the compliance requirements applicable to a specific organisation, but in general, certain aspects of DevOps are an unmistakable benefit, while other aspects require a detailed review and fine-tuning in order to comply with security regulations.

We selected ISO 27002 and NIST SP 800-53 as the reference security standards for this research effort, since both standards are widely known an applied within European as well as US organisations. Furthermore, both are sufficiently detailed in terms of the controls covered to be able to relate them to DevOps controls. For each standard, a study of the controls is conducted to evaluate the impact of De-vOps. At the same time, an extensive literature review of almost 100 scientific papers provided a good view on the relevant DevOps control objectives and the corresponding controls. It allowed to perform the mapping between the impacted security controls and (Sec)DevOps controls objectives. From this mapping we learned which (Sec)DevOps control objectives impacted the security compliance in either positive (Opportunity) or negative (Risk) way. Furthermore, the literature provided a good overview of (Sec)DevOps controls that can mitigate the abovementioned security risks. The results of the literature review are validated against the results of a number of interviews with subject matter experts, leading to the formalisation of the conclusions in SecDevOps Capability Artifact. The capabilities highlighted in the artefact are shown in Figure 10. The figure indicates the relationship between the major capabilities and the phase in the SecDevOps Gartner's toolchain where they belong. Finally, DoD Enterprise DevSecOps Reference Design recommendations are reviewed in the light of SecDevOps Capability Artifact to verify that they effectively confirm our findings.

What is true for many transformations, also applies in the case of SecDevOps - it impacts the security compliance from People, Processes and Technology perspectives:

- **People**. SecDevOps significantly changes the way in which security is integrated into the development process, which requires the introduction of new types of security roles within the organisation. Also the focus of the existing roles is shifting and, therefore, there is a need for training at different levels: from generic security training for team members to dedicated specialist training for technical profiles (e.g. developers, architects, operations, etc.).
- **Process**. SecDevOps is often introduced in the organisation together with the move to Agile software development. Therefore, it is not always easy to distinguish the process impact of SecDevOps as opposed to the impact of Agile. In general, it impacts the traditional segregation of roles, the rights each role gets within the software development and deployment chain, as well as the way in which software is designed and released.
- **Technology**. The most significant contribution of SecDevOps to the changing way we do security is due to the extensive use of automation. Automation and technology create almost endless possibilities to speed up and to improve the quality of the traditional time consuming processes such as documentation, change management and control, capacity management, event logging, monitoring and reporting. SecDevOps allows to automate many of these steps through CI/CD, build intelligent controls and alerts within the deployment pipeline and allow for automatic actions in case of failure (e.g. automatic release roll-back).

The major findings of our research suggest the controls that allow to incorporate SecDevOps into the organisation, which traditionally builds its processes around compliance requirements. The controls suggest how to satisfy these requirements without sacrificing too much on the flexibility and speed that form the major advantage of SecDevOps in first place. The most important research findings can be summarized as follows:

- Part of the tasks of the Security department should be moved to the SecDevOps teams. The latter must be able to take the responsibility for Security Gating, by using security testing automation, as well as for the iterative risk assessments and threat modelling. It will allow to position the security activities earlier in the development pipeline and to guarantee an iterative review and adjustment. A Security Champion role within the team should be able to assist in this new activities, with the support of the Security department. Obviously, this shift in responsibilities requires the acquirements of necessary technical skills by the Security people and of security knowledge by SecDevOps team members.

- The segregation of duties becomes less "a people job", and more an automated process. At the end, the CI/CD pipeline should be able to perform all the necessary checks to assure the quality of the released code and to avoid any potential fraud. However, automation will often happen gradually, where in intermediary phases manual check are still required (e.g. peer code review, Change Advisory Board (CAB)). These manual controls can be removed step-wise. For example, CAB may decide which changes are allowed to be pushed automatically through the pipeline, based on the available automated controls, the severity of the change and on the previous experience with similar releases. On the other hand, it is important to recognize that a different form of the segregation of duty will be put into place: if the pipeline is becoming the crucial/only form of controls developers should be prohibited from tampering with the pipeline code and configuration.

- SecDevOps requires new standards for software design: instead of releasing changes, products will be released instead. This leads to a new form of workload management, where features are grouped around the release of a product. It means that the dependencies between different products should be minimize as well. Service oriented architectures and micro-services are very suitable for decoupling large systems into smaller independent units that can be deployed separately in multiple environments.

- Finally, automation is the core of many recommendations, meaning that the full gain of SecDevOps can only be achieved when the majority of tasks can be done within the appropriate tooling, linked into an automatic CI/CD pipeline. The automation varies from access control (e.g. RepoMan), to automatic documentation (e.g. JIRA, GIT), to compliance check (e.g. Sonar, Fortify), monitoring and alerting (e.g. Splunk).

LIMITATIONS

The research covered in this book has a number of limitations. First of all, due to time and geographical constraints, the number of participants at expert interviews is limited to less than a dozen companies in Belgium and The Netherlands. The background of experts varied from Security Architects to Delivery Managers and provided us with the advantage of having varying points of view on the addressed problem. On the other hand, the responses on some questions provided sometimes contradictory results, which were filtered out in the Artifact by relying upon the opinion of the majority of the experts, as well as on the literature references.

Furthermore, as suggested by the Design Science Research approach (see Research Methodology), the artifact should be developed, and its effectiveness demonstrated and evaluated in practice. These steps were not completed due to the lack of time and the research was limited to the requirements collection phase.

OVERVIEW PRACTICES FOR CI/CD AND DEVOPS SECURITY COMPLIANCE

Based on this research project and the expert interviews we have derived the following key practices for CI/CD and DevOps Security Compliance:

1. Information security roles and responsibilities
   - Part of the tasks from the Security department should be moved to the SecDevOps teams
   - Define the role of the Security Champion within the agile organisation
   - Work according to "5 amigo's" principle, stimulating collaboration between Business, Development, Testing, Security & Operations
2. Segregation of duties
   - Automate the production deployment process so no person can execute the deployment without passing the automated controls first
   - Foresee full governance for sensitive phases in software development and deployment
   - A selection of (critical) code can still be peer-reviewed, but be aware of the delay it may introduce within the agile process
3. Secure development policy
   - Risk assessments should be performed from the first planning stage and continuously before every iteration
   - Make the engineer accountable for the code in production he created
   - Use automated activities to simplify and to improve software development
4. Information security awareness, education and training
   - Train Security Champion
   - Train Security Team
   - Train software developers
5. Management of privileged access rights
   - Any privileged accounts (such as root and the local administrator accounts) should be monitored very closely or, ideally, be disabled completely
   - Use timed access rights by allowing developers to request timed passwords
   - Grant employees dynamic access rights per sprint by assigning them specific responsibilities and access rights every time
6. Documented operating procedures
   - Use tooling to generate documentation
   - Use logging for documentation and traceability purposes
   - Documentation should not be shared through a medium which cannot be traced
7. Change management
   - Integrate Change Management into the Release Management process

- Changes should only be applied to production using a process that forms part of a deployment pipeline
- Integrate automatic change checks into the deployment pipeline that halt the process, if necessary

8. Capacity management
   - Operations tools should provide key insights into the performance of business transactions before the system goes live
   - Maintain a dynamic infrastructure environment

9. Event logging, monitoring & reporting
   - Integrate process monitoring tools into the deployment pipeline
   - An effective, dynamic inventory must quickly and continuously discover and validate new assets, or changes in existing assets, as soon as they appear online
   - Have a log-driven, log-specific architecture in the middle of the applications

10. Restrictions on software installation
    - In a contained environment the tooling choice can be delegated to the development team to improve the software delivery performance
    - Define policies for free tooling download
    - Teams in a common technology stack and common domains should standardize on the same tooling and configuration of CI/CD
      10. Segregation in networks
    - Use micro-segmentation for network isolation and policy control in the cloud environment
    - Development environment should be separated from Production
    - Consider moving from standard segmentation towards more flexible networking zones for business production, infrastructure testing and application factory

11. Outsourced development
    - Foresee service level agreements with hosting providers that include security
    - With off-shore development, the 3rd party needs to be on the same platform, using the same tooling
    - With cloud, we rely on the configuration of the supplier, who provides the assurance

12. Compliance with security policies and standards
    - Integrate into SecDevOps the following set of best practices: automating tests to detect non-compliance, tracking compliance breaches through automated reporting of violations, continuous monitoring and maintenance of a service catalogue with tested and certified services
    - The compliance tests should run on every commit and during deployment to the acceptance test environments
    - Introduce item tracking from requirement to the final product as a standard practice in DevOps

13. Configuration management
    - Always integrate version control into DevOps processes. This can be done by using version control systems such as Git or Subversion
    - Start each new build from an up-to-date image housing the latest patched operating system and middleware
    - Ensure that the state of production systems can be reproduced (with the exception of production data) in an automated fashion from information in version control

FUTURE RESEARCH

SecDevOps is still a relatively new paradigm and many commercial organisations are in the process of discovering the best way to adopt it to their requirements and limitations. Therefore, a periodic review of the status and the evolution in the domain of SecDevOps is required. The following elements may definitely become interesting subjects for future investigation and research:

- **Extend the SecDevOps Capability Artifact with a Capability Maturity Model.** The artefact as specified in this study is defining a number of controls that are crucial for the integration of the SecDevOps capability into a regulated organisation. However, there is no information of which controls are required for which level of SecDevOps capability maturity. Presumably, not all organisations will require the same controls que to different background and complexity of the organisational structure.

Figure 11 shows a Capability Level model described in ISO 15504 standard, which can perfectly serve security maturity measurements. This model can be used for the future research to determine which SevDevOps controls belong at which level of organisational capability maturity. The classification of controls, can, for instance, be validated through additional expert interviews.

- **Expansion of the scope of the best practices for the implementation of SecDevOps outside the boundaries of large regulated organisations.** It may provide interesting input and insight into how the SecDevOps implementation varies across different companies, depending on industry, sector (e.g. non-profit, public, commercial) and the size of the organisation.
- **Investigation of practical aspects of the implementation of the SecDevOps Capability Artifact, converting the high-level recommendation provided in this study into actual operational models.** This study is limited to the high-level control description and recommendations. However, there are plenty of opportunities to refine these suggestions from the point of view of the practical implementation. For instance, for each of the covered aspects a model / framework can be proposed on how exactly they can be integrated into the organisational processes.
- **Implementation and validation of the SecDevOps Capability Artifact within one or several reference organisations.** These findings will allow to finetune the suggestions summarised in this study and to improve it with practical examples.

Figure 1. ISO 15504 capability levels

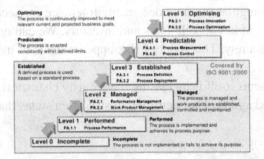

Chapter 10
Challenges and Opportunities for Security Assurance in DevOps

Dennis Verslegers
Orange Cyberdefense, Belgium

ABSTRACT

This chapter contains background information on the concepts that form the basis for the research. The author provides an overview of the integration of security activities in traditional procedural approaches and discusses the potential differences in agile, iterative, contexts. Concepts such as "continuous integration," "continuous deployment," and "shift left" are introduced. The challenges and opportunities to achieve security assurance in DevOps environments are discussed, and the objective of the research project is presented.

INTRODUCTION

The digital world is a vital playing field in this era of innovation and business transformation. Meeting market demands requires a continuous change to digital platforms. The speed with which we deliver these changes is detrimental for the competitive advantage of an organisation. To satisfy the demand for speed and agility new ways of organising work are explored and gaining in popularity rapidly. As software becomes an integral element of business growth, the focus on fast delivery of features with a tangible business value has increased. Organisations want the ability to seize opportunities without being stopped in their tracks.

Increasing pressure from regulators and a decreasing tolerance for security breaches by customers are reducing the risk appetite of key stakeholders and investors. We can only achieve business value if we can do so reliably and securely. This reduced risk appetite leads to increasing attention for secure business value creation.

The challenge which presents itself is increasing the security characteristics of our digital platforms without sacrificing speed and agility. The security industry often states that rapid change increases se-

DOI: 10.4018/978-1-7998-7367-9.ch010

curity risk. This does not necessarily hold true as security itself benefits from the ability to implement changes quickly. It allows a system to react to newly discovered vulnerabilities and the ever-changing threat landscape. What we consider secure today may not necessarily be the case tomorrow. The only way to safeguard business value is through rapid change.

The outcome of this research aims to provide a framework of validated security activities for DevSecOps environments ranked by their characteristics to improve security without sacrificing speed and agility. This framework allows organisations to build a "lean" approach to security.

This part of the book provides an overview of the drivers and approach for the research project.

Background

Software application development and IT infrastructure operations have undergone rapid change over the past decade (Tallon et al., 2019). One of the most discussed changes is the gradual shift from traditional, longer running, procedural approaches to dynamic and iterative processes (Forsgren & Humble, 2015).

An important driver for this change is the agile development approach as proposed by the manifesto for agile software development. The Agile Manifesto (Martin & Jim, 2001) defines the following core principles:

a) people and interactions over processes and tools,
b) working software instead of detailed documentation,
c) active customer participation and involvement rather than time and effort expended on negotiating contracts, and
d) willingness and ability to take on changes over steadfast commitment to a static plan.

As stated by Abrahamsson et al. (Abrahamsson et al., 2017) the focus of light and agile methods are simplicity and speed. A simplified comparison between procedural also referred to as waterfall, and agile software development practices are depicted below.

The impact of these simple tenants is not limited to the development teams. As the development approach changes, new demands are placed on the operations teams (Shahin et al., 2017a). The paradigm shift to focus on small, iterative software deliveries at a high pace adds dynamic complexity to IT operations. Speed in terms of software development translates into a need for fast deployments with a short delta time between release and availability in the production environment (Forsgren & Humble, 2015).

The introduction of agile methodologies in an organisation brings, besides the numerous benefits, also some challenges. From a process perspective, some of these challenges include areas such as functional and non-functional requirements identification, tracking of projects, quality management and risk management (Almeida & o, 2017). Traditional decision-making strategies such as hierarchical approval or review by a technical board no longer fit the increased speed with which development and operations are moving resulting in the delegation of authority and end-to-end responsibility for the various teams (Moe et al., 2008).

At the same time, technological advances such as infrastructure-as-code (Artač et al., 2017), containerisation and micro-services require specific skills and knowledge to realise their full potential (Kang et al., 2016).

Figure 1. Comparison between the typical process steps in waterfall and agile development as explained by Abrahamsson et al. (2017)

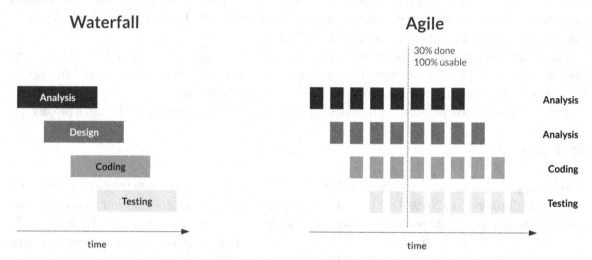

The close integration of the operations team with the development team, allowing them to collaborate early in the development cycle, is placed forward as an approach to meet these challenges and reduce friction (Ebert et al., 2016)(Smeds et al., 2015). We refer to this approach as DevOps.

Finding a complete and accurate definition of the term DevOps is difficult; this book adopts the definition as provided by Jabbari et al. (Jabbari et al., 2016):

"DevOps is a development methodology aimed at bridging the gap between Development and Operations, emphasising communication and collaboration, continuous integration, quality assurance and delivery with automated deployment utilising a set of development practices."

During a survey of Computer Associates, 88% of executives across 1425 organisations indicated to have adopted DevOps or to plan adoption in the next five years (Rahman & Williams, 2016).

DevOps, being a combination of people, process and technology, (Shahin et al., 2017b) employs a set of technical practices referred to as continuous deployment (CD). CD practices include a few key processes: version control, continuous integration, test automation and deployment automation (Forsgren & Humble, 2015).

Adoption of these Continuous Deployment practices plays an essential role in the software delivery process and may result in increased organisational performance through improved IT performance (Forsgren & Humble, 2015). An essential objective of Continuous Delivery is to reduce the risk of failure in production and gather fast feedback by integrating the various software components early in the cycle to produce deployable results with a high level of certainty (Duvall et al., 2007).

With the evolution described above, where agile iterative approaches are implemented in an increasingly rapidly evolving technological landscape aimed at increasing speed of deployment, the question rises how to integrate information security.

How can aspects such as regulatory compliance and security assurance on critical characteristics such as confidentiality, integrity, availability be considered in this fast-paced approach to enable reliable value creation with a managed level of risk? Simply put, one might argue that it does not matter how fast an organisation can develop and deploy software if security and compliance cannot keep up. Eventually, the organisation is expected to face security and compliance-related challenges or may have to accept an increased level of risk.

Figure 2. Relationship between continuous integration, delivery and deployment according to M. Shahin et al. (2017b)

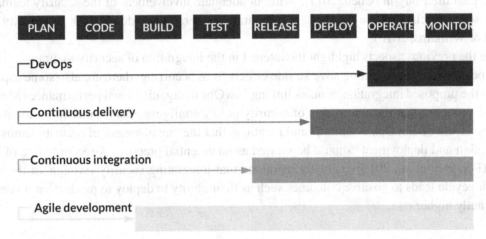

PROBLEM STATEMENT

Software security assurance aims to provide confidence in the security-related properties and functionalities, as well as the operation and administration procedures, of a given piece of software. Assurance methods are essential to achieve confidence that a solution meets its security requirements. Conventional, sequential, security assurance approaches rely on a third party's objectivity and expertise, resulting in a side-effect, documentation-focused development, which conflicts with agile development methods (Beznosov & Kruchten, 2004). In this context, third party security specialists coming "after the show" are expected to analyse, validate, test and then certify if a finished product meets its security requirements (Beznosov & Kruchten, 2004). Research points out that approximately half of the conventional assurance methods and techniques directly clash with the principles and practices of agile development. Most of these techniques create a mismatch due to their reliance on extensive documentation served as a subject of analysis, verification, and validation activities (Beznosov & Kruchten, 2004).

"The (conventional) assurance methods, applied naïvely, would create deterring delays between critically short iterations as well as prohibitively inflate the development budgets. "

Beznosov & Kruchten (2004)

Many organisations want to apply DevOps, but they are concerned by the security aspects of the produced software (Mohan & Othmane, 2016). The results of a survey on DevOps adoptions and expectations (*DevOps: The Worst-Kept Secret to Winning in the Application Economy*, 2014), where 28% of the respondents indicate that security and compliance are obstacles to adoption, reflects the importance of embedding security assurance in the DevOps process. Increasing concerns in the area of security assurance is expected due to the emergence of regulatory compliance requirements (Tashi, 2009) and the potential impact of breaches on the market value (Goel & Shawky, 2009) and reputation (Sinanaj et al., 2015) of organisations.

In organisations with a strong emphasis on security, the security teams are essential stakeholders in the adoption of DevOps methodologies. Integrating security assurance aspects into the DevOps processes is vital to gain their buy-in (Chen, 2017). Without adequate involvement of the security team, rapidly deployed software changes are more likely to contain vulnerabilities due to lack of security reviews (Rahman & Williams, 2016).

Where the previous aspects highlight the interest in the integration of security assurance in DevOps from the perspective of preventing adverse side effects from occurring, there are also some aspirational aspects to the proposed integration. Studies linking DevOps to organisational performance (Mann et al., 2018) also point out that the automation of security policy configurations is "mission-critical to reach the highest levels of DevOps evolution" and mentions that the "involvement of security teams in technology design and deployment" should be viewed as an essential practice. A recent "State of DevOps Report" (Forsgren et al., 2019) explicitly mentions that integrating security throughout the software delivery lifecycle leads to positive outcomes such as their ability to deploy to production on demand at a significantly higher rate.

Figure 3. Mapping of IATF Information Security Engineering (ISSE) to waterfall and agile process steps

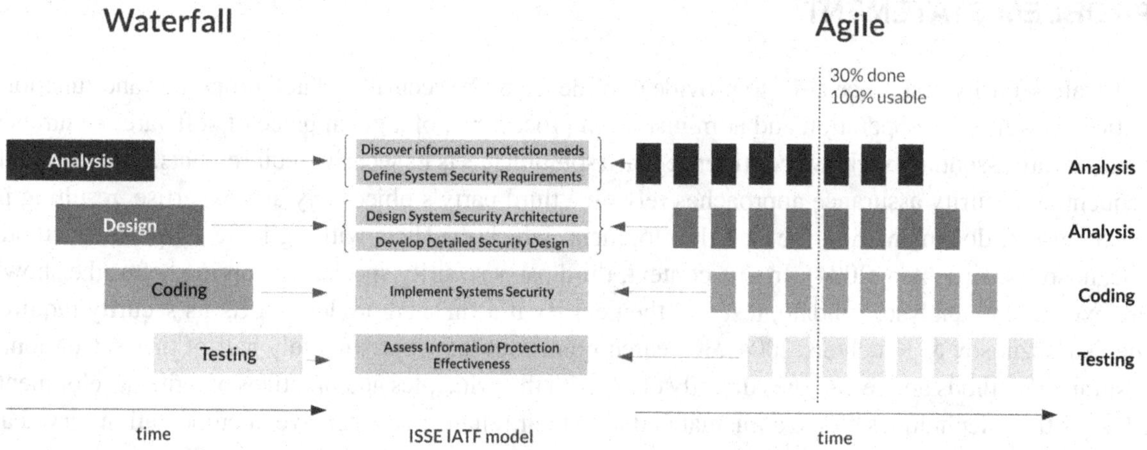

There is also an inverse effect where doing DevOps well enables to do security well (Forsgren et al., 2019), the same principles that drive good outcomes for software development such as culture, automation, measurement and sharing are like those that drive good security outcomes. Automation enables organisations to respond more swiftly to security issues (Forsgren et al., 2019). Therefore, we argue that a strong DevOps culture also supports more robust security.

These aspects drive the tight integration of security into DevOps; a practice commonly referred to as "DevSecOps" or "SecDevOps" (Mohan & Othmane, 2016). DevSecOps proposes solutions to improve security on the three dimensions of DevOps, namely people, process and technology (McCarthy et al., 2015).

An example of the mindset found in DevSecOps is expressed in the Rugged Software manifesto (Corman et al., n.d.):

"Rugged" describes software development organisations that have a culture of rapidly evolving their ability to create available, survivable, defensible, secure, and resilient software.

A concept frequently cited in the context of DevSecOps is "shift left on security" (Jiménez et al., 2018)(Airaj, 2016) (Mansfield-Devine, 2018). It refers to the principle of integrating security-related activities as early as possible in the process because the cost and time required to identify and remediate issues as applications advance through the various development stages increase significantly (Westland, 2004). At its core "shift left on security" advocates delegating authority and coaching of all contributors in the process to integrate security in their work, allowing the security team to focus on support in specific complex cases.

Figure 4. Visualization of the "shift left on security" concept based on the premises of (Jiménez et al., 2018) and (Westland, 2004)

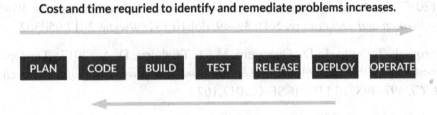

Cost and time requried to identify and remediate problems increases.

Shift application security to the left

The question which arises is how to integrate security assurance activities in DevOps, thereby leveraging its strengths to increase the security posture of the produced software and providing increased speed of response (Forsgren et al., 2019), without introducing delays between critically short iterations (Beznosov & Kruchten, 2004).

To achieve this, organisations need a framework of relevant security activities enabling an organisation to integrate security assurance in DevOps with a clear view on the effectiveness as well as the delay caused in the process of continuous delivery.

OBJECTIVE OF THIS RESEARCH

The focus of this research is to establish a framework to effectively integrate security assurance activities in DevOps without sacrificing speed and agility. The framework, therefore, needs to factor in the delay caused to the development and operations processes of the various activities. The main research question is defined as follows:

Which set off security activities are applicable to DevOps and how do they relate to the perspectives of effectiveness and delay caused in the process of continuous deployment?

The outcome of this question allows the design of a framework for security assurance in DevOps by integrating security in DevOps without sacrificing speed and agility. The framework aims to provide an overview of the relevant security activities and their characteristics in terms of effectiveness and delay caused to the process of continuous deployment.

REFERENCES

Abrahamsson, P., Salo, O., Ronkainen, J., & Warsta, J. (2017). *Agile Software Development Methods: Review and Analysis*. Academic Press.

Airaj, M. (2016). Enable cloud DevOps approach for industry and higher education: Enable cloud DevOps approach for industry and higher education. *Concurrency and Computation*, *29*(5), e3937. doi:10.1002/cpe.3937

Almeida, F. (2017). Challenges in migration from waterfall to agile environments. *World Journal of Computer Application and Technology*, *5*(3), 39–49. doi:10.13189/wjcat.2017.050302

Artač, M., Borovšak, T., Nitto, E. D., Guerriero, M., & Tamburri, D. A. (2017). DevOps: Introducing Infrastructure-as-Code. *2017 IEEE/ACM 39th International Conference on Software Engineering Companion (ICSE-C)*, 497–498. 10.1109/ICSE-C.2017.162

Beznosov, K., & Kruchten, P. (2004). *Towards agile security assurance*. doi:10.1145/1065907.1066034

Chen, L. (2017). Continuous Delivery: Overcoming adoption challenges. *Journal of Systems and Software*, *128*, 72–86. doi:10.1016/j.jss.2017.02.013

Corman, J., Rice, D., & Williams, J. (n.d.). *Rugged Software manifesto*. https://ruggedsoftware.org/

DevOps: The Worst-Kept Secret to Winning in the Application Economy. (2014). https://www.ca.com/content/dam/ca/us/files/msf-hub-assets/research-assets/devops-winning-in-application-economy.pdf

Duvall, P. M., Matyas, S., & Glover, A. (2007). *Continuous integration: improving software quality and reducing risk*. Academic Press.

Ebert, C., Gallardo, G., Hernantes, J., & Serrano, N. (2016). DevOps. *IEEE Software*, *33*(3), 94–100. doi:10.1109/MS.2016.68

Forsgren, N., & Humble, J. (2015). The Role of Continuous Delivery in it and Organizational Performance. SSRN *Electronic Journal*. doi:10.2139srn.2681909

Forsgren, N., Smith, D., Humble, J., & Frazelle, J. (2019). *2019 Accelerate State of DevOps Report*. Academic Press.

Goel, S., & Shawky, H. A. (2009). Estimating the market impact of security breach announcements on firm values. *Information & Management*, *46*(7), 404–410. doi:10.1016/j.im.2009.06.005

Jabbari, R., Ali, N., Petersen, K., & Tanveer, B. (2016). *What is DevOps? A Systematic Mapping Study on Definitions and Practices.* doi:10.1145/2962695.2962707

Jiménez, M., Rivera, L. F., Villegas, N. M., Tamura, G., Müller, H., & Gallego, P. (2018). *DevOps' Shift-Left in Practice: An Industrial Case of Application.* doi:10.29007/lh14

Kang, H., Le, M., & Tao, S. (2016). Container and Microservice Driven Design for Cloud Infrastructure DevOps. *2016 IEEE International Conference on Cloud Engineering (IC2E)*, 202–211. 10.1109/IC2E.2016.26

Mann, A., Brown, A., Stahnke, M., & Kersten, N. (2018). *Puppet - State of DevOps Report 2018.* https://puppet.com/resources/whitepaper/state-of-devops-report

Mansfield-Devine, S. (2018). DevOps: Finding room for security. *Network Security, 2018*(7), 15–20. doi:10.1016/S1353-4858(18)30070-9

Martin, F., & Jim, H. (2001). Agile-Manifesto.pdf. *Software Development, 9*(8), 28–35.

McCarthy, M. A., Herger, L. M., Khan, S. M., & Belgodere, B. M. (2015). Composable DevOps: Automated Ontology Based DevOps Maturity Analysis. *2015 IEEE International Conference on Services Computing*, 600–607. 10.1109/SCC.2015.87

Moe, N. B., Dings, T., & Dyb, T. (2008). Understanding Self-Organizing Teams in Agile Software Development. *19th Australian Conference on Software Engineering (aswec 2008)*, 76–85. 10.1109/ASWEC.2008.4483195

Mohan, V., & Othmane, L. B. (2016). SecDevOps: Is it a Marketing Buzzword? *2016 11th International Conference on Availability, Reliability and Security (ARES)*, 542–547. 10.1109/ARES.2016.92

Rahman, A. A. U., & Williams, L. (2016). *Software security in DevOps: synthesizing practitioners' perceptions and practices.* doi:10.1145/2896941.2896946

Shahin, M., Babar, M. A., & Zhu, L. (2017). Continuous Integration, Delivery and Deployment: A Systematic Review on Approaches, Tools, Challenges and Practices. *IEEE Access: Practical Innovations, Open Solutions, 5*, 3909–3943. doi:10.1109/ACCESS.2017.2685629

Sinanaj, G., Muntermann, J., & Cziesla, T. (2015). How Data Breaches Ruin Firm Reputation on Social Media!-Insights from a Sentiment-based Event Study. *Wirtschaftsinformatik, 2015*, 902–916.

Smeds, J., Nybom, K., & Porres, I. (2015). *DevOps: a definition and perceived adoption impediments.* Academic Press.

Tallon, P. P., Queiroz, M., Coltman, T., & Sharma, R. (2019). Information technology and the search for organizational agility: A systematic review with future research possibilities. *The Journal of Strategic Information Systems, 28*(2), 218–237. doi:10.1016/j.jsis.2018.12.002

Tashi, I. (2009). Regulatory Compliance and Information Security Assurance. *2009 International Conference on Availability, Reliability and Security*, 670–674. 10.1109/ARES.2009.29

Westland, J. C. (2004). The cost behavior of software defects. *Decision Support Systems, 37*(2), 229–238. doi:10.1016/S0167-9236(03)00020-4

Chapter 11
Research Findings in the Domain of Security Assurance in DevOps

Dennis Verslegers
Orange Cyberdefense, Belgium

ABSTRACT

This chapter of the book provides an overview of the approach and outcomes of the research project. It consists of the following sections: Research approach and deliverables provides an overview of the rationale behind the research methods selected for the project and the outcomes of the literature review. A detailed overview of the security activities is identified during the research project. This chapter contains an in-depth analysis performed for each of the security activities and design factors identified during the previous research steps.

RESEARCH APPROACH AND DELIVERABLES

Selection of Research Methods

The topics of DevOps and DevSecOps are on the vanguard of information technology and therefore are subject to constant and fast evolution. Recent academic papers refer to DevOps as being a novel concept (Jabbari et al., 2016a) and point out that the number of relevant publications on DevSecOps is low (Mohan & Othmane, 2016b). An exploratory approach was chosen for this research project, allowing to gain familiarity with the concept of lean security, to generate new insights and return these insights to the body of knowledge.

Design science combines behavioural science and design science based on the premise that this combination is most suited to address fundamental problems faced in the productive application of information technology. Technology and behaviour are considered inseparable in IS research (Hevner et al., 2004). This premise aligns with a consensus in the proposed research domain; the combination

DOI: 10.4018/978-1-7998-7367-9.ch011

of process, technology and people (Martin & Jim, 2001)(McCarthy et al., 2015) is essential for both DevOps and Information Security to become effective.

This research project develops the specifications for an artefact detailing a set of security activities applicable to DevOps and their characteristics in terms of effectiveness and delay caused in the process of continuous deployment.

This artefact proposes an initial solution to the problem of integrating security in DevOps and as such contributes to the body of knowledge. Therefore, design science methods are selected as the basis for this research. Design science methods are described by Hevner et al. as follows (Hevner et al., 2004):

"A research paradigm in which a designer answers questions relevant to human problems via the creation of innovative artefacts, thereby contributing new knowledge to the body of scientific evidence. The designed artefacts are both useful and fundamental in understanding that problem."

As described by Recker (Recker & Recker, 2012), design science starts from the existing knowledge base to provide the material through which design science research is accomplished, thereby achieving rigour. This knowledge may consist of existing theories from science and engineering, specifications of currently known design, useful facts about currently available products, lessons learned from the experience of researchers in earlier design science projects, and plain common sense (Wieringa, 2014).

Subsequently, the researcher engages in a relevance cycle to bridge the environment of the research project with the design science activities, thereby providing relevance in the application domain. At the heart of design science is the design cycle which iterates between the core activity of building and evaluating the design artefacts.

Wieringa indicates that additional research may be required in cases where current scientific research does not provide an answer (Wieringa, 2014). The research domain selected for this project is on the vanguard of technological and cultural practices in the field of secure information systems development and operations in agile environments. As a result, knowledge on this subject is constantly evolving and therefore encourages us to employ techniques that facilitate knowledge gathering based on expert experience to generate new insights.

An approach to generating new insights is the use of Group Support Sessions. They facilitate the effective collection, organisation, evaluation, cross-impact analysis and reporting of data (Bobbert, 2017a). Group Support Sessions have been proposed as a qualitative research method in the decision-making process within the domain of Business Information Security (BIS) due to the stimulation of free-flowing discussion and sharing of experiences eliciting the views of all participants. Previous studies state that GSS provides a solution to the problem of capturing and sharing knowledge and as such can be used to feed decision making (Bobbert, 2017a).

A similar combination of design science research methods and Group Support Sessions for exploratory research has been applied in previous research to derive prioritised lists of items in the field of information security (Bobbert & Mulder, 2013).

Based on the need for this research project to elicit the views of practitioners and a similar need to provide a prioritised list of items, the choice was made to apply GSS. This research employs a combination of techniques to interact with experts: (a) surveys, (b) interviews and (c) group support sessions.

Figure 1. Overview of design science as defined by Hevner (2007)

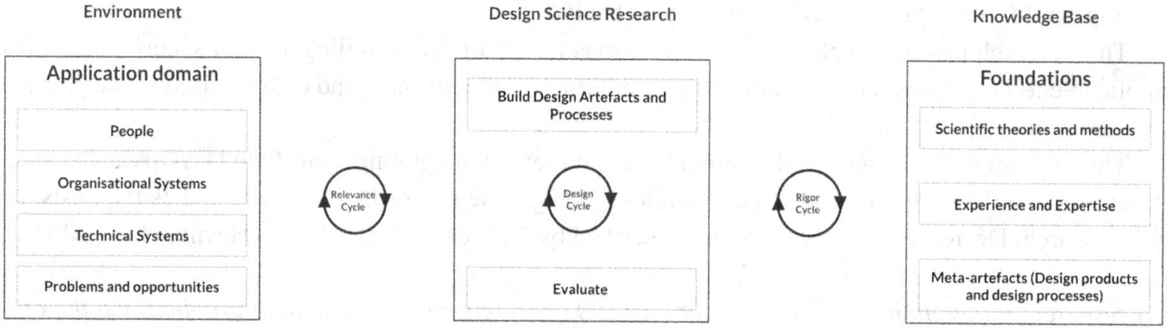

Design Problems and Knowledge Questions

Wieringa (Wieringa, 2014) emphasises the need to distinguish design problems from knowledge questions. Design problems call for a change in the real world and are solved through a design which is evaluated by its utility with respect to the stakeholder goals. There is not one single best solution. Knowledge questions, on the other hand, ask for knowledge about the world as it is. The answers to knowledge questions are evaluated by truth, independent of stakeholder goals.

Therefore, Wieringa emphasises the importance to make a clear distinction between knowledge questions and design problems as they need to be treated, and more importantly evaluated, differently. To this end, a goal structure for design science research projects is proposed (Wieringa, 2014).

This proposed structure was applied to the research design for this project. As stated earlier, this research addresses the need to integrate security activities in DevOps. The objective of the project is to specify an artefact enabling the integration of security activities in DevOps with little friction (a1).

To this end the following knowledge goals were identified: (k1) definition of the terms DevOps and DevSecOps, (k2) identification of relevant activities and design factors for DevSecOps, (k3) framework of security activities and design factors considered relevant in DevSecOps according to DevSecOps experts and (k4) prioritised framework of security practices, activities and design factors in DevSecOps according to experts. Two instrument design goals are required to facilitate the gathering of information for the knowledge goals: (i1) construct survey to perform expert validation and elaboration on the list identified in K2 and (i2) construct survey and GSS prerequisites to perform ranking of the framework established in K3.

Research Approach

A clear definition of the terms DevOps and DevSecOps contribute to the reusability and comparability of this research project. The starting point for the research therefore consists of establishing a clear definition for the core concepts of DevOps and DevSecOps (k1). These definitions are gathered from the literature and will be presented to a group of security experts through a survey for validation. The related research questions for this goal are:

Figure 2. Schematic overview of the structure for the research

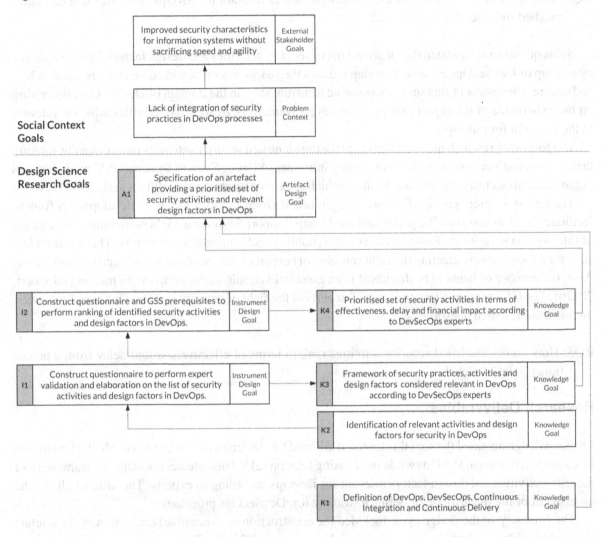

RQ1a: What is the definition of DevOps?
RQ1b: What is the definition of DevSecOps?

A two-step process was applied to gather a list of activities and design factors which are considered relevant for DevSecOps (k2). The first step consists of the identification of security activities applicable in the context of DevOps through a literature review of academic literature. During the literature review, publications from a period of multiple years were included to ensure a broad coverage (2015-2020). In a subsequent step, the findings of this literature review were listed in a structured way by grouping the security activities that exhibit similar purposes and advocate similar notions using thematic analysis as described by Nowell, Norris, White et al. (Nowell et al., 2017). This activity enables us to answer research question 2:

RQ2: Which set of security activities and design factors relevant to DevOps processes can be distinguished from academic literature?

Subsequently, we validated the identified list of security activities and design factors through a survey by a group of DevSecOps experts. This step reduces the risk of inaccuracy introduced by researcher bias and ensures the results of this study incorporate a current state in the domain of DevSecOps. By relying on the experience of the expert group, we achieve a framework of security activities which is relevant to the field (fit for purpose).

The following research question aims to place the identified security activities into context by ranking them based on effectiveness and the friction they introduce. A Group Support System (GSS) workshop was organised with security experts to determine which security activities are useful for security in DevOps.

The necessary prerequisites (I2), such as agenda and group composition, were developed up-front to facilitate the GSS sessions. We performed the Group Support Sessions using a sufficiently sized group of information security professionals to ensure reliability and replicability of results. There is no 'ideal size' for a focus group; selecting the right (number of) experts is key to obtaining collective intelligence. Also, the number of items to be discussed is an essential variable in the setup of the meeting (Bobbert, 2017b). The GSS session elicits the view of experts in the field of security and DevOps thereby answering the third research question:

RQ3: How do the identified security activities rank in terms of effectiveness and delay from a practitioner point of view?

Research Deliverables

The knowledge obtained through this research allowed the design of an artefact to enable the integration of security activities in DevOps while minimising friction (a1). This artefact consists of a framework of security activities and design factors relevant for DevOps according to experts. The artefact allows the assessment of the maturity of, or define a strategy for, DevSecOps processes.

The final step of the design cycle included the construction and (iterative) evaluation of the artefact to provide evidence of utility in the stated problem context. This final step as considered out of scope for this research project due to time constraints, it is proposed as a topic for future research.

ACADEMIC LITERATURE REVIEW

Searching the Knowledge Base

During this research emphasis was placed on the generation of progressive insights into security activities relevant in the context of DevOps. The exploratory nature of this research was reflected in the choice to apply the forward search method for the academic literature review. A forward search attempts to identify relevant papers to the research domain and to follow their citations to discover newer insights.

The preliminary investigation in the scope of producing the research proposal pointed out that finding a clear definition for the terms DevOps and DevSecOps through literature review is challenging (RQ1) (Jabbari et al., 2016b). It was decided to retain the definitions identified during the preparatory phase of this project and to request validation by the DevSecOps experts during the validation and elaboration step. The definitions are:

Figure 3. Schematic overview of the research design

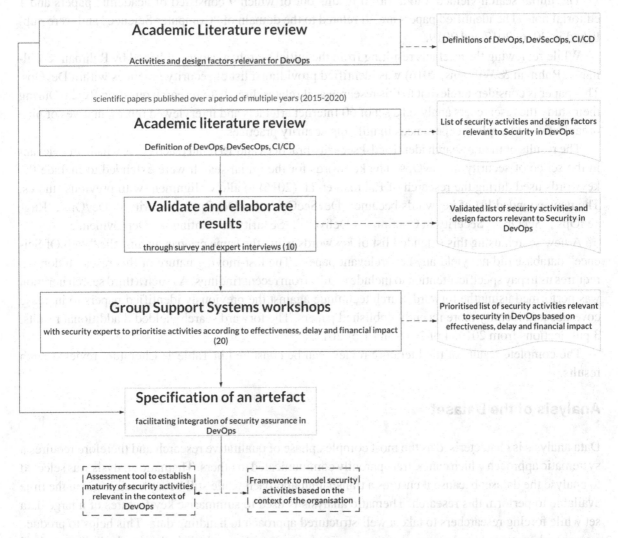

§ DevOps: "DevOps is a development methodology aimed at bridging the gap between Development and Operations, emphasising communication and collaboration, continuous integration, quality assurance and delivery with automated deployment utilising a set of development practices." (Jabbari et al., 2016b)

§ RuggedOps (DevSecOps): "Rugged" describes software development organisations that have a culture of rapidly evolving their ability to create available, survivable, defensible, secure, and resilient software (Corman et al., 2010)

To answer the second research question (RQ2), and achieve knowledge goals K1 and K2, an initial search against the "Web of Science" database was performed based on the keywords 'DevSecOps', 'SecDevOps' and 'Security in DevOps'. A query against these keywords in the title, written in English and published between 2015 and 2020 was constructed to identify relevant papers.

The initial search yielded a total of 10 results out of which 9 consisted of academic papers and 1 editorial note. The identified papers are all related to the domain of "Computer Science" and were published in 2016, 2017 and 2019.

While reviewing the artefacts resulting from the initial search, a paper published by Rahman & Williams (Rahman & Williams, 2016) was identified providing a list of security practices within DevOps. This paper is considered relevant to this research as it relates directly to research question RQ2a. During their study the researchers analysed a set of 60 Internet artefacts and interviewed representatives of nine organisations to explore experiences in utilising security practices.

The results of their research identified 4 security practices and 15 related activities which are relevant in the scope of security in DevOps. The keywords for the initial search were extended to include the keywords used during the research of Rahman et. al. (2016) to allow alignment with previous studies. The new extended list of keywords became: 'DevSecOps', 'SecDevOps', 'Security in DevOps', 'RuggedOps', 'SecOps', 'Security in Continuous Delivery', 'Security in Continuous Deployment'.

A new search using this extended list of keywords using the same criteria against the "Web Of Science" database did not yield any new relevant papers. The fast-moving nature of this research domain requires us to pay specific attention to include results from recent findings. As such a third search iteration was performed using the forward search technique against the previously identified papers to increase coverage and identify more recently published papers. The forward search yielded 8 additional results, 3 publications from 2019, 4 of 2018 and 1 of 2017.

The complete results of the literature review can be consulted in Table 1: Literature review search results.

Analysis of the Dataset

Data analysis is characterised as the most complex phase of qualitative research and therefore requires a systematic approach which can be transparently communicated to others. Thematic analysis was selected to analyse the dataset because it ensures a transparent approach while still being feasible given the time available to perform this research. Thematic analysis is used to summarise key features of a large data set while forcing researchers to take a well-structured approach to handling data. This helps to produce a clear and organised final report. A risk when applying thematic analysis is that its flexibility can lead to inconsistency and a lack of coherence when developing themes derided from the research data. To improve the trustworthiness of this study the decision was taken to apply the structured approach for thematic analysis as proposed by Lorelli et al. (Nowell et al., 2017).

The procedure is both practical and effective and consists of six steps which may be performed in an iterative fashion: (1) familiarising yourself with your data, (2) generating initial codes, (3) searching for themes, (4) reviewing themes, (5) defining and naming themes and (6) producing the report. In this section a detail is provided on how the different steps of this process are applied to this research.

As a first step the researcher is encouraged to read through the complete data set at least once prior to starting the data analysis. This step aims to ensure the researcher is sufficiently familiar with the information available in the dataset. During this step all identified papers are loaded and categorised in a research tool of choice and read through actively. A short summary of the contents and the rationale for inclusion or exclusion of the paper was written down. A allowing traceability. The papers identified through the fourth iteration were not included in the initial data analysis exercise due to their large number. Instead they were used to verify the completeness of the mapping exercise by reading through them and comparing to the activities and design factors produced through the thematic analysis.

The next step in the approach involved the definition of an initial set of codes based on the previously defined research questions. These codes would then be used to label various sections of the papers allowing a relationship to be identified between the research questions and the information available in the selected set of papers.

The initial set of codes (Table 2: Overview of codes used for thematic analysis) was confirmed by reading through the complete data set to ensure it allowed to cover the breadth and depth of the information required for this research. In a second iteration five codes were added to allow capturing and structuring of additional information.

With the codes for thematic analysis defined the next step, where these codes are applied to the contents of the papers in the dataset, was started. Each relationship between a code and a statement was added to a mapping table allowing further analysis.

Based on this mapping table the work of reviewing and naming themes starts. As a starting point all excerpts were loaded into a mind mapping software and grouped by code so that all activities, design factors, etc would be grouped together. Subsequently common themes were identified by grouping the various excerpts into a first level of themes. For some activities the range of activities belonging to a theme was still too broad necessitating a second level grouping. This leads to the definition of "level 1" and "level 2" activities where the latter is a more detailed subset of the former.

For example, a theme which quickly became apparent within the activity codes was "manual security testing". All excerpts belonging to this theme were then further grouped based on their content leading to (a) performing manual penetration testing and (b) performing manual security reviews. This approach was applied to all the excerpts mapped to the "activity" and "design factor" codes resulting in a final set of activities and design factors.

Interim Conclusion

The results obtained through this activity, as represented in Table 3: Results of thematic analysis of security activities and Table 4: Results of thematic analysis of design factors, allowed us to complete the second knowledge goal (K2) and as such serves as a starting point to answer the second research question (RQ2):

Which set of security activities and design factors relevant to DevOps processes can be distinguished from academic literature?

VALIDATION AND PRIORITISATION OF THE RESEARCH FINDINGS

Preparation

This research requires capturing the current state-of-the-art approaches in the fast-evolving domain of DevSecOps. As such this research cannot remain limited to literature review which mainly provides retrospective insights and often fails to capture new perspectives. To close this gap the choice was made to include a panel of DevSecOps experts. The insights of DevSecOps experts were leveraged in two different phases.

In a first phase a group of experts was consulted to validate and elaborate the findings of the literature research. This allowed capturing state-of-the-art knowledge and insights while also reducing researcher bias thereby contributing in answering research question 2 (RQ2). This was achieved through a survey and optional structured interview of these experts. The protocol used for this survey and the results are detailed here.

In a second phase a group of experts have been requested to perform a ranking of the validated activities in terms of effectiveness, financial impact and delay thereby contributing to answering research question 3 (RQ3). The ranking-type Delphi method was applied to develop group consensus about the relative importance in terms of these three aspects for the security practices which were refined in the previous step.

The Delphi method is described by Linstone and Turoff (Linstone et al., 1975) as follows:

Delphi may be characterised as a method for structuring a group communication process so that the process is effective in allowing a group of individuals to deal with a complex problem. To accomplish this "structured communication" is provided: some feedback of individual contributions of information and knowledge; some assessment of the group judgment or view; some opportunity for individuals to revise views; and some degree of anonymity for the individual responses.

Due care has been applied to safeguard the validity of the results increase the confidence in this study and improve reusability of the results. Validity as defined by Saunders, Lewis and Thornhill (Saunders et al., 2007) refers to (1) the appropriateness of the measure used, (2) the accuracy of the analysis of the results and (3) generalisability of the findings.

When looking specifically at the aspect of measurement validity it is important to ensure the use of appropriate measures to gather information so that the data obtained gives an accurate reflection of the actual situation in practice. The exploratory nature of this research combined with the broad subject domain makes it a complex issue requiring the knowledge from people who understand different aspects of security (Saunders et al., 2007). Therefore, it is important to leverage group knowledge, a group of experts will be able to provide more appropriate answers compared to individual expert's opinions.

Okoli and Pawlowski (Okoli & Pawlowski, 2004)point out the importance of selecting appropriate experts as a very important aspect of the measurement validity of the Delphi method. Their guidelines were closely followed while preparing the solicitation of qualified experts for both phases of this research. When drafting the selection criteria special attention was payed to accept experts with a broad set of skills to ensure inclusion of various roles across a wide range of organisations. This is important for research in such a broad domain due to the wide range of stakeholders and disciplines involved. As such the background knowledge and the role of the experts in their respective organisations is expected to play an important part in their perceptions. For instance, people who come from a development background may have a different view compared to someone from networking or operations teams. Likewise, someone being part of a security architecture team will have a different view compared to application security engineers.

When looking at the target organisations some differentiators may be hypothesised to result in different views on Dev(Sec)Ops. These differentiators could be the size, the variation in the level of regulatory requirements and the amount of legacy technology in the organisation. Therefore, a wide range of potential organisations was included in the selection criteria.

An overview of the selection criteria for experts is provided in Table 5: Selection criteria for DevSec-Ops experts

The knowledge nomination worksheet (KRNW was iteratively populated with the names of potential experts for participation in this study. In a first iteration potential experts from personal networks of the researchers and promotors were leveraged to establish an initial set. Gradually this list was extended by experts proposed by people who were contact for participation in this survey. Potential academic experts were selected from the list of authors whose papers were identified during the literature review phase of this research. All experts on the KRNW were contacted through a formal invitation to request their collaboration for the first and second phase. The invitations and responses were also included in the KRNW.

Prioritisation by Expert Panel

With the validated and elaborated list of security activities established the research project continued to prioritise the activities on aspects which are relevant to DevSecOps. To achieve this a second expert panel was established to leverage the knowledge of security experts to rank them in terms of effectiveness, delay and financial impact. To facilitate this a Group Support Session (GSS) was organised bringing together the experts while stimulating free-flowing discussions and sharing of experiences. The session was performed on 15th of April 2020 from 16:00 to 18:00 GMT+1 through an online video conferencing platform and leveraging group support systems software (MeetingWizard).

The target group size for the GSS session was set at 10 participants to ensure an appropriate group size allowing enough qualified data without introducing a high level of noise. The number of items to be discussed was limited to the 29 security activities to make the best out of the limited time and "processing power" of the group (Bobbert, 2017b).

A total of 8 experts participated in the ranking session. These experts received an invitation beforehand together with an overview of the outcome of this research so far allowing time to prepare should they wish to do so.

The security activities were grouped by category (collaboration, non-automated and automated security activities) and treated in these groups to avoid long uninterrupted ranking steps. The experts where explicitly asked not to perform the ranking by category but to perform the ranking holistically. The experts were also invited to provide comments to express their thoughts and reflections, which they did extensively.

Each security activity was ranked on a Likert scale from 1 to 5 for each of the following three aspects:

§ **effectiveness**: the degree to which an activity contributes to the security of the resulting software
§ **delay**: the degree to which an activity is expected to cause delays in the development and operations processes
§ **financial**: the degree to which an activity is expected to have financial impact such as effort or licensing costs

This scale was explained to the experts by providing two examples on how the ranking could be applied to "Performing penetration testing" and "Performing continuous assurance". It was clearly expressed that these were only meant as illustration of the ranking scales. From this point on the session was strictly moderated by prof. Dr. Yuri Bobbert to avoid any influence from the researcher. The participants had the opportunity to ask questions when the context of a certain activity was unclear. The raw result set was downloadable for the experts upon completion of the session.

Descriptive statistics were applied to the dataset to determine the qualitative scores for each of the three aspects for which a consensus exists between most of the experts. This was achieved by calculating the median and Inter-Quartile Range (IQR) for each item. The median is a measure of the central tendency and the IQR provides a measure of the spread. A full SPSS analysis was not performed on the dataset due to the limited time available. The results of this analysis can be found in Table 6: Results of the prioritisation of security activities by the expert panel.

The comments provided by the participants were read and analysed to place the scoring in context. In most cases where the scoring displayed a large spread (IQR), indicating dissonance of opinion, the comments provided enough insight to explain the differences. A write-up of the conclusions for each activity was drafted and provided to the experts for feedback. The resulting design factors were used to enrich the overview of security activities in DevSecOps as presented in the Detailed overview of the security activities identified during the research project chapter of this book.

Interim Conclusion

The results of this analysis were used to construct the prioritised list of security activities relevant in DevSecOps thereby providing an answer to the third and last research question RQ3 and the related knowledge goal K3.

DETAILED OVERVIEW OF THE SECURITY ACTIVITIES IDENTIFIED DURING THE RESEARCH PROJECT

Collaboration

Performing Continuous Feedback from Production to Development

Various researchers mention performing continuous feedback between production and development teams as an essential factor for effective DevSecOps. Such feedback loops allow establishing a security culture and provide learning opportunities based on real-world scenarios and incidents. These learning opportunities also form the basis for continuous improvement throughout the development and operations chain (Jaatun et al., 2017).

Swift feedback from production to development teams enables organisations to make quick changes in the code if a vulnerability is discovered (in operations). Establishing an emergency code base response (Jaatun et al., 2017) allows organisations to react to security incidents through timely modification of applications. The continuous deployment pipeline introduced earlier acts as an enabler to ensure the modification is deployed in the production environment without delays.

Design Factors

A cornerstone to enable emergency code base response is a process to feed the discovered flaws or defects in operations to the development team. The number of flaws and defects discovered in production can be leveraged as a metric to measure the level of security in each piece of software. An interface between developers and incident response operators needs to exist to facilitate the timely exchange of information.

A suggested approach is to create an internal forum to discuss attacks and exchange attack patterns (Jaatun, 2018). Another recommended practise is to assign a contact point for the software security group outside of office hours.

Feeding back information regarding security defects and incidents is only one part of the approach. As mentioned by researchers (Jaatun et al., 2017) and confirmed by various members of the DevSecOps expert panel, it remains challenging to ensure the prioritisation of security defects against functional requirements. Potential approaches to overcome this is to set time limits for security defects (Jaatun et al., 2017) or dedicating a fixed percentage of effort or sprints to implementing security-related features and fixing defects. Weaknesses detected in production must be included in the QA processes to reduce the risk of regressions (Carter, 2017).

Expert Ranking

Within DevSecOps feedback loops are considered crucial to share both successes and issues to foster collaboration and improve the quality of the results. Continuous feedback from production to development contributes to this objective while also ensuring that issues do not go unnoticed for weeks. There is an agreement within the expert panel that performing continuous feedback from production to development is highly effective (mdn= 4.5, IQR= 1.25), does not cause significant delays (mdn= 4.0, IQR= 1.50) and does not have a significant financial impact (mdn= 4.0, IQR= 1.00). This scoring assumes that appropriate stakeholder management is applied to include only relevant roles and leverages automation to the extent possible.

Frameworks

§ OWASP SAMM v2.0: Operations: Incident Management
§ BSIMM v10: Deployment: Configuration Management & Vulnerability Management

Design Factors

Table 1. Design factors for performing continuous feedback from production to development

Design factor	Perceived as relevant by experts
Set a time limit for all lower-priority security defects	5 of 7
Provide feedback on attack patterns to your developers	7 of 7
Establish an internal forum to discuss attacks	5 of 6
Organise emergency code base response	5 of 6
Incorporate security tests as part of QA for detected incidents to avoid regression	5 of 6

Quick Facts

Table 2. Quick facts for performing continuous feedback from production to development

Why should you do this?	Enables faster collaboration
	Ensure that issues are detected and resolved faster
	Enables a security culture to be established
What to avoid?	Require manual intervention to perform the communication
	Involve too many or incorrect roles in the organisation

Provide Security Training

Software security education is an import aspect of security, regardless of the chosen development approach. Some researchers argue that security training gains in importance in agile development teams due to the decentralisation of responsibilities and the increased speed of iterations.

How much education is required remains a matter of debate, as it is not possible to teach every developer to be a software security expert (Jaatun et al., 2017) (Torkura et al., 2017). The expert panel has also mentioned this limitation.

Design Factors

A potential alternative approach to security training is to involve developers in building security tools. Doing so provides them with learning opportunities and gives the added benefit of improving DevSecOps processes(Mohan & Othmane, 2016a). The DevSecOps expert panel indicated that security training should be hands-on.

The experts suggest approaching training with a broader perspective and targeting all the relevant roles, including product owners. Some examples of topics to consider are educating teams how to refactor software and explaining to product owners that this is very important, teaching approaches to think from a risk perspective, perform risk prioritisation and explaining that cloud-native services are not secure by default.

Tailoring the level of security training based on the criticality of the role and the current level of a security mindset is recommended. Training on specific topics only needs to be performed once; however, it is crucial to ensure that people put their new skills and knowledge into practice.

Expert Ranking

Sharing and increasing knowledge within the DevSecOps teams is perceived as an essential enabler for DevSecOps. The expert panel agrees that providing security training is a useful activity (mdn=4.0, IQR=1.00) and that it does not cause a significant delay (mdn=4.0, IQR=1.50). The experts strongly agree that moderate financial impact is to be expected (mdn=3.0, IQR=0.50). High-quality external training material can be expensive; the panel expects financial consequences to be moderate.

Frameworks

§ OWASP SAMM v2.0: Governance: education and guidance
§ BSIMM v10: Governance: training

Table 3. Design factors for providing security training

Design factor	Perceived as relevant by experts
Teach every developer enough to enable them to identify areas where they would benefit from the advice of an expert	7 of 7
Security training should be hands-on	Added by an expert
Educate teams how to refactor software	Added by an expert
Educate product owners on the importance of security	Added by an expert
Ensure that team members can put their new skills and knowledge into practice	Added by an expert
Organise centralised training sessions and workshops	Added by an expert

Design Factors

Quick Facts

Table 4. Quick facts for providing security training

Why should you do this?	Increase the security knowledge and mindset in the organisation
What to avoid?	Not tailoring the contents of the trainings to the specific roles

Establish Security Satellites

Various studies point out the importance of establishing security satellites as part of the development teams. The principle is that the security team teaches one of the developers about security, and then [that person] disseminates the information to the rest of the team. It is really about knowledge sharing (Carter, 2017).

The importance of embedding security knowledge and responsibility within each development team originates from the idea that short-term helicopter-style incursions of a centralised security team are inevitably perceived as an outside agent hindering progress (Jaatun et al., 2017). The presence of security-minded developers across the organisation will aid in resolving quick fixes in case of incidents. The role of security satellites is often also referred to as security champions (Jaatun, 2018).

Within the expert panel, there was some debate on the definition of the term where experts indicated a difference between security satellites and security champions. Some experts in the panel pointed out that people who are a member of an external security team who can be contacted by DevOps teams should be referred to as "security satellites" whereas security champions are people embedded in the teams with some security knowledge. During this research, we use both terms interchangeably.

Design Factors

Experts from the DevSecOps panel point out that establishing security satellites is a starting point to implement the principle of 'Security to the left'. However, they feel that the most effective way to implement security champions is to allow the teams to discover the usefulness of this role by themselves and not having it forced upon them.

Expert Ranking

Establishing security satellites allows injecting the DevSecOps teams with security knowledge and best practices. It also provides a feedback loop from the teams towards the central security team on their needs and helps to foster a security mindset in the organisation. Giving appropriate levels of authority to security satellites increases their efficiency. The expert panel agrees that this activity is highly effective (mdn=5.0, IQR=1.00), even more so in large organisations where there may be boundaries between the development and centralised information security teams. Incorporating this activity in the approach is not believed to cause significant delays in the DevSecOps process (mdn=4.0, IQR=0.50). Different views exist on the expert panel regarding the financial consequences as they expect these to be in direct relation to the time assigned to security duties (mdn=3.0, IQR=3.0).

Frameworks

§ OWASP SAMM v2.0: Governance: education and guidance
§ BSIMM v10: Governance: training

Design Factors

Table 5. Design factors for establish security satellites

Design factor	Perceived as relevant by experts
Allow people to discover the role of security satellites, do not assign these roles top-down	Added by an expert
Ensure security satellites are provided enough autonomy and authority	Added by an expert

Quick Facts

Table 6. Quick facts for establish security satellites

Why should you do this?	Improve the security mindset	
	Share knowledge and best practices	
	Gather feedback from teams on security aspects	
What to avoid?	Not providing enough autonomy and authority to security satellites	

Practice Incident Response

DevSecOps teams should organise red-team exercises performing security drills on the deployed software (Myrbakken & Colomo-Palacios, 2017). Such preparedness exercises should involve both incident responders and developers (Jaatun, 2018). The objective of these exercises is to find and exploit vulnerabilities in the system, thereby not only assisting in the identification of security flaws but also providing metrics on incident response capabilities. These metrics can be leveraged to find solutions and improvements. An additional benefit to these exercises is that they break silos and promote collaboration(Carter, 2017).

Expert Ranking

Practising incident response from DevSecOps perspective aims at involving the development and operations team in the exercise. Doing so allows not only more effective response to incidents in the future but also to improve the security mindset across the various roles involved. The expert panel agrees that this activity can be useful but to varying degrees (mdn=4.0, IQR=1.25). There is more dissonance on the delay (mdn=4.0, IQR=1.25) and financial consequences (mdn=3.0, IQR=1.50) of this activity. The differences in scoring originate from the when and who is performing this activity. The expert panel advises performing incident response exercises during a low time.

Frameworks

§ OWASP SAMM v2.0: Incident Management: Mature incident management
§ BSIMM v10: Deployment: Configuration management & vulnerability management

Quick Facts

Table 7. Quick facts for practice incident response

Why should you do this?	Increase effectiveness of incident response
	Increase security mindset across various roles

Establish a Security Mindset Across the Organisation

Several of the interviewed experts referred to the "feature first" mentality in their organisation. They pointed out that this approach is harmful to the security posture of the platform and creates friction between security aspects and functional aspects throughout the development and operations cycles. The DevSecOps expert panel indicated the importance of training management on security concepts to overcome this obstacle. Product owners and project managers are directly involved in balancing security with functionality while considering business value. Product owners and manager should be considered essential stakeholders.

Expert Ranking

The experts mostly agree that this activity can be highly effective (mdn=4.0, IQR=1.25) while they do not expect this activity to cause significant delays (mdn=4.0, IQR=1.50). There is a debate on the financial consequences. Some experts indicated that implementing this activity requires a lot of effort (N=1, 28%), while others indicated that this does not necessarily have to be the case (N=4, 42%). Changing the mindset is expected to be a slow and gradual process, regardless of the effort invested.

Frameworks

§ OWASP SAMM v2.0: Governance: Education and guidance
§ BSIMM v10: Governance: Training

Design Factors

Table 8. Design factors for establish a security mindset across the organisation

Design factor	Perceived as relevant by experts
Train management on security concepts	Added by an expert
Reward the teams who fix the most vulnerabilities	Added by an expert
Share best practices across the organisation	Added by an expert
Establish a code quality standard for each programming language	Added by an expert

Quick Facts

Table 9. Quick facts for establish a security mindset across the organisation

Why should you do this?	Obtain a good balance between security and functionality from a business value perspective
What to avoid?	Limiting the "security mindset" to technical roles

USE OF NON-AUTOMATED ACTIVITIES

Performing Security Requirements Analysis

Software development is usually performed based on requirements defined by stakeholders. In many cases, these requirements focus on functional requirements and omit focused security requirements. Studies call for the definition of security requirements to include then from the start of a project (Oyetoyan et al., 2016).

Design Factors

Security requirements should be testable, clear, consistent, complete, unambiguous, measurable and accompanied by acceptance criteria (Derksen et al., 2018). A commonly suggested approach to identify these security requirements is using 'abuse cases' and 'misuse cases' which provides an inverse perspective leading to the identification of security requirements. Another approach for security requirements analysis is to leverage goal-oriented requirements analysis (GORE) (Tuma et al., 2018). A detailed explanation of goal-oriented requirements analysis is not feasible in the content of this research; however, various papers and case studied are available in the literature.

The threat modelling and risk analysis steps described in this book can also provide input for the security requirements analysis phase. Moreover, the security requirements identified during this activity provide the ability to compare design options to select the most appropriate approach from a security perspective (Jaatun et al., 2017).

The expert panel involved in this research agreed that security requirements form an essential aspect of DevSecOps. A recurring remark during the survey and interviews is that security requirements business stakeholders often set back security requirements in favour of functional requirements. Various experts wonder how to strike the right balance between functional and non-functional requirements. An approach applied by one of the respondent organisations was to dedicate a fixed percentage of effort during each sprint to security-related requirements.

Expert Ranking

Useful security requirements analysis requires a tailored approach; the effort spent defining security requirements needs to be in line with the level of sensitivity of the application. The definition of high-level security requirements itself is not problematic; however, detailing them towards technical requirements is complicated and requires significant security knowledge. Therefore, it makes sense to stop at the definition of high-level security requirements for less sensitive applications while going into greater technical detail for sensitive applications.

Overall security requirements analysis is perceived as moderately effective (Mdn=3.0, IQR=0.50) which is counterintuitive seeing that experts throughout this research indicated that security requirements form a cornerstone in the process from design over implementation to testing. Delays caused by performing security requirements analysis and the financial consequences are directly related to the depth of the analysis and its level of integration with sprints.

These variables may explain the different views of experts. A suggested approach would be the creation of proper baselines acting as boilerplates to speed up requirements analysis for standard applications.

Table 10. Design factors for performing security requirements analysis

Design factor	Perceived as relevant by experts
Treat security requirements as non-functional requirements	9 of 9
Leverage metrics gathered during the security requirements analysis phase to evaluate the security level of alternative designs	5 of 7
Leverage goal-oriented requirements analysis (GORE) to perform security requirements analysis	5 of 5
Leverage process metrics	Added by expert
Establish rules to balance functional and non-functional requirements	Added by expert
Leverage model-based systems engineering	Added by expert

Frameworks

§ OWASP SAMM v2.0: Intelligence: Security features and design
§ BSIMM v10: Design: Security requirements

Design Factors

Quick Facts

Table 11. Quick facts for performing security requirements analysis

Why should you do this?	Ensure security aspects are considered during development and testing
	Identify business logic errors
What to avoid?	Applying the same rigour in terms of security requirements analysis to every application regardless of their level of sensitivity

Performing Threat Modelling

Threat modelling aims at identifying potential weaknesses in each design and the related threats which could be posed by them. We suggest considering threat modelling as an input for risk analysis during the design phase of an information system.

Design Factors

Performing threat modelling from the initial planning stage while placing the results in context through a risk analysis is paramount to establish security by design. By applying this process iteratively throughout the development cycle, an approach to identify and prioritise weaknesses and examine controls already in place allows an effective risk treatment strategy to be defined (Myrbakken & Colomo-Palacios, 2017).

Threat modelling starts by creating an abstraction of the application under design or development, usually employing a Data Flow Diagram. A threat modelling approach, such as STRIDE (Hernan et al., 2006) is applied by iterating over the model elements to identify all potential security threats. The approaches to identifying potential threats are numerous ranging from the use of mnemonics, which is more suitable for experienced security professionals, to the use of attack trees or serious games such as "Elevation of Privilege" (Shostack, 2014) or "OWASP Cornucopia" (Watson, 2012) empowering development and operations teams to perform threat modelling autonomously. Results obtained from threat modelling provide a basis to develop Quality Assurance (QA) tests (Oyetoyan et al., 2016).

Researchers (Tuma et al., 2018) distinguish several categories of threat modelling techniques: risk-centric, attack-centric or software-centric. It is not possible to unambiguously categorise each threat analysis methods in one of these groups. In the context of DevSecOps analysis, velocity is preferred over analysis systematicity. A strong focus on the most critical assets is vital to allow timely results.

Table 12. Design factors for performing threat modelling

Design factor	Perceived as relevant by experts
Perform threat modelling from a risk-centric perspective	10 of 10
Perform threat modelling from an attack-centric perspective	10 of 10
Perform threat modelling from a software-centric perspective	8 of 9
Ensure compatibility of threat modelling outcomes from a scope and result perspective	6 of 6
Introduce abuse cases and problem frames to perform threat modelling	10 of 10
Implement traceability of threat modelling results in the code base	6 of 6
Leverage automated threat impact analysis	7 of 7
Threat modelling should be performed through an iterative approach	Added by expert
Leverage both quantitative and qualitative approaches	Added by expert

Shostack (Shostack, 2008)argues that software-centric threat modelling techniques are most suitable for identifying threats in information systems. He also mentions that it makes sense to perform threat modelling without (security) experts due to their short supply and that by including the people involved in building the system they obtain a sense of ownership and an understanding of the security model.

The DevSecOps panel consulted during this research confirmed that the various threat modelling approaches (risk-, attack- and software-centric) could prove valuable depending on the objective and scope of the threat modelling activity. They recommend the use of different techniques such as attack trees, abuse cases and problem frames during the identification of potential weaknesses. However, the expert panel emphasises selecting a threat modelling approach which is iterative and where possible assisted by automation to reduce the time required to perform a threat modelling exercise. Annotating the outcome of the threat modelling directly in the codebase can be an excellent method to allow traceability. However, according to experts, these techniques have not yet matured to the level where they can be leveraged broadly.

Teams are more likely to consistently perform threat modelling at the start of each iteration if the time required to perform the activity is reduced. Consistently performing threat modelling at the start of an iteration leads to broader coverage of the threat models. A challenge related to such an iterative approach is to ensure that the full breadth of the system is covered. Maintain oversight on the compatibility of the threat modelling outcomes from both a scope and a result perspective is therefore challenging.

Expert Ranking

Overall experts on the panel perceive threat modelling as moderately effective (Mdn=3.0, IQR=0.50). A wide range of approaches to threat modelling exist. Each of these approaches provides different outcomes depending on the technique, goal and stage in the DevSecOps cycle where they are performed. These differences make it challenging to achieve a shared view on the topic. According to the experts, threat modelling allows the detection of weaknesses before the actual implementation is started and propose leveraging it as the basis for risk analysis and security testing. Threat modelling also allows the identification of the potential blast radius of an attack by analysing the potential chains of weaknesses. Some experts point out that manual threat modelling is only moderately useful for these purposes; instead, they recognise a significant advantage in using threat modelling to increase the security mindset in developers. Most experts agree that automation of threat modelling and focusing on technical security components are vital characteristics of effective threat modelling. Threat modelling is considered to have a moderate impact in terms of delay and financial consequences. In principle, it can be done reasonably fast by experienced people with appropriate tooling. Licenses for these tools and the required training to do it will have a cost associated however

Frameworks

§ OWASP SAMM v2.0: Design: Threat assessment
§ BSIMM v10: Intelligence SSDL touchpoints: Attack Models Architecture Analysis

Design Factors

Quick Facts

Table 13. Quick facts for performing threat modelling

Why should you do this?	As a starting point for risk analysis
	To identify areas for security testing
	To identify the blast radius of potential attacks
	To increase the security mindset (when performed manually)
What to avoid?	To apply threat modelling at the business process level

Performing Risk Analysis

Risk analysis aims to identify risks and quantify them as a function of likelihood and impact. The results of a threat modelling exercise can be placed in context through a risk assessment by applying a calculation to the identified weaknesses based on attributes such as likelihood and impact. The outcomes allow a risk factor to be determined while considering existing countermeasures.

The DREAD framework, which defines the attributes Damage, Reproducibility, Exploitability, Affected Users and Discoverability, is sometimes used as a methodology in (technical) risk assessments. Based on experiences at Microsoft Shostack (Shostack, 2008) states that DREAD is not well suited for software-centric risk analysis as it seems to add numbers without defining their scales, generating a risk of making a risk assessment appear algorithmic when it is not.

Design Factors

The outcome of a risk assessment enables stakeholders to decide on risk treatment options considering effectiveness, cost and impact. All stakeholders, including business, should be part of the risk assessment to ensure that they consider security during product development and to provide visibility continuously (Derksen et al., 2018).

Previous studies on multi-cloud DevOps implementations (Rios et al., 2017) indicate that continuous risk assessment is also vital to select relevant security controls and metrics for inclusion in Security SLA's for cloud services.

Expert Ranking

Performing risk analysis is perceived by the expert panel as a beneficial approach (mdn=4.0, IQR=0.50) to prioritise security efforts such as the rigour applied in defining security requirements and performing security testing. When performing risk analysis, it is essential to ensure the approach remains practical and yields tangible results. A practical approach prevents risk analysis from becoming a theoretical exercise which provides little benefits. The experts expect a relatively low direct impact on delay in the DevSecOps process (mdn=3.5, IQR=1.00). For this to be the case, the process must be performed in a parallel track and should be completed before the actual technical activities start. However, selecting a practical approach with the appropriate level of intensity is critical to ensure those team members with the required security knowledge to complete this activity can keep up with the pace of development. The financial impact of performing risk analysis is presumed to be limited (mdn=4.0, IQR=0.25) assuming that, as recommended, a practical and pragmatic approach and intensity is selected.

Frameworks

§ OWASP SAMM v2.0: Design: Threat assessment
§ BSIMM v10: Intelligence SSDL touchpoints: Attack Models Architecture Analysis

Table 14. Design factors for performing risk analysis

Design factor	Perceived as relevant by experts
Performing risk analysis continuously before each iteration	Not validated
Performing risk analysis during the design phase	Not validated
Include a broad range of stakeholders including the business owner when setting security goals	Not validated
Establish clear rules regarding information exchange across teams and maintain a log for every access to sensitive data	Not validated
Provide security knowledge and tools and encourage the Development and Operations teams to integrate themselves	Not validated
Consider gamification for finding vulnerabilities or bugs	Not validated

Design Factors

Quick Facts

Table 15. Quick facts for performing risk analysis

Why should you do this?	To prioritise security related efforts
What to avoid?	Performing risk analysis to accurately measure risk

Establishing Security SLA's for Cloud Providers

Various cloud technology-oriented research papers propose approaches to the definition of security service level agreements for cloud providers. The challenge resides in generating the required Security SLA's for across various cloud providers and maintaining oversight of the applications consuming the cloud services related to these security SLAs (Rios et al., 2019). An area of research related to this is the automation of Cloud Security SLA generation.

Design Factors

The DevSecOps expert panel mentioned that it is essential for a DevSecOps team to be involved in determining the required and desired cloud provider from a security control capability perspective. They also mention that education of the decision-makers with regards to the retention of responsibility for security in cloud environments is vital.

Expert Ranking

Expert panel members point out the importance of availability and integrity aspects of information systems residing in the cloud. Establishing security SLAs is a measure to achieve certainty regarding these two crucial aspects. Other experts point out that these security SLAs only have minimal effect on the actual security levels and that the statements in these SLAs are defined on such a high level that they do not necessarily provide tangible benefits. These different viewpoints explain part of the variability in the effectiveness scoring, the experts agree on the importance but see varying levels of actual benefits to investing effort in establishing these SLAs. About one-third of the experts perceives this activity as very effective (N=4, 37%) whereas another this of the experts indicate low effectiveness (N=2, 37%) The experts do also not seem to have a shared view on how much delay establishing these security SLAs introduces in the DevSecOps process (mdn=4.5, IQR=2.25), some point out that in most cases this activity is not organised in-line with the pipeline. Therefore, there should not be any delay. When looking at the financial impact of this activity the panel agrees that in general establishing the SLAs does not necessarily introduce significant costs (mdn=3.5, IQR=1.75), in some cases premium services may be required to obtain reasonable SLAs which in turn would represent a considerable financial consequence.

Frameworks

§ OWASP SAMM v2.0: Design: Security requirements
§ BSIMM v10: Intelligence: Security features and design

Design Factors

Table 16. Design factors for establishing security SLA's for cloud providers

Design factor	Perceived as relevant by experts
Ensure that the DevSecOps team is involved in determining the required cloud provider security controls	Added by expert
Educate decision makers on the retention of responsibility for security in cloud environments	Added by expert

Quick Facts

Table 17. Quick facts for establishing security SLA's for cloud providers

Why should you do this?	To obtain additional assurance regarding availability and integrity of cloud services

Performing Continuous Assurance

Performing continuous assurance as an activity refers to continuously checking if a system satisfies internal and external requirements. One such example is proving compliance with the requirements set by regulatory bodies (Rahman & Williams, 2016). One can expect a significant efficiency gain when leveraging some of the metrics provided by the automated activities described in this book for continuous assurance. In literature, this concept is sometimes referred to as compliance-as-code. Performing continuous assurance requires significant preparation and analysis to identify the various controls and metrics and providing a framework to correlate them.

Design Factors

Most experts on the DevSecOps panel agreed with the relevance of this activity and pointed in the direction of "policy-as-code" whereby requirements would be expressed in code and automatically enforced through the infrastructure and the CI/CD pipeline.

Expert Ranking

Expert panel members believe that compliance-as-code is key to highly regulated industries; however, there is some disagreement on its bottom-line effectiveness. Over half of the experts rated this activity as very effective (N=4/N=5, 60%) while some rated it as very ineffective. These experts warn that compliance should not be confused with security; an insecure application may still be compliant. Performing continuous assurance may increase security awareness, thereby helping the establishment of a security mindset within the organisation. The experts expect this activity to create a significant delay in the DevSecOps process due to the coordination efforts between the large number of stakeholders involved (CISO, Compliance, Business, Dev and Ops). Judging the financial impact of continuous assurance is complex when offsetting the cost against the potential gains in risk avoidance. The experts point out that for this reason, the business should always be in the driver seat to decide on the balance between the cost of non-compliance (mainly determined by fines) and the cost of compliance (including effort but also the cost of delaying a potentially lucrative business feature). Overall, the inherent financial impact of performing continuous assurance is perceived as significant.

Frameworks

§ OWASP SAMM v2.0: Governance: Policy & Compliance
§ BSIMM v10: Governance: Compliance & Policy

Table 18. Design factors for performing continuous assurance

Design factors	Perceived as relevant by experts
Leverage compliance-as-code	Added by expert

Table 19. Quick facts for performing continuous assurance

Why should you do this?	Potentially part of the barrier to entry for your industry
	To avoid impact from sanctions and fines
	To establish a security mindset
What to avoid?	Not putting the business stakeholders in the driver's seat when determining compliance objectives

Design Factors

Quick Facts

Performing Manual Security Testing

DevOps in general and DevSecOps precisely have placed a strong emphasis on the automation of activities. Nevertheless, some manual security testing activities are referenced both in literature and during interviews with members of the expert panel. The two main types of activities which were distinguished consist of "manual penetration testing" and "manual security reviews". Whether they are part of DevSecOps is under debate; however, the experts agree on their effectiveness.

Frameworks

§ OWASP SAMM v2.0: Verification: Security testing
§ BSIMM v10: SSDL touchpoints: Security testing

Design Factors

Table 20. Design factors for performing manual security testing

Design factor	Perceived as relevant by experts
Limit manual penetration testing to critical components or perform the manual penetration testing in parallel to reduce impact on deployment lead times	5 of 8
Automated security testing is an important enabler to increase the value of manual testing	Added by expert
Performing manual penetration testing should depend on the criticality of the application and the level of customisation	Added by expert

Performing Manual Penetration Testing

Manual penetration testing provides a view of the potential attack surfaces and methods to which an application is vulnerable through real-world testing by a security professional. The results of manual penetration testing are proportional to the knowledge of the person performing the test and the time attributed to the engagement. In general, manual penetration testing allows for more accurate results because a human can try to examine context which is otherwise challenging to automate (Siewruk et al., 2019). Several research studies mention this activity (Jaatun, 2018; Jaatun et al., 2017).

An issue mentioned by researchers and members of the DevSecOps expert panel is that performing manual penetration testing in environments with frequent deployments of changes is challenging to implement due to the time constraints.

A proposed approach to overcome this limitation is to limit manual penetration testing to critical components or perform them in parallel to the continuous deployment process to reduce the impact on deployment lead times (Siewruk et al., 2019).

The DevSecOps expert panel points out that performing automated security testing is an enabler for valuable manual security testing. The penetration tester can be allowed to focus on aspects which cannot be automated by using automated security testing to perform low-complexity activities. They also confirm that the necessity to perform manual security testing could be determined based on the criticality of the application and the level of customisation the application has gone through.

Expert Ranking

Manual penetration testing is rated very effective by the expert panel (mdn=4.0, IQR=1.50); however, there are some differing views. The experts expect many facets to having a direct impact on the effectiveness of this activity. These aspects include the skillset of the person performing the penetration test, the process to handle the results of the penetration test, and the frequency of testing. Experts who scored the activity lower in terms of effectiveness point out that for these reasons it is more beneficial to to focus on automated penetration testing (N=2, 25%). Experts who rated the activity as highly effective state that a skilled penetration tester with proper preparation (e.g. access to security requirements and threat modelling outcomes) is very effective in identifying vulnerabilities in applications. The experts agree that penetration testing is very time consuming (mdn=2.0, IQR=1.50) and therefore must be organised outside of the DevSecOps pipeline to avoid introducing delays. The time required combined with the expert skills required to perform this activity results in a significant financial impact (mdn=2.0, IQR=1.25).

Quick Facts

Table 21. Quick facts for performing manual penetration testing

Why should you do this?	Obtain direct results in terms of vulnerabilities and exploitability in an application
What to avoid?	Perform a penetration test as a one off prior to going to production
	Perform a penetration test without providing output of requirements analysis and threat modelling activities
	Perform penetration testing in-line with the deployment pipeline

Performing Manual Security Review

The activity of performing a manual security review is viewed as an approach to increase the security of software (Mohan & Othmane, 2016a) (Jaatun, 2018). The activity consists of a manual review considering the security requirements and outcomes of the threat modelling phases to ensure the correct implementation of the required security measures. Inviting stakeholders to perform security tests during the product review, also referred to as a demo, provides the opportunity to break the system's security and try things that intruders or deceptive users would do to see how the system reacts (Derksen et al., 2018).

Expert Ranking

The expert panel generally perceives manual security reviews to only be moderately effective (mdn=3.0, IQR=1.50). Some experts indicated that it could act as an effective way to get a quick insight in the actual security level of an application given that this is not just performed based on documentation but consists of an actual review of the implementation together with the development and operations teams. The expert panel expects manual security reviews to introduce significant delays (mdn=2.0, IQR=1.00) in the DevSecOps process because it requires time from the development resources. A moderate financial impact is expected (mdn=3.0, IQR=1.25).

Quick Facts

Table 22. Quick facts for performing manual security review

Why should you do this?	To gain insight in the way security requirements are implemented in an application
What to avoid?	Performing a manual security review purely based on documentation

Secure the CI/CD Pipeline

The CI/CD pipeline, forming the technical backbone of DevSecOps activities, may itself be vulnerable to security attacks or misconfigurations (Gruhn et al., 2013). The security of a deployment pipeline may be threatened by malicious code being deployed through the pipeline or by allowing direct communication between components in the testing and production environments (Rimba et al., 2015). Therefore, it is crucial to ensure the CI/CD pipeline itself is adequately secured and that proper role-based access permissions and auditing is performed against the automated activities included in the pipeline.

When looking at the security of the pipeline three distinct scenarios play a role: (a) the pipeline may deploy an artefact which has not been validated, (b) an artefact may be deployed without going through the complete pipeline or (c) the production environment may be accessible from a different environment.

The objective of securing the CI/CD pipeline should be to assure that the pipeline is secure against attacks and cannot behave in a way that is not intended.

Expert Ranking

There is some debate about the effectiveness, delay and financial cost of securing the CI/CD pipeline within the expert panel. Overall, they believe this activity to be very effective (mdn=4.0, IQR=1.50), some experts pointed out that the CI/CD pipeline is the central axe along which DevSecOps activities are performed and therefore securing it is vital. A base level of security is relatively easy to achieve as most CI/CD related tools and products have some protections already built-in to them. Delay to the DevSecOps process is expected to be low (mdn=4.0, IQR=1.25) because securing the pipeline is a parallel process and is not subject to continuous change. The complexity of the pipeline and the chosen depth of security controls contribute to the financial cost of this activity. The expert panel expects a moderate financial impact (mdn=3.5, IQR=1.25). Costs to be account for include the time required to implement the selected controls and potential associated licensing costs.

Frameworks

§ OWASP SAMM v2.0: Implementation: Secure build & deployment
§ BSIMM v10: Deployment: Configuration & Vulnerability management

Quick Facts

Table 23. Quick facts for secure the CI/CD pipeline

Why should you do this?	To ensure the CI/CD pipeline provides reliable assurance the security processes is followed when developing or deploying applications.
What to avoid?	Not leveraging the default security controls build into the tooling

USE OF AUTOMATED ACTIVITIES

Performing Automated Security Testing

The objective to automate security controls and verifications where possible is a core concept in DevSec-Ops (Myrbakken & Colomo-Palacios, 2017). To achieve these objectives, organisations integrate a wide range of tools into the automated deployment pipeline (Carter, 2017). An option to speed up implementation and adoption is to leverage self-managed, automated and scalable security services provided by SecaaS platforms (Torkura et al., 2017).

The DevSecOps expert panel points out the importance of the "Fail fast" principle when evaluating efficiency gains through automated security testing. These automated security tests should be integrated early in the development cycle, allowing short feedback loops on security.

Selecting tools with flexible APIs is essential to their ability to integrate with existing organisational processes tightly. The approach needs to support the existing processes as opposed to imposing changes to them.

The implementation of automated security tests should be understandable to developers, and the expected outcomes included in the definition of done. The developer and operation buy-in will increase if they understand the objectives of the implemented tests and the guardrails they provide.

Three detailed security activities are derived from this security activity to cover its significant breadth and depth.

Frameworks

§ OWASP SAMM v2.0: Verification: Security testing
§ BSIMM v10: SSDL touchpoints: Security testing

Design Factors

Table 24. Design factors for performing automated security testing

Design factor	Perceived as relevant by experts
Leverage SecaaS by using cloud provided self-managed, automated and scalable security services	7 of 8
Integrate the security tools in an automated deployment pipeline	10 of 10
Automate as many security controls and verifications as possible	10 of 10
Ensure the team and management understands and supports the security validations integrated in the automated deployment pipeline	Added by expert
Fail fast when security validations do not pass	Added by expert
Integrate the validations in the Definition of Done	Added by expert
Ensure APIs (of security verifications) align with organisational processes allowing the implementation to be easy to understand	Added by expert
Automated testing is geared towards finding implementation bugs but generally not suited to spot design flaws.	Added by expert

Performing Automated Run-Time Testing

Automated run-time testing, also referred to as Dynamic Application Security Testing (DAST) is a testing methodology where an application is tested from the outside. It requires the application to be running and functional in order to be able to scan for security vulnerabilities.

An essential drawback to this technique is the limited scope which it can cover (Siewruk et al., 2019). As pointed out by an expert on the DevSecOps panel, automated run-time testing has issues supporting newer (web) technologies leading to a severely reduced coverage.

Design Factors

To compensate for this limitation, one could leverage functional tests by running them through automated run-time testing tooling allowing the tool to cover a broader set of the application function. An important note is that in this case, the coverage of the scans is dependent on the coverage of the functional tests (Siewruk et al., 2019).

Experts on the DevSecOps panel point out that implementing useful automated run-time testing is challenging when DevOps is not at a proper level (of automation). The recommended approach is to start small and extend later by focusing on the most relevant applications.

Expert Ranking

The expert panel mostly agrees that automated run-time testing is moderately effective (mdn=3.0, IQR=1.25). Experts who rated the effectiveness lower have done so indicating that this technique has become less effective in detecting vulnerabilities in modern technologies (Single Page Applications, SSO, MFA) and the potential number of high false positives. Those who rated it as very effective indicated that it provides good direct feedback and can cover a large volume of applications. The expert panel expects limited delays in the DevSecOps process (mdn=4.0, IQR=0.50) assuming the time required to perform a scan is maintained at a low level, and the number of false positives is limited. Implementing these tools to block the pipeline on failures, however, is expected to cause significant delays. There is debate regarding the financial consequences of automated run-time testing. Some experts rated this as very low impact, whereas others rated it very high. These different ratings can be explained by choice of tools and licensing models. The financial impact is considered low if one opts for using some of the open-source tools available; however, it may rise quickly if top-of-the-line commercial products are selected.

Design Factors

Table 25. Design factors for performing automated run-time testing

Design factors	Perceived as relevant by experts
Perform automated run-time testing at four levels: (1) pre-authentication scanning, (2) post-authentication scanning, (3) independent backend scanning and (4) complete workflows	8 of 8
Ensure automated run-time testing is implemented for a broad scope of test scenarios	8 of 8
Ensure proper unit tests are in place to optimise run-time testing efficiency	7 of 8
Start small, extend later	Added by expert
Get DevOps to a proper maturity level before implementing automated run-time testing	Added by expert

Quick Facts

Table 26. Quick facts for performing automated run-time testing

Why should you do this?	If you need to cover many web applications
What to avoid?	Implement block on fail

Performing Automated Static Testing

During automated static testing, also referred to as Static Application Security Scanning (SAST), the application code is analysed to identify potential security or other quality-related errors. These errors are usually related to coding mistakes or insecure coding practices which may introduce vulnerabilities in the application under development.

Ideally, this verification is performed during the software development to allow shorter feedback loops in the development sprints (Derksen et al., 2018). By integrating an automated code review into the development process, appropriate feedback can be provided to the software developers of interest, using open source and commercial static analysis tools (Rahman & Williams, 2016). The extent of static code analyses in the CI chain is recognised in the DevSecOps maturity model as one of the four essential axes to achieve security aspects (Mohan & Othmane, 2016a).

Design Factors

An important design factor for automated code review is the speed and accuracy of the feedback provided to developers. If the feedback loop created through automated code review is slow developers are less likely to include it in their workflow or will use it less frequently. Additionally, if the results of the automated code review contain too many false positives, the developer will need to spend much time sifting through the findings and is more likely to miss essential findings or to dismiss the results all together (Siewruk et al., 2019).

According to the DevSecOps expert panel, it is essential to ensure enough code coverage by automated static testing. Static testing performed on modules may not correctly follow complete flows when applications are not scanned. It is also important to require independent justification when marking findings as false positives and to perform adequate follow-up to ensure fixes genuinely address the root cause of the issue. Building a good process around static testing is vital, one approach which has proven useful is to implement so-called build breakers which stop the CI/CD pipeline upon discovery of new severe issues.

One should consider prioritising efforts on static code analysis for the most relevant applications due to the time and effort required to get it right.

Expert Ranking

There is debate among experts regarding the effectiveness of performing static security testing (mdn=3.5, IQR=2.0). The majority believes, however, that it is very (N=5, 36%) to highly effective (N=4, 14%) while one expert finds it highly ineffective (N=1, 12%). The main limitation to the effectiveness of static testing is the potentially high number of false positives. These same false positives are also the primary source of delay, in general running the tooling can be fast however managing the false positives can become time-consuming when used in blocking mode where the pipeline is interrupted upon detecting issues. Reduction of the impact of false positives is possible by performing a baseline scan to suppress false positives and creating global lists of accepted false positives. These factors explain the slight disagreement between the experts (mdn=3.0, IQR=1.50). The expert panel expects a significant financial impact (mdn=2.5, IQR=1.00) as most experts point out that there is a lack of good open-source tooling, and the commercial products can become quite expensive.

Design Factors

Table 27. Design factors for performing automated static testing

Design factor	Perceived as relevant by experts
Minimise the number of false positives resulting from static testing	8 of 8
Independent justification of whether a false positive is false and whether the fix genuinely addresses the cause	Added by expert
Ensure code coverage (applications should be tested as a whole, not scans run on separate modules, otherwise the static testing will not properly follow flows).	Added by expert
Should prioritise the most relevant applications first since this takes a lot of time.	Added by expert
Good process is key, build breakers are important.	Added by expert

Quick Facts

Table 28. Quick facts for performing automated static testing

Why should you do this?	Provide fast feedback loops for developers on common secure code violations
What to avoid?	Not tuning the rulesets by performing a baseline scan or establishing lists of globally accepted false positives

Integrate Security Tests in Unit Testing

Security testing should start at the feature or component level before performing any system integration. Tests should cover both unauthorised misuse and violations of the assumptions on which the component was built (Rahman & Williams, 2016).

Design Factors

Including tests to verify the correct implementation of security requirements for a given feature is recommended. These tests should target not only specific security features, but also non-functional security requirements of other features used in security contexts. Several experts on the DevSecOps panel indicated that security testing should start as early as possible in the development cycle (shift left) and that adequate coverage of security aspects in unit testing allows, in combination with other activities, the security experts to focus on the problematic aspects of security.

Expert Ranking

Security experts strongly disagree on the effectiveness of this activity. Half of the expert panel indicated that this activity is highly effective; however, some other experts rated it very low, explaining that it only works in theory and is not realistic in practice. A prerequisite to this activity is detailed security requirements from the start of the project and adequate security knowledge on the part of the developers. Experts are also divided on the delays caused by this activity (mdn=3.0, IQR=3.50); however, they mostly agree that it has severe financial consequences (mdn=1.0, IQR=1.50).

Design Factors

Table 29. Design factors for Integrate security tests in unit testing

Design factor	Perceived as relevant by experts
Consider the team needs to have special security related knowledge	Added by expert
Start early on to avoid bottlenecks when deploying to production	Added by expert

Quick Facts

Table 30. Quick facts for Integrate security tests in unit testing

Why should you do this?	Gain deep assurance for very sensitive software features or functions which have detailed security requirements

Performing Continuous Monitoring

Performing continuous monitoring can be applied to a variety of system and application aspects and therefore covers a wide range of detailed activities. Researchers suggest monitoring system-related data (Rahman & Williams, 2016) but also software vulnerabilities (Fraile et al., 2018), security metrics (Rahman & Williams, 2016) and security SLA's (Rios et al., 2017). Gathering security metrics during the SDLC (Myrbakken & Colomo-Palacios, 2017) to maintain an overview of application behaviour (Jaatun, 2018) is recommended.

Besides, continuous monitoring is cited as relevant in the context of demonstrating compliance with policies (Mohan & Othmane, 2016a). It is essential (when automating security controls) to be able to generate evidence on demand that controls are working and that they are effective (Myrbakken & Colomo-Palacios, 2017). To that end, ensuring complete coverage of all assets and resources is vital.

Metrics gathered through continuous monitoring are preferably exposed to development and operations teams through so-called self-service monitoring and alerting (Diaz et al., 2019). According to the DevSecOps expert panel, this aspect is critical in supporting the concept of shifting security to the left" and plays a role in preventing unnecessary escalations while allowing teams to become self-organised when it comes to fixing issues.

Continuous monitoring itself should be implemented as code, allowing a versioned and repeatable deployment of the required monitoring infrastructure (monitoring-as-code) (Diaz et al., 2019).

Members of the DevSecOps expert panel pointed out that automated deployment is a crucial enabler to become successful at automated continuous monitoring. Without it, the challenge to ensure the correct monitoring configuration is applied to each resource rises.

Frameworks

§ OWASP SAMM v2.0: Governance: Strategy & Metrics
§ BSIMM v10: Governance: Strategy & Metrics

Design Factors

Table 31. Design factors for Performing continuous monitoring

Design factor	Perceived as relevant by experts
Ensure continuous monitoring covers a wide range of resources and metrics including logical security, availability and intrusions	9 of 9
Leverage monitoring as code to establish a versioned and repeatable deployment of monitoring infrastructure	8 of 8

Performing Continuous Monitoring of Security SLAs

Continuous monitoring of security service level agreements once components are deployed, and running, allows reaction measures to be taken in case of potential or actual violations (Rios et al., 2017). These results can be fed back to development to provide learning opportunities. There are also tools being developed which allow dynamic adaptation of multi-cloud applications to ensure the security status included in security service level agreements are maintained, allowing a swift reaction to possible security incidents. The DevOps team should verify that the metrics defined for the security controls are reaching the target levels, thereby ensuring that security and privacy levels of multi-cloud applications are attained (Rios et al., 2019).

Expert Ranking

The expert panel has not reached agreement on the ranking of this activity. Most experts agree that monitoring the security SLAs only makes sense if one can do something with this information from a security perspective. Therefore, the expert rate this activity as moderately effective (mdn=3.0, IQR=2.00). The activity is only seen as useful in organisations with multi-cloud platforms where the ability exists to move resources from one provider to another in case of a security failure. Delays on the DevSecOps process are considered limited, assuming the monitoring is automated and running in parallel. The financial consequences are expected to be moderate (mdn=3.5, IQR=1.75).

Table 32. Quick facts for Performing continuous monitoring of security SLAs

Why should you do this?	To leverage the capabilities offered by multi-cloud platforms to react to security failures by moving resources

Quick Facts

Performing Continuous Monitoring of Security Metrics Throughout the SDLC Using CI/CD Tooling

Leveraging Continuous Integration and Continuous Deployment (CI/CD) tools to gather security relevant metrics to identify issues such as coding mistakes or vulnerable dependencies (Myrbakken & Colomo-Palacios, 2017) is an important activity within DevSecOps. Doing so enables organisations to track threats and vulnerabilities in real-time and allows continuous evaluation of required versus achieved security levels (Jaatun et al., 2017).

Expert Ranking

Gathering security metrics by levering CI/CD tooling is considered a highly effective security activity by most of the security experts (mdn=4.0, IQR=1.00) assuming careful selection of the metrics. The expert panel considers the delays caused by this activity limited (mdn=4.0, IQR=1.50) as most of the heavy lifting is part of other activities in the area of continuous monitoring. This activity leverages the outputs generated by other tools to provide indicators of security and knowledge levels in teams. The experts believe there is a moderate financial impact related to this activity which is mainly driven by the effort to implement and maintain it (mdn=3.0, IQR=1.50).

Table 33. Quick facts for Performing continuous monitoring of security metrics throughout the SDLC using CI/CD tooling

Why should you do this?	To identify security issues in software
	To measure security knowledge of teams and identify blind spots
What to avoid?	Collect metrics without a clear strategy

Table 34. Design factors for Performing continuous monitoring of system metrics using automated tools

Design factor	Perceived as relevant by experts
Ensure that other enablers such as automated deployments and configuration management are implemented	Added by expert

Quick Facts

Performing Continuous Monitoring of System Metrics Using Automated Tools

Continuously monitoring metrics such as resource usage and reaction times allows potential detection of malicious activity. System-related information, such as CPU usage and memory usage, can be gathered and stored for further analysis using automated tools (Rahman & Williams, 2016).

Design Factors

Experts point out that automated deployments and configuration management are required to make effective use of this activity.

Expert Ranking

There is some debate regarding the effectiveness of this activity however most of the experts perceive it as being very effective (mdn=4.0, IQR=1.50) without cause a significant delay in the process (mdn=4.0, IQR=1.50). The experts agree that there are moderate financial consequences to be expected mainly driven by efforts to implement and maintain it (mdn=3.0, IQR=0.75).

Table 35. Quick facts for Performing continuous monitoring of system metrics using automated tools

Why should you do this?	To gain security knowledge from system behaviour

Table 36. Design factors for Performing continuous monitoring of security controls

Design factor	Perceived as relevant by experts
Ensure that formal controls are tracked	Added by expert
Especially important in highly regulated environments	Added by expert

Design Factors

Quick Facts

Performing Continuous Monitoring of Security Controls

This activity refers to implementing specific monitoring solutions to determine the effectiveness of security controls for a given application such as SSL settings, access control mechanisms and vulnerable dependencies. This allows the generation of evidence on demand that controls are working and that they are effective (Myrbakken & Colomo-Palacios, 2017).

Design Factors

Experts on the panel pointed out that this activity is especially relevant in highly regulated environments and where formal controls can and should be tracked.

Expert Ranking

The expert panel strongly agrees that performing continuous monitoring of security controls is highly effective (mdn=4.0, IQR=0.50) and agree that it introduces limited delays (mdn=4.0, IQR=1.00). Some debate exists on the financial consequences however a moderate financial consequence, mainly driven by the effort to implement and maintain the solution, is expected (mdn=3.0, IQR=1.50).

Table 37. Quick facts for Performing continuous monitoring of security controls

Why should you do this?	To gain insights on the effectiveness of security controls

Design Factors

Quick Facts

Performing Continuous Monitoring of Application Behavior

Continuous monitoring of application behaviour, such as input and output, assists in detecting and preventing malicious activity by identifying changes in patterns (Jaatun, 2018). This activity is commonly implemented through tools such as Web Application Firewalls (WAF) or run-time application security protection (RASP).

Design Factors

Experts on the panel point out that the DevOps team needs to have a good understanding of normal application behaviour before attempting to identify deviations.

Expert Ranking

The expert panel strongly agrees on the effectiveness of performing continuous monitoring of application behaviour (mdn=4.0, IQR=0.25). There is some debate on the delay and financial impact caused by this activity. Experts on the panel believe the delay caused will be limited (mdn=4.0, IQR=1.50) while the financial consequences will be considerable (mdn=2.0, IQR=1.50).

Design Factors

Table 38. Design factors for Performing continuous monitoring of application behavior

Design factor	Perceived as relevant by experts
A DevOps team should be aware of the normal application behaviour allowing identification of deviations	Added by expert

Quick Facts

Provide Self-Service Monitoring Capabilities to dev and ops

Providing a flexible monitoring infrastructure allowing teams to configure their monitoring and alerting services according to their criteria allows 'fast and continuous feedback from Ops to Dev (Diaz et al., 2019). This concept is referred to as 'you build, you run, and now you monitor it'. Such self-service monitoring and alerting solution allow breaking silos between dev, ops and security teams by opening access to critical security metrics, enabling a sharing culture and continuous improvement. When looking at the implementation of such a solution, one should strive for a large extend of automation, leading to the concept of 'monitoring as code'.

Table 39. Quick facts for Performing continuous monitoring of application behavior

Why should you do this?	To detect abnormal activity

Design Factors

Self-service monitoring capabilities should allow teams to be self-organised when it comes to fixing issues. Self-service capabilities prevent unnecessary escalations and form an essential enabler for shifting security to the left.

Expert Ranking

Implementing self-service monitoring capabilities is perceived as moderately effective by experts (mdn=3.0, IQR=1.25). The activity is not expected to cause significant delays (mdn=4.0, IQR=1.50) nor significant financial impact (mdn=4.0, IQR=1.50).

Design Factors

Table 40. Design factors for Provide self-service monitoring capabilities to dev and ops

Design factor	Perceived as relevant by experts
Enable teams to be self-organised when it comes to fix security issues	Added by expert

Quick Facts

Implement Centralized Dashboards

This activity was identified through the DevSecOps expert panel and refers to establishing centralised dashboards which provide a clear view on security metrics related to secure development and operations.

Expert Ranking

Building centralised dashboards is considered a moderately effective security activity (mdn=3.5, IQR=1.50) and is expected to cause minimal delays in the DevOps process (mdn=4.0, IQR=1.50).

Table 41. Quick facts for Provide self-service monitoring capabilities to dev and ops

Why should you do this?	To establish a sharing culture and allow teams to become self-organised with respect to fixing security issues
	To reduce the number of escalations

The experts fully agree that this activity is not expected to cause any significant financial consequences (mdn=4.0, IQR=0.00). This rating is under the assumption that the process of gathering and reporting on the information is fully automated.

Quick Facts
Table 42. Quick facts for Implement centralized dashboards

Why should you do this?	To provide visibility into security related aspects to a wide variety of stakeholders

Implement Automated Remediation

Dynamic adaptation and reaction to security incidents have been mentioned in the context of multi-cloud environments (Rios et al., 2017). Maintaining security levels across multi-cloud applications can be achieved by automatically taking necessary remediation measures on security degradation. Our DevSecOps expert panel indicated that they have never seen it being useful in practice.

Design Factors

Experts point out that manual verification remains essential to validate the completeness and appropriateness of automated remediation.

Expert Ranking

The expert panel considers this activity to be moderately effective (mdn=3.0, IQR=1.50). Some experts point out that getting automated remediation up and running requires attention to potential breaking changes and that effort is potentially better spent in preventing the issue from occurring. Automated remediation can only become effective when many other aspects are fully automated. Experts do not expect that this activity will cause significant delays in the DevSecOps process (mdn=5.0, IQR=0.50) due to the emphasis on automation. The activity can represent a moderate financial impact (mdn=3.0, IQR=1.50) driven by licensing costs and effort to implement and maintain.

Frameworks

§ OWASP SAMM v2.0: Operations: Incident management
§ BSIMM v10: Deployment: Configuration management & Vulnerability management

Design Factors

Table 43. Design factors for Implement automated remediation

Design factor	Perceived as relevant by experts
Always manually verify results from time to time	Added by expert

Quick Facts

Table 44. Quick facts for Implement automated remediation

Why should you do this?	To provide visibility into security related aspects to a wide variety of stakeholders

Performing Security Configuration Automation

Security configuration automation is mentioned several times throughout academic literature, for example, in the context of automating software-defined firewalls to allow the deployment of consistent policies across firewalls in an organisation (Rahman & Williams, 2016). Experts also refer to security configuration automation as security-as-code allowing the definition of security policies which can be embedded and enforced throughout the development and operations processes from the project get-go. Such codified security policies can be activated automatically or manually and stored in a central repository for reuse on a new project (Myrbakken & Colomo-Palacios, 2017).

Expert Ranking

Performing security configuration automation is seen as an effective approach (mdn=4.0, IQR=0.25) while causing minimal delays (mdn=4.0, IQR=1.50). The concepts behind configuration automation and potential best practices are proven, well known and can easily be integrated into existing solutions such as system configuration automation tooling. There is debate regarding the financial consequences, mainly driven by the effort to implement and maintain the solution. A moderate financial impact should be expected (mdn=3.0, IQR=2.00).

Frameworks

§ OWASP SAMM v2.0: Implementation: Secure build
§ BSIMM v10: Deployment: Configuration management & Vulnerability management

Quick Facts

Table 45. Quick facts for Performing security configuration automation

Why should you do this?	To consistently manage security configurations across a wide range and volume of devices

Implement Secrets Management

This activity was identified through the DevSecOps expert panel and referred to automating the management of secrets used by applications and infrastructure components.

Expert Ranking

Implementing secrets management is perceived as an effective security activity, with half of the experts (N=5, 50%) rating is as highly effective. They do point out that it may sometimes become complicated or even theoretically impossible to secure all secrets in an environment. It is essential to pay appropriate attention to ensure the availability of this system and to implement a proper authorisation model to increase its effectiveness. This activity It is not expected to cause significant delays (mdn=4.0, IQR=1.50) in the DevSecOps process, however, depending on the choice of product, is expected to have some financial consequences (mdn=3.5, 1.75).

Frameworks

§ OWASP SAMM v2.0: Implementation: Secure build
§ BSIMM v10: Deployment: Configuration management & Vulnerability management

Quick Facts

Table 46. Quick facts for Implement secrets management

Why should you do this?	To consistently manage security configurations across a wide range and volume of devices

Manage Digital Supply Chain

During this research activities were identified which relate to managing the risks originating from using external components in the software development process. These components are not under the direct control of the organisation, and any weaknesses in them may significantly impact the security of the systems build on top of them. Therefore, we decided to group these activities under the umbrella of 'managing the digital supply chain'.

Frameworks

§ OWASP SAMM v2.0: Implementation: Secure build
§ BSIMM v10: Deployment: Configuration management & Vulnerability management

Establish Artefact and Source Code Registries Which are Automatically Scanned for Vulnerabilities

An expert of the expert panel added the activity of establishing artefact and source code repositories with automated scanning for vulnerabilities. This activity refers to the actions taken to continuously scan and approve entries in artefact and source code registries, thereby preventing the use of vulnerable resources in development or operations.

Expert ranking

Experts perceive this activity to be very effective (mdn=4.0, IQR=1.00) and is expected to cause moderate delays (mdn=3.0, IQR=1.50) specifically if manual verifications are required. There are no significant financial consequences anticipated (mdn=4.0, IQR=1.50) beyond some licensing costs for the tooling.

Quick Facts

Table 47. Quick facts for Manage digital supply chain

Why should you do this?	Prevent developers from using vulnerable components in their applications
What to avoid?	Manual verification of each component

Implement Automated Container Security Scanning

A member of the devsecops expert panel added the activity of automated container security scanning. This activity refers to implementing automated security scanning of container images and configuration for known vulnerabilities and weaknesses preventing vulnerable images or insecure configurations from being used as a basis for development or operations.

Expert Ranking

Automated container security scanning has been rated as a very effective security activity by the expert panel (mdn=4.0, IQR=1.25). However, experts do point out that container scanning is only a part of the puzzle. In principle, container scans are very fast to execute, not causing significant delays (mdn=4.0, IQR=0.00). Some financial consequences are anticipated (mdn=3.5, IQR=1.00).

Quick Facts

Table 48. Quick facts for Implement automated container security scanning

Why should you do this?	To detect vulnerabilities or vulnerable configurations in container images

Performing Automated Software Composition Analysis

Automated software composition analysis is a technique where the third-party libraries and code on which an application depend are verified for known vulnerabilities (Mohan & Othmane, 2016a). This verification is achieved by comparing the components and their versions against a database of known vulnerabilities. It is generally accepted that the ecosystem of dependencies on which applications are build may introduce significant risk to an application throughout its entire lifespan. Therefore, automated software composition analysis should be performed continuously over time and not only during the development cycle. This activity is closely related to the concept of controlling open source risks (Jaatun, 2018).

Design Factors

Several experts on the DevSecOps panel mentioned that ensure that only approved libraries can be used during development time is an approach to reduce the problem at the source. However, this does not provide coverage for vulnerabilities which are detected while an application is active in a production environment and not subject to development activities. A process to handle new vulnerabilities is required to ensure proper prioritisation and sign-off. Experts recommend starting with the most critical applications first as the potential fall-out may be significant.

Expert Ranking

Performing automated software composition analysis is a moderately effective security activity (mdn=3.0, IQR=1.25), mainly due to the potential overload of information. The technique is very capable of findings security weaknesses however verification of the applicability to the context of the application is required and may often indicate that the vulnerable function in the dependency is not used in the context of the application. It may introduce a moderate delay in the DevSecOps process (mdn=3.0, IQR=1.00) if applications have many dependencies and a full scan is expected on every build. A moderate financial impact is expected (mdn=3.0, IQR=1.50) mainly due to effort related to verifying false positives and placing results in the context of the application. Treating the identified issues late in the development or operations process may, however, represent considerable effort.

Design Factors

Table 49. Design factors for Performing automated software composition analysis

Design factor	Perceived as relevant by experts
Make use of approved libraries	Added by expert
Start with the most critical (end-user facing) applications first	Added by expert
Establish a process to handle new vulnerabilities	Added by expert

Quick Facts

Table 50. Quick facts for Performing automated software composition analysis

Why should you do this?	To detect vulnerabilities in libraries and dependencies used by software
What to avoid?	Requiring a full scan of all dependencies on every build in continuous deployment environments

REFERENCES

Bobbert, Y. (2017). On Exploring Research Methods for Business Information Security Alignment and Artefact Engineering. *International Journal of IT/Business Alignment and Governance, 8*(2), 28–41. doi:10.4018/IJITBAG.2017070102

Bobbert, Y., & Mulder, H. (2013). *Group Support Systems Research in the Field of Business Information Security: A Practitioner's View*. doi:10.1109/hicss.2013.244

Carter, K. (2017). Francois Raynaud on DevSecOps. *IEEE Software, 34*(5), 93–96. doi:10.1109/MS.2017.3571578

Corman, J., Rice, D., & Williams, J. (2010). *Rugged Software manifesto*. Academic Press.

Derksen, B., Neggers, M., Onwezen, D., & Zelen, S. (2018). *Agile Secure Software Lifecycle Management Secure by Agile Design*. Secure Software Alliance.

Diaz, J., Perez, J. E., Lopez-Pena, M. A., Mena, G. A., & Yague, A. (2019). Self-Service Cybersecurity Monitoring as Enabler for DevSecOps. *IEEE Access: Practical Innovations, Open Solutions, 7,* 100283–100295. doi:10.1109/ACCESS.2019.2930000

Fraile, F., Flores, J. L., Anaya, V., Saiz, E., & Poler, R. (2018). *A Scaffolding Design Framework for Developing Secure Interoperability Components in Digital Manufacturing Platforms*. doi:10.1109/is.2018.8710510

Gruhn, V., Hannebauer, C., & John, C. (2013). *Security of public continuous integration services.* doi:10.1145/2491055.2491070

Hernan, S., Lambert, S., Ostwald, T., & Shostack, A. (2006). Threat modeling-uncover security design flaws using the stride approach. *MSDN Magazine*, 68–75.

Hevner, M., March, Park, & Ram. (2004). Design Science in Information Systems Research. *Management Information Systems Quarterly*, *28*(1), 75. doi:10.2307/25148625

Jaatun, M. G. (2018). *Software Security Activities that Support Incident Management in Secure DevOps.* doi:10.1145/3230833.3233275

Jaatun, M. G., Cruzes, D. S., & Luna, J. (2017). *DevOps for Better Software Security in the Cloud.* Advance online publication. doi:10.1145/3098954.3103172

Jabbari, R., Ali, N., Petersen, K., & Tanveer, B. (2016). *What is DevOps? A Systematic Mapping Study on Definitions and Practices.* doi:10.1145/2962695.2962707

Linstone, H. A., Turoff, M., & others. (1975). *The Delphi method.* Academic Press.

Martin, F., & Jim, H. (2001). Agile-Manifesto.pdf. *Software Development*, *9*(8), 28–35.

McCarthy, M. A., Herger, L. M., Khan, S. M., & Belgodere, B. M. (2015). Composable DevOps: Automated Ontology Based DevOps Maturity Analysis. *2015 IEEE International Conference on Services Computing*, 600–607. 10.1109/SCC.2015.87

Mohan, V., & Othmane, L. (2016a). *SecDevOps: Is It a Marketing Buzzword? Mapping Research on Security in DevOps.* doi:10.1109/ares.2016.92

Mohan, V., & Othmane, L. B. (2016b). SecDevOps: Is it a Marketing Buzzword? *2016 11th International Conference on Availability, Reliability and Security (ARES)*, 542–547. 10.1109/ares.2016.92

Myrbakken, H., & Colomo-Palacios, R. (2017). *DevSecOps: A Multivocal Literature Review.* doi:10.1007/978-3-319-67383-72

Nowell, L. S., Norris, J. M., White, D. E., & Moules, N. J. (2017). Thematic Analysis. *International Journal of Qualitative Methods*, *16*(1). doi:10.1177/1609406917733847

Okoli, C., & Pawlowski, S. D. (2004). The Delphi method as a research tool: An example, design considerations and applications. *Information & Management*, *42*(1), 15–29. doi:10.1016/j.im.2003.11.002

Oyetoyan, T. D., Cruzes, D. S., & Jaatun, M. G. (2016). An Empirical Study on the Relationship between Software Security Skills, Usage and Training Needs in Agile Settings. *2016 11th International Conference on Availability, Reliability and Security (ARES)*, 548–555. 10.1109/ares.2016.103

Rahman, A. A. U., & Williams, L. (2016). *Software Security in DevOps: Synthesizing Practitioners' Perceptions and Practices.* doi:10.1145/2896941.2896946

Recker, J., & Recker, J. (2012). *Scientific Research in Information Systems, A Beginner's Guide.* doi:10.1007/978-3-642-30048-6_1

Rimba, P., Zhu, L., Bass, L., Kuz, I., & Reeves, S. (2015). Composing Patterns to Construct Secure Systems. *2015 11th European Dependable Computing Conference (EDCC)*, 213–224. 10.1109/edcc.2015.12

Rios, E., Iturbe, E., Larrucea, X., Rak, M., Mallouli, W., Dominiak, J., Muntes, V., Matthews, P., & Gonzalez, L. (2019). Service level agreement-based GDPR compliance and security assurance in (multi) Cloud-based systems. *IET Software, 13*(3), 213–222. doi:10.1049/iet-sen.2018.5293

Rios, E., Iturbe, E., Mallouli, W., & Rak, M. (2017). *Dynamic security assurance in multi-cloud DevOps*. doi:10.1109/cns.2017.8228701

Saunders, M., Lewis, P., & Thornhill, A. (2007). *Research methods*. Business Students.

Shostack, A. (2008). *Experiences Threat Modeling at Microsoft*. Academic Press.

Shostack, A. (2014). *Elevation of privilege: Drawing developers into threat modeling*. Academic Press.

Siewruk, G., Mazurczyk, W., & Karpinski, A. (2019). Security Assurance in DevOps Methodologies and Related Environments. *International Journal of Electronics and Telecommunications, 65*(2), 211–216.

Torkura, K. A., Sukmana, M. I. H., Cheng, F., & Meinel, C. (2017). *Leveraging Cloud Native Design Patterns for Security-as-a-Service Applications*. doi:10.1109/SmartCloud.2017.21

Tuma, K., & Calikli, G. (2014). Threat analysis of software systems: A systematic literature review. *ACM SIGPLAN Notices, 49*(6), 275–294. doi:10.1016/j.jss.2018.06.073

Watson, C. (2012). *OWASP Cornucopia Ecommerce Website Edition*. https://www.owasp.org/images/7/71/Owasp-cornucopia-ecommerce_website.pdf

Wieringa, R. J. (2014). *Design Science Methodology for Information Systems and Software Engineering*. doi:10.1007/978-3-662-43839-8_1

APPENDIX

Table 51. Literature review search results

ID	Title	Author(s)	Year	Iteration
P1	Self-Service Cybersecurity Monitoring as Enabler for DevSecOps	Jessica Diaz, Jorge E. Perez, Miguel A. Lopez-Pena, Gabriel A. Mena, Agustin Yague	2019	1
P2	Software Security Activities that Support Incident Management in Secure DevOps	Unknown, Martin Gilje Jaatun	2019	1
P3	Security Assurance in DevOps methodologies and related environments	Grzegorz Siewruk, Wojciech Mazurczyk & Andrzej Karpiński	2019	1
P4	Software Security in DevOps: Synthesizing Practitioners' Perceptions and Practices	Akond Ashfaque Ur Rahman, Laurie Williams	2016	1
P5	DevSecOps: A Multivocal Literature Review	Myrbakken, H; Colomo-Palacios, R	2017	1
P6	SecDevOps: Is It a Marketing Buzzword? Mapping Research on Security in DevOps	Mohan, V; ben Othmane, L	2016	1
P7	Security Practices in DevOps	Rahman, AAU; Williams, L	2016	1
P8	Dynamic Security Assurance in Multi-Cloud DevOps	Rios, E; Iturbe, E; Mallouli, W; Rak, M	2017	1
P9	Francois Raynaud on DevSecOps	Carter, K	2017	1
P10	DevOps for Better Software Security in the Cloud	Jaatun, MG; Cruzes, DS; Luna, J	2017	1
P11	A systematic mapping study of infrastructure as code research	Rahman, A; Mandavi-Hezaveh, R; Williams, L	2019	3
P12	Threat analysis of software systems: A systematic literature review	Tuma, K; Calikli, G; Scandariato, R	2018	3
P13	DevOps in practice: A multiple case study of five companies	Lucy Ellen Lwakatare, Terhi Kilamo, Teemu Karvonen, Tanja Sauvola, Ville Heikkil‰o, Juha Itkonen, Pasi Kuvaja, Tommi Mikkonen, Markku Oivo, Casper Lassenius	2019	3
P14	Service level agreement-based GDPR compliance and security assurance in (multi)Cloud-based systems	Rios, E; Iturbe, E; Larrucea, X; Rak, M; Mallouli, W; Dominiak, J; Muntes, V; Matthews, P; Gonzalez, L	2019	3
P15	A Scaffolding Design Framework for Developing Secure Interoperability Components in Digital Manufacturing Platforms	Fraile, F; Flores, JL; Anaya, V; Saiz, E; Poler, R	2018	3
P16	Scaling Agile Software Development to Large and Globally Distributed Large-scale Organizations	Fraile, F; Flores, JL; Anaya, V; Saiz, E; Poler, R	2018	3
P17	A Multivocal Literature Review on the use of DevOps for e-learning systems	Sanchez-Gordon, M; Colomo-Palacios, R	2018	3
P18	Leveraging Cloud Native Design Patterns for Security-as-a-Service Applications	Torkura, KA; Sukmana, MIH; Cheng, F; Meinel, C	2017	3
P19	A systematic mapping study on security in agile requirements engineering	Curcio, K.; Navarro, T.; Malucelli, A.; Reinehr, S.	2018	A
P20	Security requirements Engineering in the Agile Era: How does it work in Practice?	Daneva, M.; Wang, C.	2018	A
P21	Effectiveness of using card games to teach threat modelling for secure web application development	Thompson, M.; Takabi, H.	2016	A
P22	Continuous Integration, Delivery and Deployment: A Systematic Review on Approaches, Tools, challenges and Practices	Mojtaba, S.; Muhammad A.B.; Liming Z.	2017	A

Table 52. Overview of codes used for thematic analysis

RQ	Code	Description	Iteration
RQ2	Principle	A fundamental truth or proposition that serves as the foundation for DevSecOps	1
RQ2	Practice	A security practice is a collection of activities can be grouped based on existing similarities within activities	1
RQ2	Activity	A security activity focuses on achieving a small, well-defined goal that has a tangible output	1
RQ2	Design factor	A specific way a security activity can be performed to increase its effectiveness or efficiency	1
	Challenge	An obstacle which may impede or complicate the implementation of a security activity or practice in DevSecOps	2
	Opportunity	An advantage or improvement related to the implementation of a security activity or practice in DevSecOps	2
	Methodology	A system of methods applicable to DevSecOps	2
	Frameworks	A defined and agreed approach that intends to improve security relevant to DevSecOps	2
	Best practices	A repeated method used by people to improve sec in DevSecOps	2
	Tool category	A category of security tools which can be used to implement or facilitate a security activity	2

Table 53. Results of thematic analysis of security activities

Activity	Description	Level
Performing continuous feedback from production to development	This activity refers to continuously feeding security metrics and information on security incidents from production back to development.	1
Provide security training	This activity refers to training a wide range of stakeholders such as developers, architects and product owners on security aspects.	1
Establish security satellites	This activity refers to creating a network of security savvy people throughout the various teams involved in software development. These people are regularly referred to as security champions.	1
Practice incident response	This activity refers to practicing incident response through red-team exercises and security drills.	1
Performing automated security testing	This activity refers to aspects of security testing which can be automated thereby providing actionable information.	1
Performing automated run-time testing	This activity refers to the dynamic, interactive testing of a deployed application using automated tools (DAST)	2
Performing automated static testing	This activity refers to code review using automat-ed tools to detect common vulnerability patterns (SAST).	2
Integrate security tests in unit testing	This activity refers to leveraging unit testing to perform security-oriented tests such as boundary testing.	2
Performing automated software composition analysis	This activity refers to the verification of all dependencies (e.g. third-party libraries) for known vulnerabilities (SCA)	2
Implement automated remediation	This activity refers to the dynamic adaptation and reaction to security incidents.	1
Implement automation of software licensing	This activity refers to ensuring that users are purchasing, installing, and using software as per the conditions set by the software vendor of interest, using automated tools.	1
Performing security configuration automation	This activity refers to automating security configurations (hardening) throughout the lifecycle of an environment.	1
Performing security requirements analysis	This activity refers to the definition of security requirements.	1
Performing threat modelling	This activity refers to performing threat modelling to establish a common model of an application and subsequently identifying potential threats.	1
Performing risk analysis	This activity refers to analysing the threats to an application in the context of the business impact and likelihood to establish a risk score.	1
Establishing security SLA's for cloud providers	This activity refers to establishing security service level agreements for cloud providers based on the security requirements of a given application.	1
Performing continuous monitoring	This activity refers to actions enabling a continuous view on various security aspects of development and operations activities.	1
Performing continuous monitoring of security SLA's	This activity refers to performing continuous monitoring to confirm compliance with security service level agreements for cloud providers.	2
Performing continuous monitoring of security metrics throughout the SDLC using CI/CD tooling	This activity refers to leveraging the Continuous Integration and Continuous Deployment tools to gather security relevant metrics which can be monitored to identify risks such as coding mistakes or vulnerable dependencies.	2
Performing continuous monitoring of system metrics using automated tools	This activity refers to continuously monitoring metrics such as resource usage and reaction times. Based on patterns in these metrics malicious activity could potentially be detected.	2
Performing continuous monitoring of security controls	This activity refers to implementing specific monitoring solutions to determine the effectiveness of security controls for a given application such as SSL settings, access control mechanisms and vulnerable dependencies.	2
Performing continuous monitoring of application behaviour	This activity refers to continuous monitoring of application behaviour such as input and output to determine changes in patterns which may indicates malicious activity. This activity is commonly implemented through tools such as Web Application Firewalls (WAF).	2
Provide self-service monitoring capabilities to dev and ops	This activity refers to building monitoring capabilities so that they allow dev and ops to define the collection of metrics, definition of thresholds and alerts themselves making it a shared responsibility.	2
Performing continuous assurance	This activity refers to continuously validating if the software of interest is compliant with relevant regulatory requirements.	1
Performing manual security testing	This activity refers to aspects of security testing activities which cannot be automated and need to be performed manually.	1
Performing manual penetration testing	This activity refers to performing manual penetration tests.	2
Performing manual security review	This activity refers to performing manual security reviews which is usually a combination of manual code analysis combined with documentation review and stakeholder interviews.	2
Performing automated security testing of the CI/CD pipeline	This activity refers to performing automated security testing of the CI/CD pipeline to identify weaknesses.	1

Table 54. Results of thematic analysis of design factors

Activity	Design factor
Performing automated Security Testing	Leverage SecaaS by using cloud provided self-managed, automated and scalable security services
Performing automated Security Testing	Integrate the security tools in an automated deployment pipeline
Performing automated Security Testing	Automate as many security controls and verifications as possible
Perform automated run-time testing	Perform automated run-time testing at four levels: (1) pre-authentication scanning, (2) post-authentication scanning, (3) independent backend scanning and (4) complete workflows
Perform automated run-time testing	Ensure automated run-time testing is implemented for a broad scope of test scenarios
Perform automated run-time testing	Ensure proper unit tests are in place to optimise run-time testing efficiency
Performing automated static testing	Minimise the number of false positives resulting from static testing
Performing security requirements analysis	Treat security requirements as nonfunctional requirements
Performing security requirements analysis	Leverage metrics gathered during the security requirements analysis phase to evaluate the security level of alternative designs
Performing security requirements analysis	Enable the evaluation of alternative designs through suitable metrics during security requirements analysis to determine variations in security levels of a given design and make appropriate choices
Performing security requirements analysis	Leverage goal-oriented requirements analysis (GORE) to perform security requirements analysis
Performing threat modelling	Perform threat modelling from a risk-centric perspective
Performing threat modelling	Perform threat modelling from a attack-centric perspective
Performing threat modelling	Perform threat modelling from a software-centric perspective
Performing threat modelling	Ensure compatibility of threat modelling outcomes from a scope and result perspective
Performing threat modelling	Introduce abuse cases and problem frames to perform threat modelling
Performing threat modelling	Make use of attack or threat trees to perform threat modelling
Performing threat modelling	Implement traceability of threat modelling (results) in the code base
Performing threat modelling	Automated threat impact analysis
Performing risk analysis	Performing risk analysis continuously before each iteration
Performing risk analysis	Performing risk analysis during the design phase
Performing risk analysis	Include a broad range of stakeholders including the business owner when setting security goals
Performing risk analysis	Establish clear rules regarding information exchange across teams and maintain a log for every access to sensitive data
Performing risk analysis	Provide security knowledge and tools and encourage the Development and Operations teams to integrate themselves
Performing risk analysis	Consider gamification for finding vulnerabilities or bugs
Performing continuous monitoring	Ensure continuous monitoring covers a wide range of resources and metrics including logical security, availability and intrusions
Performing continuous monitoring	Leverage monitoring as code to establish a versioned and repeatable deployment of monitoring infrastructure
Performing manual security testing	Limit manual penetration testing to critical components or perform in parallel to reduce impact on deployment lead times
Performing continuous feedback from production to development	Set a time limit for all lower-priority security defects
Performing continuous feedback from production to development	Give attack patterns to your developers
Performing continuous feedback from production to development	Build an internal forum to discuss attacks
Performing continuous feedback from production to development	Establish emergency code base response
Performing continuous feedback from production to development	Incorporate security tests as part of QA for detected incidents
Providing security training	Teach every developer enough to enable them to identify areas where they would benefit from the advice of an expert
Performing risk analysis	Establish clear rules regarding information exchange across teams and maintain a log for every access to sensitive data

Table 55. Selection criteria for DevSecOps experts

Category	Criteria
Disciplines and skills	Academic: • Published article on the subject matter of agile security, DevSecOps or CI/CD in peer reviewed academic magazines; OR • Master or PhD thesis on the subject matter of agile security, DevSecOps or CI/CD Practitioner: • min 5 years of professional experience; AND • Holding a relevant certification in information security from certifying bodies such as SANS, ISC2 or ISACA; AND • Technical experience in a DevSecOps environment and have an affiliation with security; OR • Organisational experience structuring people, process or tool aspects of a DevSecOps environment.
Target Organisations	• Large sized organisations (+1000 employees) in a highly or moderately regulated sector; • Medium sized organisations (+250 employees) in a highly or moderately regulated sector; • Digital native organisations; • Government institutions; • Organisations providing information security services

Table 56. Results of the prioritisation of security activities by the expert panel

Activity	Effectiveness		Delay		Financial	
	Mean	IQR	Mean	IQR	Mean	IQR
Collaboration						
Performing continuous feedback from production to development	4.5	1.25	4.0	1.50	4.0	1.00
Provide security training	4.0	1.00	4.0	1.50	3.0	0.50
Establish security satellites	5.0	1.00	4.0	0.50	3.0	1.50
Practice incident response	4.0	1.25	4.0	1.50	3.0	1.50
Establish a security mindset across the organisation	4.0	1.25	4.0	1.50	3.0	2.00
Use of automated activities						
Performing automated security testing						
Performing automated run-time testing	3.0	1.25	4.0	0.50	3.0	1.50
Performing automated static testing	3.5	2.00	3.0	1.50	2.5	1.00
Integrate security tests in unit testing	4.5	1.25	4.0	0.50	3.0	1.50
Performing continuous monitoring						
Performing continuous monitoring of security SLA's	3.0	2.00	4.0	2.00	3.5	1.75
Performing continuous monitoring of security metrics throughout the SDLC using CI/CD tooling	4.0	1.00	4.0	1.50	3.5	1.00
Performing continuous monitoring of system metrics using automated tools	4.0	1.50	4.0	1.50	3.0	0.75
Performing continuous monitoring of security controls	4.0	0.50	4.0	1.00	3.0	1.50
Performing continuous monitoring of application behaviour	4.0	0.25	4.0	1.50	2.0	1.50
Provide self-service monitoring capabilities to dev and ops	3.0	1.25	4.0	1.50	4.0	1.50
Implement centralised dashboard	3.5	1.50	4.0	1.50	4.0	0.00
Implement automated remediation	3.0	1.50	5.0	0.50	3.0	1.50
Performing security configuration automation	4.0	0.25	4.0	1.50	3.0	2.00
Implement secrets management	4.5	2.00	4.0	1.50	3.5	1.75
Manage digital supply chain						
Establish artefact and source code registries which are automatically scanned for vulnerabilities	4.0	1.00	3.0	1.50	4.0	1.50
Implement automated container security scanning	4.0	1.25	4.0	0.00	3.5	1.00
Performing automated software composition analysis	3.0	1.25	3.0	1.00	3.0	1.50
Use of non-automated activities						
Performing security requirements analysis	3.0	0.50	3.0	1.50	3.5	2.00
Performing threat modelling	3.0	0.50	3.0	1.00	3.0	0.25
Performing risk analysis	4.0	0.50	3.5	1.00	4.0	0.25
Establishing security SLA's for cloud providers	3.0	2.00	4.5	2.25	3.5	1.00
Performing continuous assurance	4.0	2.00	3.0	2.00	3.0	0.25
Performing manual security testing						
Performing manual penetration testing	4.0	1.50	2.0	1.50	2.0	1.25
Performing manual security review	3.0	1.50	2.0	1.00	3.0	1.25
Securing the CI/CD pipeline	4.0	1.50	4.0	1.25	3.5	1.25

Chapter 12
Findings and Core Practices in the Domain of Security Assurance in DevOps

Dennis Verslegers
Orange Cyberdefense, Belgium

ABSTRACT

This chapter of the book provides an overview of the results obtained during this research project and presents the prioritised list of security practices to integrate security assurance in DevOps. It includes answers to the questions formulated at the start of the research project and contains an overview of the findings for collaboration, automated, and non-automated security activities. The authors discuss the overall conclusions based on the outcomes of the research. Finally, an overview of the limitations and possibilities for future research are presented.

SUMMARY OF THE RESEARCH RESULTS

At the start of this research project three research questions were formulated to provide an answer to the challenge on how to integrate security assurance activities in DevOps without creating friction.

As a foundation for this research the first question aimed to provide an agreed definition on DevOps and DevSecOps. The proposed definitions were derived from academic papers (Jabbari et al., 2016) and presented to a panel of 10 DevSecOps experts out of which 7 agreed on the proposed definitions. These results allow us to make the following statements:

Definition of DevOps: DevOps is a development methodology aimed at bridging the gap between Development and Operations, emphasising communication and collaboration, continuous integration, quality assurance and delivery with automated deployment utilising a set of development practices.

Definition of DevSecOps: RuggedOps (DevSecOps): "Rugged" describes software development organisations that have a culture of rapidly evolving their ability to create available, survivable, defensible, secure, and resilient software.

DOI: 10.4018/978-1-7998-7367-9.ch012

Subsequently this research project set out to identify a set of security activities and design factors which are relevant in the context of DevSecOps. This set of relevant security activities and design factors was established through a literature review followed by an expert validation and elaboration session involving 10 DevSecOps experts. The result is a set of 29 validated security activities on which there is strong agreement between the experts on their relevance to DevSecOps.

A total of 31 security activities and 34 design factors identified through literature research are considered relevant in the context of security for DevOps. During the validation and elaboration, the expert panel indicated one activity to be irrelevant to DevSecOps (-1) and added 3 additional activities (+3) bringing the total on 33 security activities. The expert panel also added an additional 53 design factors bringing the total number of design factors to 87.

The third and final objective for this research project aims to rank the identified security activities based on effectiveness and delay caused in the process of continuous deployment by a group of practitioners. The results of the previous research question were prioritised by a group of 8 DevSecOps experts during a Group Support Session (GSS). The security activities, grouped into the categories of collaboration, use of automated activities and use of non-automated activities, have been ranked by the experts as presented below (prioritised list of security activities) according to the following three aspects:

§ **Relevance**: an indication of the number of experts who perceived the activity as relevant to DevSecOps. This score was obtained during the second phase of the process

§ **Effectiveness**: an indication of the degree to which the expert panel believes the activity is contributing to the security of the software under development

§ **Delay**: an indication of the degree to which the expert panel believes the activity causes delays in the development or operations process

CONCLUSION

The identified security activities can be categorized into three distinct practices: (a) collaboration, (b) use of non-automated activities and (c) use of automated activities.

Collaboration

Experts and scholars place a strong emphasis on the People aspect for DevOps and DevSecOps. Fostering a security culture throughout the organisation is a crucial enabler to improve security assurance. Every stakeholder in the process, from ideation to operations, contribute to security within their respective areas of responsibility. Loss of contextual knowledge and delays can be achieved by avoiding the need for frequent hand-over to a dedicated security team.

Table 1. Prioritised list of security activities

		Relevance	Effectiveness		Delay		Financial	
			mdn	iqr	mdn	iqr	mdn	iqr
Collaboration								
Performing continuous feedback from production to development	This activity refers to continuously feeding security metrics and information on security incidents from production back to development.	9 of 10	4.5	1.25	4	1.50	4	1.00
Provide security training	This activity refers to training a wide range of stakeholders such as developers, architects and product owners on security aspects.	9 of 10	4	1.00	4	1.50	3	0.50
Establish security satellites	This activity refers to creating a network of security savvy people throughout the various teams involved in software development. These people are regularly referred to as security champions.	10 of 10	5	1.00	4	0.50	3	1.50
Practice incident response	This activity refers to practicing incident response through red-team exercises and security drills.	7 of 10	4	1.25	4	1.50	3	1.50
Establish a security mindset across the organisation	This activity refers to actions taken to increase the attention and awareness on security related aspects of secure development and operations across both operational and managerial levels of an organisation.	Added by expert	4	1.25	4	1.50	3	1.50
Use of automated activities								
Performing automated security testing	This activity refers to aspects of security testing which can be automated thereby providing actionable information.	10 of 10						
Performing automated run-time testing	This activity refers to the dynamic, interactive testing of a deployed application using automated tools (DAST)	8 of 10	3	1.25	4	0.50	3	1.50
Performing automated static testing	This activity refers to code review using automat-ed tools to detect common vulnerability patterns (SAST).	8 of 10	3.5	2.00	3	1.50	2.5	1.00
Integrate security tests in unit testing	This activity refers to leveraging unit testing to perform security-oriented tests such as boundary testing.	9 of 10	4.5	2.25	3	3.50	1	1.50
Performing continuous monitoring	This activity refers to actions enabling a continuous view on various security aspects of development and operations activities.	9 of 10						
Performing continuous monitoring of security SLAs	This activity refers to performing continuous monitoring to confirm compliance with security service level agreements for cloud providers.	8 of 9	3	2.00	4	2.00	3.5	1.75
Performing continuous monitoring of security metrics throughout the SDLC using CI/CD tooling	This activity refers to leveraging the Continuous Integration and Continuous Deployment tools to gather security relevant metrics which can be monitored to identify risks such as coding mistakes or vulnerable dependencies.	7 of 9	4	1.00	4	1.50	3.5	1.00
Performing continuous monitoring of system metrics using automated tools	This activity refers to continuously monitoring metrics such as resource usage and reaction times. Based on patterns in these metrics malicious activity could potentially be detected.	5 of 9	4	1.50	4	1.50	3	0.75
Performing continuous monitoring of security controls	This activity refers to implementing specific monitoring solutions to determine the effectiveness of security controls for a given application such as SSL settings, access control mechanisms and vulnerable dependencies.	8 of 9	4	0.50	4	1.00	3	1.50
Performing continuous monitoring of application behaviour	This activity refers to continuous monitoring of application behaviour such as input and output to determine changes in patterns which may indicates malicious activity. This activity is commonly implemented through tools such as Web Application Firewalls (WAF).	8 of 9	4	0.25	4	1.50	2	1.50
Provide self-service monitoring capabilities to dev and ops	This activity refers to building monitoring capabilities so that they allow dev and ops to define the collection of metrics, definition of thresholds and alerts themselves making it a shared responsibility.	8 of 9	3	1.25	4	1.50	4	1.50
Implement centralised dashboards	This activity refers to establishing centralised dashboard which provide a clear view on security metrics related to secure development and operations.	Added by expert	3.5	1.50	4	1.50	4	0.00

continued on following pa

ed

		Relevance	Effectiveness		Delay		Financial	
			mdn	iqr	mdn	iqr	mdn	iqr
Implement automated remediation	This activity refers to actions enabling a continuous view on various security aspects of development and operations activities.	10 of 10	3	1.50	5	0.50	3	1.50
Performing security configuration automation	This activity refers to automating security configurations (hardening) throughout the lifecycle of an environment.	9 of 10	4	0.25	4	1.50	3	2.00
Implement secrets management	This activity refers to automating the management of secrets used by applications and infrastructure components.	Added by expert	4.5	2.00	4	1.50	3.5	1.75
Manage digital supply chain	This activity refers to actions taken to manage the risks introduced using external components which are not under the direct control of the organisation.	Added by expert						
Establish artefact and source code registries which are automatically scanned for vulnerabilities	This activity refers to the actions taken to continuously scan and approve entries in artefact and source code registries thereby preventing the use of vulnerable resources in development or operations.	Added by expert	4	1.00	3	1.50	4	1.50
Implement automated container security scanning	This activity refers to implementing automated security scanning of container images and configuration for known vulnerabilities and weaknesses preventing vulnerable images or insecure configurations to be used as a basis for development or operations.	Added by expert	4	1.25	4	0.00	3.5	1.00
Performing automated software composition analysis	This activity refers to the verification of all dependencies (e.g. third-party libraries) for known vulnerabilities (SCA)	8 of 10	3	1.25	3	1.00	3	1.50
Use of non-automated activities								
Performing security requirements analysis	This activity refers to the definition of security requirements.	9 of 10	3	0.50	3	1.50	3.5	2.00
Performing threat modelling	This activity refers to performing threat modelling to establish a common model of an application and subsequently identifying potential threats.	10 of 10	3	0.50	3	1.00	3	0.25
Performing risk analysis	This activity refers to analysing the threats to an application in the context of the business impact and likelihood to establish a risk score.	10 of 10	4	0.50	4	1.00	4	0.25
Establishing security SLA's for cloud providers	This activity refers to establishing security service level agreements for cloud providers based on the security requirements of a given application.	10 of 10	3	2.00	4.5	2.25	3.5	1.00
Performing continuous assurance	This activity refers to continuously validating if the software of interest is compliant with relevant regulatory requirements.	9 of 10	4	2.00	3	2.00	3	0.25
Performing manual security testing	This activity refers to aspects of security testing activities which cannot be automated and need to be performed manually.	8 of 10						
Performing manual penetration testing	This activity refers to performing manual penetration tests.	8 of 8	4	1.50	2	1.50	2	1.25
Performing manual security review	This activity refers to performing manual security reviews which is usually a combination of manual code analysis combined with documentation review and stakeholder interviews.	8 of 8	3	1.50	2	1.00	3	1.25
Secure the CI/CD pipeline	This activity refers to using a set of tools to prevent any changes such as access control or build street definitions which may impact the security of the CI/CD pipeline.	8 of 10	4	1.50	4	1.25	3.5	1.25

Shifting this responsibility to the various stakeholders requires them to be offered the necessary skills and knowledge addressing the specific concerns within their area of responsibility. The intention is not to create security experts of every person in the organisation but rather to infuse contextual knowledge which allows the actors to contribute to the security of the product or platform. To identify the required skills and knowledge, we can leverage security performance metrics collected on a team level throughout the development cycle. By pinpointing the specific areas of concerns, we can limit the provided knowledge base to those topics which make a real difference—thereby avoiding information overload for the teams by including broad scoped security knowledge which does not apply to their jobs.

Sharing of incidents and security weaknesses clearly and understandably contributes to the security culture. These occasions provide learning opportunities while reinforcing the message that the organisation takes security seriously.

A practical approach should include appointing the required security contact points for the various teams. Motivated people with security interest can be identified in the various teams and leveraged as contact points within the teams themselves. Furthermore, security experts can be seconded to the teams to provide expert knowledge and support when required.

The objective is to establish a culture where solid engineering practices, such as a focus on quality and reliability, open communication and learning opportunities, have a place.

The objective is to establish a culture where solid engineering practices, such as a focus on quality and reliability, open communication and learning opportunities, have a place.

Use of Non-Automated Activities

The identified non-automated security activities contribute significantly to the security mindset described in the collaboration section of this book. The activities that are part of this area tend to be very similar compared to procedural approaches as they cannot be automated and tend to require some degree of expert security skills and knowledge. Nevertheless, we consider these activities essential to achieve security by design and must receive appropriate attention.

Integrating these activities in the DevOps process in such a way that they do not cause significant friction and delays requires specific approaches and, in most organisations, are subject to experimentation. Achieving scalability of non-automated activities is possible by focusing efforts on high(er) risk applications and components. As everything in DevOps, an incremental approach is advised. The objective of completeness is often subordinate to consistent execution of the activity. One way to achieve this is to perform these activities in a timeboxed approach so that the teams remain motivated and committed to performing them regularly, e.g. at the start of every sprint.

Increased efficiency for these activities is possible by leveraging the knowledge of security experts to create models, frameworks and blueprints which help the teams perform the job. These resources must be understandable and usable by development and operations teams.

A point of attention when leveraging such an incremental approach is the overall coherence and overview of the results from partial activities. This area is expected to remain the primary focus of specialised security teams.

Use of Automated Activities

Automation of security activities is vital to approach security activities in a scalable way in DevSecOps. Often these automated activities are integrated into the CI/CD pipelines we discussed earlier in the book. By doing so, one creates technical guard rails which assist teams in meeting security objectives and assures the security teams. The automation capabilities also provide the unique potential to embed decisions taken at the governance and management level into the operational domain and as such, is an avenue to close the traditional gap between governance and operational levels.

When looking at DevSecOps in literature, we notice that this area of DevSecOps receives the most attention. Some experts argue that this domain is the only domain which genuinely belongs to DevSecOps, the other practices such as collaboration and non-automated activities have strong resemblances to their counterparts in a procedural security approach. In the previous sections, we have already demonstrated that the approach for the activities belonging to these practices differs significantly. The practices of collaboration and non-automated security activities are vital to ensure that the selected automated activities are performed with the appropriate goals and context.

General Findings

The premise of this research project revolved around the concept that DevOps and security assurance can reinforce each other when implemented appropriately. Integrating security assurance activities enables organisations to gain confidence in their security posture (Forsgren et al., 2019), improve security characteristics (Rahman & Williams, 2016) and may remove an obstacle for the adoption of DevOps in an organisation (*DevOps: The Worst-Kept Secret to Winning in the Application Economy*, 2014) (Chen, 2017). It also allows organisations to increased speed of response to security issues (Forsgren et al., 2019) and open the path to leverage concepts such as quality culture and automation for the benefit of security (Forsgren et al., 2019). However, this integration must be performed appropriately to avoid the introduction of significant delays (Beznosov & Kruchten, 2004) which could potentially prohibit the materialisation of increased organisational performance (Forsgren & Humble, 2015).

Therefore, this research set out to define a framework of relevant security activities enabling an organisation to integrate security assurance in DevOps with a clear view on the effectiveness but also the delay caused to the process of continuous delivery.

Based on the results of this research project we conclude that for the most part traditional security assurance activities remain relevant in DevOps. There was a strong agreement among the security experts that the identified activities remained relevant in the context of DevSecOps. Furthermore, the activities, while having varying rating in terms of effectiveness and financial impact, display a similar level of delay.

This leads us to the conclusion that in order to avoid introducing delays in DevSecOps it is not so much about doing different things as it is about doing things differently. These differences are visible in the design factors gathered throughout this research and represented in part three of this book. Overall the following tenant expresses the difference in approach quite eloquently:

"We prefer automated over manual, repeatable over one-off. "

Some of the trade-offs which are made in DevSecOps are counterintuitive and radically different from traditional approaches to information security. Some examples of these are:

§ Reduce the detection rate in favour of reducing the number of false positives;
§ Prefer short iterative threat modelling exercises at the beginning of each sprint at the expense of threat model coverage;
§ Prefer limiting the available libraries during development to pre-approved versions over pre-production validations;
§ Foster engineering spirit and peer review over quality gates with validation steps;
§ Continuously feed information from production to development

§ Parallelise slower processes and invest in blue green deployments to roll back when needed.

Figure 1. Map of DevSecOps activities identified during this research

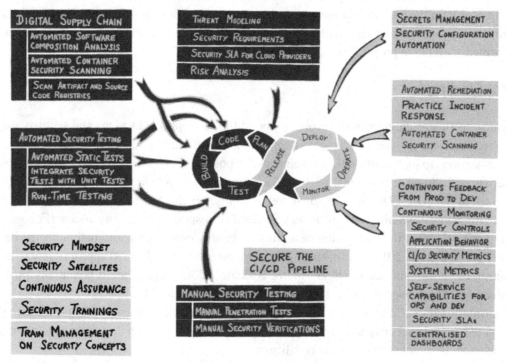

These trade-offs and differences in approaches are translated into design factors for each of the security activities and presented in part three of this book.

All the involved security experts strongly emphasised the aspects of collaboration and knowledge for people working in DevSecOps teams. One could say that this is an instantiation of solid engineering practices, DevSecOps is about teams of engineers who gather around a problem to understand and solve it through mutually agreed approaches. It is built on the concept of combining disciplines and creating a learning culture. The main theme in DevSecOps is the creation of feedback loops for learning purposes and ensuring you have the right measurements in place to get the most learnings out of each investment thereby gradually finding out what works and what doesn't, while taking into account the relevant security aspects.

Overall the following tenants hold true:

§ Share security learning experiences and create a security engineering mindset;
§ Shift security responsibility to the teams and create supporting mechanisms to get the job done;
§ Leverage security automation whenever possible;
§ Establish security measurements to gain insight and learning opportunities;
§ When selecting activities and approaches one should favour reducing delay over reducing financial
    impact

The identified set of security activities, their ranking in terms of effectiveness, delay and financial impact and the in-depth design factors presented in this research project can be leveraged by organisations to model their DevSecOps approach to include security activities based on their characteristics in terms of effectiveness, delay and financial consequences. The design factors represented in this research project provide insights in suitable implementations of these activities. The activities and design factors combined can be used to establish a roadmap for integrating security activities in a DevSecOps environment and as a benchmarking tool to assess the existing implementation.

As a closing summary a map of DevSecOps activities as identified during this research linked to the various stages of the DevSecOps process is presented in

CORE PRACTICES IN DEVSECOPS

During our research on DevSecOps, several activities were identified and subsequently ranked by an influential DevSecOps practitioners' group. We had the opportunity to have in-depth conversations with several experts who gave insight into the reasoning behind their rankings. They shared their views on the elements which are crucial to establish a successful DevSecOps program. The following ten core practices for DevSecOps were derived from the insights they provided.

Maintain a Business Perspective Business Context

Contribution from a wide range of stakeholders is required to integrate security into digital platforms. Business stakeholders must value security to ensure that appropriate resources are available and able to focus on the implementation of the selected measures. Assigning resources such as time and budget is inevitable if we want the security activities to affect the bottom-line security of the platform. One can argue that building security into the early stages of the process ensures a more efficient path to reach the security objectives. The need to increase budget or delay delivery of business functionality is to be expected. Business stakeholders must be enabled to make these trade-offs. Therefore, a genuine understanding of the consequences and opportunities from a security perspective are required. The translation of security initiatives into business impact reduction is a must. This translation is possible by establishing a clear relationship between the proposed security measures, the relevant threats and related business impact.

An added benefit by placing things into the context of business risk is that the outcomes provide clear objectives, a "true north", for all stakeholders including the technical oriented roles. The product teams gain the required flexibility to identify and experiment with potential approaches by placing initiatives in a business context.

During our research, the activities identified supporting this practice include **performing risk assessment** and **training management on security concepts**.

Implement Mechanisms for Security Trade-Offs

Closely related to maintaining a business perspective is the concept of implementing mechanisms to enable the business stakeholders to make security trade-offs. In many conversations, the experts involved in our research indicated that their organisations had a "functionality first" approach when deciding on priorities for their product teams. The implicit de-prioritisation of security investments can be explained partly them being labelled "non-functional requirements" and the the difficulty to understand them without the necessary technological background knowledge. Many product owners have a primary objective of driving the evolution of their product in terms of features, thereby favouring speed of delivery over other aspects of software products.

Placing security requirements in a business context helps to attract increased attention of business stakeholders and product owners. The next step is to ensure that security trade-offs are made explicit. This can be achieved by introducing security prioritisation mechanisms, one example being the introduction of a taxation system whereby part of each sprint's capacity for security-related improvements. Another approach consists of establishing a model where a certain level of "security defect density" acts as a barrier to entry for any new functionality.

During our research, the activities identified supporting this practice include **performing risk assessment** and performing **security requirements**.

Establish a Common Shared Security Perspective

Bringing people with the required skillsets together around a common objective is considered a prerequisite for high performing teams. It allows them to focus and orchestrate their efforts thereby achieving the objective efficiently and effectively. This concept also applies to security objectives, security objectives require a multidisciplinary approach and often benefit from a shared understanding. Providing clarity and fostering collaboration allows all contributors to participate in the decision-making process and are likely to be more supportive throughout the process. Investing in this level of collaboration may lead to significant gains in implementation speed. The contributors will be more likely to look for solutions towards the common goals instead of creating obstacles.

During our research, the activities identified supporting this practice include **performing threat modelling** and performing **security requirements**.

Empower Teams[1]

Gathering teams around a common shared security perspective and placing the objectives into a business context are essential steps. Another aspect is to empower the contributors and provide them the freedom to identify the optimal path to achieve the objectives. The required levels of experimentation and learning to achieve this are only possible with autonomy and authority. In the DevSecOps world, the balance between maintaining control and empowering teams is known as "building guardrails". These guardrails define the margins within which the teams can manoeuvre freely. Practising delegated authority enables contributors to feel responsible for the result and fosters continuous improvement initiatives.

During our research, the activity identified supporting this practice are **establishing security satellites, performing security trainings, training management on security concepts** and **performing continuous feedback from production to development**.

Ensure Security Scalability

Security requires both investments in terms of resources (time, budget) and investments in terms of focus and attention from the individuals involved. Directing the same level of attention to each project will quickly drain resources and create an inevitable "security fatigue". Closely related to this issue is that in all organisations involved in our research the security experts were significantly outnumbered and experienced difficulties sustaining high-paced distributed product team demands. Some experts indicated that the best way forward is to leverage the security team to create consumable security services and tailored methodologies that enable the development and operations teams to build security into their activities. Although not every team member can be expected to become a security expert, it is possible to enable all contributors to carry their weight regarding security. In many cases, this approach may even lead to better results; security experts themselves lack the deep technical insight available to the engineers who master a specific domain.

During our research, the activity identified supporting this practice are **establishing security satellites, performing automated security testing,** and **securing the digital supply chain**.

Establish a Security Culture by Promoting Communication and Collaboration

Ultimately, an organisation's security culture determines the success of security initiatives. Fostering such a security culture requires open communication across all layers of the organisation. Management should express the importance they place on security and communicate their expectations. Similarly, the security, development and operations teams should be encouraged to share information regarding security weaknesses, challenges, and incidents with one another to achieve continuous improvement. Far too often, security incidents or penetration test results are treated with secrecy and only distributed to a minimal number of stakeholders.

During our research, the activities identified supporting this practice include **performing security training, establishing a security mindset** and performing **continuous feedback from production to development**.

Leverage and Secure the Pipelines

One of the core premises of DevOps lies in its emphasis on automation. Automation provides numerous opportunities from a security perspective as it allows embedding security through every step of the process in a scalable way. Some of the advantages include repeatability and consistency, near immediate feedback to team members and more scalable use of the available security experts. When appropriately implemented, automated tools can also significantly streamline assurance activities and increase auditors' trust by ensuring that every output has gone through various security validations.

As these pipelines hold the potential power to take outputs from development to production, it is important to properly secure them to ensure the required security controls cannot be circumvented.

During our research, the activities identified supporting this practice include **performing security configuration automation, performing automated security testing, securing the digital supply chain,** and securing **the CI/CD pipeline**.

Establish Security Observability

Security observability requires logging and monitoring for security-related events to be available throughout a digital platform's lifecycle[2]. The logging and monitoring capabilities are preferably implemented in an automated fashion to ensure consistency across deployments. Collection of security metrics can be performed through the CI/CD pipeline and obtained from running applications. The metrics must be made available to teams through a self-service approach allowing identification of security issues as early as possible in the system development process. Early implementation and access to these systems allow the various teams to build operational knowledge, establish baselines and prepare incident response tactics, and relevant alerts early in the process. These preparations allow the teams to get a significant head start when discovering security incidents during the platform's operations phase.

During our research, the activities identified supporting this practice include **performing continuous security monitoring**, including monitoring of security controls, application behaviour, gathering metrics throughout the CI/CD pipeline and providing centralised dashboards and self-service capabilities for dev and ops, and establishing **security SLAs for cloud providers**.

Practice Incident Response

Gathering information from various sources to identify security weaknesses and incidents provide the necessary foundations for incident response. To ensure proper handling of incidents, the different teams must establish detection and response procedures. Incident response exercises should involve a wide range of competences, including operations, development, and security to put these procedures to the test and create a security culture throughout the organisation.

During our research, the activity identified supporting this practice is **practicing incident response**.

Perform Continuous Assurance

Assuring internal and external stakeholders on the security posture of digital platforms is, in many cases, a key driver to integrate security into DevOps processes. Our research's various activities produce relevant metrics on the state of security and the likelihood that security weaknesses are introduced during the design, implementation, or operations phases of the digital platform. These metrics provide significant opportunities for automated continuous assurance when their integrity is guaranteed. Approaches enforcing security policies through technical means (security-as-code, compliance-as-code) contribute to the effectiveness of the security program and at the same time, eases compliance exercises significantly.

During our research, the activity identified supporting this practice is also referred to as **performing continuous assurance**.

RESEARCH LIMITATIONS

During this research project some limitations were encountered which may influence the validity of the outcomes. They are listed here for your reference:

§ The literature review performed in the scope of this research project focused on relatively narrow search criteria to keep the number of identified papers within a workable limit given the time available for this study. Therefore, important contributions in the field of agile security and DevSecOps may be absent in the dataset used for this research;
§ The grouping of security activities and design factors was performed by the researcher thereby potentially introducing bias. This threat has been partially reduced by performing a validation step with an expert panel;

§ A mistake was made during the creation of the survey leading to the design factors for risk analysis not being validated by the expert panel;

§ The results of the elaboration by the expert panel were interpreted by the researcher to extend the list of identified security activities and design factors. The comments were interpreted by the researcher and presented to the experts for verification however this was not done through a formal survey due to time limitations and the risk of "survey fatigue" on the respondents' part;

§ The prioritisation performed by the expert panel was based on short descriptions of the security activities, differences in interpretation between the members of the expert panel may have remained unnoticed and may have influenced the scoring;

§ The mapping with OWASP SAMM and BSIMM and the mapping with the various stages of the DevSecOps process as presented in this book were not validated by an expert panel;

§ The analysis of the prioritisation by the expert panel was performed using simple methods of descriptive statistics, there was no complete SPSS analysis performed;

§ The artefact answering research question 3 was not evaluated in real world environments therefore the design cycle was not completed.

FUTURE RESEARCH

Various avenues for future activities were identified during this research project. The framework of security activities relevant for DevSecOps, developed as an answer to the third research question, should be tested in real world situations to measure and assess actual usability and to perform improvements to the framework. This is left as a potential avenue for future research.

Each of the identified security activities identified during this project represent a potential subject for future research. Such research could focus on establishing complete inventory of design factors and measure their influence on effectiveness, delay and financial consequences.

In addition, an overview of the tooling landscape for each of the identified security activities could be created to increase the real-world usability of the framework.

A final approach to extend the framework would be to create an extensive mapping between existing standards such as the Agile Secure Software Framework (SSA), OWASP SAMM and BSIMM but also to standards such as ISO27000, NIST Cybersecurity Framework or CIS controls.

REFERENCES

Beznosov, K., & Kruchten, P. (2004). *Towards agile security assurance.* doi:10.1145/1065907.1066034

Chen, L. (2017). Continuous Delivery: Overcoming adoption challenges. *Journal of Systems and Software, 128*, 72–86. doi:10.1016/j.jss.2017.02.013

DevOps: The Worst-Kept Secret to Winning in the Application Economy. (2014). https://www.ca.com/content/dam/ca/us/files/msf-hub-assets/research-assets/devops-winning-in-application-economy.pdf

Forsgren, N., & Humble, J. (2015). The Role of Continuous Delivery in it and Organizational Performance. SSRN *Electronic Journal.* doi:10.2139srn.2681909

Forsgren, N., Smith, D., Humble, J., & Frazelle, J. (2019). *2019 Accelerate State of DevOps Report.* Academic Press.

Jabbari, R., Ali, N., Petersen, K., & Tanveer, B. (2016). *What is DevOps? A Systematic Mapping Study on Definitions and Practices.* doi:10.1145/2962695.2962707

Rahman, A. A. U., & Williams, L. (2016). *Software security in DevOps: Synthesizing practitioners' perceptions and practices.* doi:10.1145/2896941.2896946

ENDNOTES

[1] When we refer to teams, we do not necessarily mean teams in the traditional sense where people are placed together under the same reporting line and authority. We perceive teams as a free collective of contributors around common objectives. These contributors may, from a management perspective, belong to different teams within the organisation.

[2] The statement that "code is eating the world" is very much applicable to the development of digital platforms. Traditionally the term code is strongly associated with traditional development of applications. We see the introduction of development approaches in other domains at an increasing rapid pace. Software defined networks, systems deployment and configuration automation and cloud stacks which expose functionality through APIs result in the infrastructure domain adopting many of the techniques (such as versioning and unit testing) that were previously situated in the development domain. This also influences the way security can be integrated in the process as demonstrated through the numerous automated activities reflected in our research. Therefore, the term secure software development lifecycle would be better interpreted as secure systems development lifecycle in the current age.

About the Contributors

Yuri Bobbert is a business scientist. He is Chief Security Officer at ON2IT, a global cybersecurity player and academic director / professor at Antwerp Management School (AMS). Bobbert holds a double PhD from both the University of Antwerp and Radboud University. He is specialised in establishing risk, security, and compliance functions within a wide variety of organisations. He advised +300 companies. Bobbert served for 10 years as CEO of B-ABLE and 6 years as CRO/CSO of UWV and NN Group. At NN Group he led the merge with DeltaLloyd. Bobbert published +100 papers, books, and blogs.

Maria Chtepen is a security practitioner with a strong background in Enterprise Architecture. Over 15 years of experience within different industries working across business and technology teams. Building collaboration and communities across varying organizational domains to achieve the optimal synergies between technology and company goals and objectives. Maria has a strong academic background and is an author of various publications in international journals and conference proceedings.

Tapan Kumar comes from India and holds a bachelor's degree in information technology from renowned university in India and did his masters from Antwerp management school in Enterprise It Architecture (2018-2020). He has worked with few of the biggest IT organizations like IBM, Cognizant and Tata Consultancy Services in his 15-year career. With his experience and specialization within Banking and Financial domain he has always been associated with a consultancy organization that worked in a distributed model, initially with the waterfall model in lately with the agile model. Tapan started his career as a mainframe engineer and gradually found interest in the field of architect and currently he the Solution/Application architect within a leading bank in The Netherlands in the domain of Payments and Cards.

Yves Vanderbeken helps clients introducing new business models based on adoption of digital strategies and technologies. He works as country CTO in Belgium for a global IT Services company. Yves manages the development of strategic Business – IT roadmaps with client leadership, focusing on the full spectrum of IT (Information, Applications, and Infrastructure). With over 25 years of experience in IT, Yves combines extensive cross industry expertise from government, finance, and manufacturing. Yves holds 2 MBA degrees from Antwerp Management School. His book chapters are based on a combination of academic research and working at some of the referenced agencies.

Dennis Verslegers is a seasoned IT Security Expert with over 20 years of experience in a wide variety of roles. His passion is to bring Business and IT together on Cyber Security and help them step up and get security right. Being active as a Strategic Advisor Application Security at a global security provider, he engages in risk management and secure development advisory services for several international organisations and institutions. Dennis holds an MSc degree from Antwerp Management School and multiple professional certifications. He loves to actively share his knowledge as an ISACA trainer and guest lecturer at Antwerp Management School.

Index

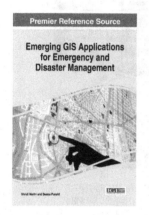

IGI Global's Transformative Open Access (OA) Model:
How to Turn Your University Library's Database Acquisitions Into a Source of OA Funding

Well in advance of Plan S, IGI Global unveiled their OA Fee Waiver (Read & Publish) Initiative. Under this initiative, librarians who invest in IGI Global's InfoSci-Books and/or InfoSci-Journals databases will be able to subsidize their patrons' OA article processing charges (APCs) when their work is submitted and accepted (after the peer review process) into an IGI Global journal.

How Does it Work?

Step 1: **Library Invests in the InfoSci-Databases:** A library perpetually purchases or subscribes to the InfoSci-Books, InfoSci-Journals, or discipline/subject databases.

Step 2: **IGI Global Matches the Library Investment with OA Subsidies Fund:** IGI Global provides a fund to go towards subsidizing the OA APCs for the library's patrons.

Step 3: **Patron of the Library is Accepted into IGI Global Journal (After Peer Review):** When a patron's paper is accepted into an IGI Global journal, they option to have their paper published under a traditional publishing model or as OA.

Step 4: **IGI Global Will Deduct APC Cost from OA Subsidies Fund:** If the author decides to publish under OA, the OA APC fee will be deducted from the OA subsidies fund.

Step 5: **Author's Work Becomes Freely Available:** The patron's work will be freely available under CC BY copyright license, enabling them to share it freely with the academic community.

Note: This fund will be offered on an annual basis and will renew as the subscription is renewed for each year thereafter. IGI Global will manage the fund and award the APC waivers unless the librarian has a preference as to how the funds should be managed.

Hear From the Experts on This Initiative:

"I'm very happy to have been able to make one of my recent research contributions *freely available* along with having access to the *valuable resources* found within IGI Global's InfoSci-Journals database."

– Prof. Stuart Palmer,
Deakin University, Australia

"Receiving the support from IGI Global's OA Fee Waiver Initiative *encourages me to continue my research work without any hesitation*."

– Prof. Wenlong Liu, College of Economics and Management at Nanjing University of Aeronautics & Astronautics, China

Printed in the United States
by Baker & Taylor Publisher Services